The Channels of Mars

The Channels of Mars

BY VICTOR R. BAKER

 UNIVERSITY OF TEXAS PRESS, AUSTIN

Library of Congress Cataloging in Publication Data

Baker, Victor R.
  The channels of Mars.
  Bibliography: p.
  Includes indexes.
  1. Mars (Planet)—Surface.
  2. Geomorphology.   I. Title.
QB641.B23   559.9'23   81-7549
ISBN 0-292-71068-2   AACR2

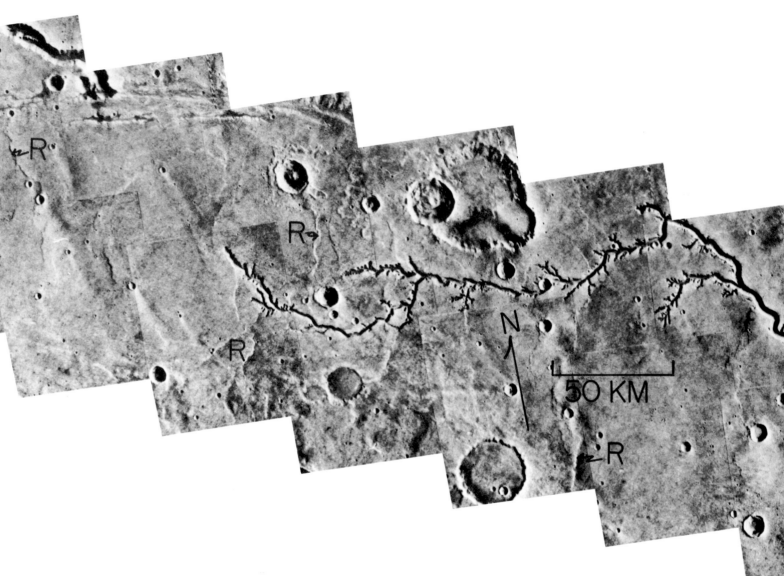

Above: Figure 4.12. (See page 66.) *Facing title page*: Mosaic of NASA Viking images showing an extensive region of outflow channeling measuring 1,400 by 1,800 kilometers and centered at 15°N, 33.75°W on Mars. Major features include Chryse Planitia (upper left), Ares Vallis (right center), Simud Vallis (bottom center), and Shalbatana Vallis (bottom left). (Mosaic assembled by the U.S. Geological Survey—U.S.G.S. Miscellaneous Investigations Series, Maps I-1343 and I-1345.)

TO SNOOKIE

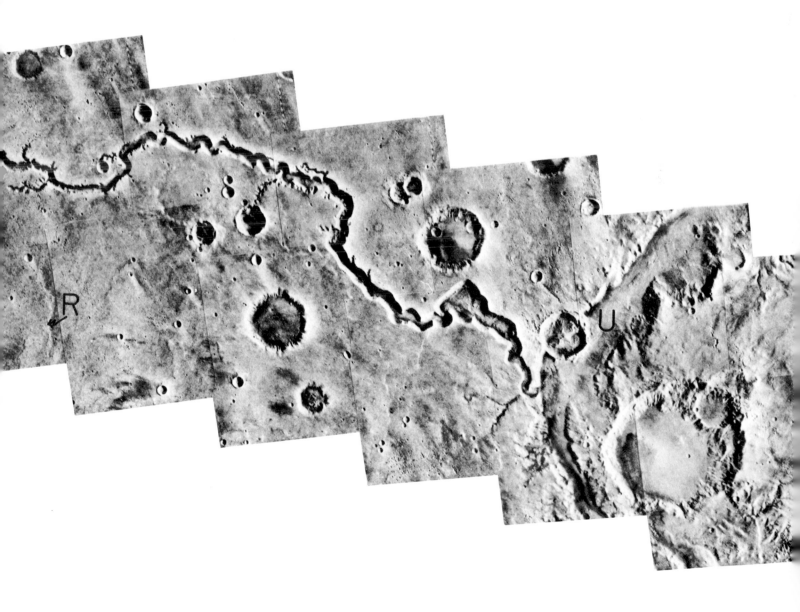

Contents

Preface and Acknowledgments   xiii

1. Telescopes, Canals, and Space Probes   3
2. The Geomorphology of Mars   14
3. Channel Types, Distribution, Ages, and Proposed Origins   34
4. Patterns and Networks of Martian Valleys   56
5. Ice and the Martian Surface   88
6. The Outflow Channels   108
7. The Channeled Scabland: An Earth Analog   140
8. Catastrophic Flood Processes   161
9. Mars: A Water Planet?   175

References Cited   185
Index   195

Tables

1.1. Mars Data 3
1.2. International Astronomical Union Names for Features on Mars 7
2.1. Geological Mapping Units for Mars 17
2.2. Ages for Some Geomorphic Surfaces on Mars 21
3.1. Large Channels on Mars 36
3.2. Ages for Some Martian Channel Surfaces 38
3.3. Martian Outflow Channels Compared to Small Valley Networks 44
3.4. Characteristic Morphological Features of Martian Outflow Channels 46
3.5. Compatibility of Various Fluid-Flow Processes with Certain Features of Martian Outflow Channels 54
5.1. Possible Periglacial and Permafrost Features on Mars 92
5.2. Comparison of Atmospheric Constituents on Earth and Mars 105
7.1. Important Bed Forms in the Channeled Scabland 147
8.1. Estimated Flow Parameters for Martian Outflow Channels and Possible Earth Analogs 162
8.2. Regression Equations for Shapes of Streamlined Forms on Earth and Mars 170
9.1. Indicators of Present or Former Water-Related Processes on Mars 176

Figures

1.1. Viking image of Phobos 4
1.2. Schiaparelli's map of Mars 5
1.3. E. M. Antoniadi's mapping of Mars 6
1.4. Mariner 4 picture of Martian surface 6
1.5. Mariner 9 frames showing Mangala Vallis 8
1.6. A portion of Mangala Vallis 8
1.7. Viking 1 landing ellipse 10
1.8. The Viking lander 10
1.9. Viking lander picture of the Viking 1 landing site 11
1.10. Viking lander picture of the Viking 2 landing site 11
1.11. Patches of white frost at the Viking 2 landing site 12
1.12. The Viking biology experiments 12
2.1. View of the Argyre impact basin 15
2.2. Geologic map of Mars 16
2.3. Martian crater Bamberg 18
2.4. Rampart craters in the Chryse Planitia 19
2.5. The crater Yuty 20
2.6. Mauna Loa shield volcano 22
2.7. Viking mosaic of Olympus Mons 23
2.8. Summit caldera of Olympus Mons 23
2.9. Appollinarsis Patera 24
2.10. Southwest rift zone of Mauna Loa 25
2.11. Lunae Planum wrinkle ridges 25
2.12. Western Valles Marineris and Noctis Labyrinthus 27
2.13. Tithonium Chasma and Ius Chasma 27
2.14. Central Valles Marineris 28
2.15. Eastern Valles Marineris 29
2.16. Chaotic terrain at Hydaspis Chaos 30
2.17. Fretted terrain in Protonilus Mensae 30
2.18. Fretted terrain 31
2.19. Martian north polar cap 32
2.20. Large dunefields 33
3.1. Outflow channels northeast of the Valles Marineris system 35
3.2. Small-scale "runoff" channels 36
3.3. Upper reaches of Mamers Valles 37
3.4. Crater counts for several Martian outflow channels 39
3.5. Viking imagery of upper Kasei Vallis 40
3.6. Small valley networks 42
3.7. Sketch map of outflow channels south of the Chryse Planitia 43
3.8. Streamlined uplands at the outflow of Ares Vallis into the southern Chryse Planitia 44
3.9. Viking mosaic of lower Kasei Vallis 45
3.10. "Scabland" topography in the southern Chryse Planitia 45
3.11. The "braided" section of Mangala Vallis 47
3.12. Oblique image of Ravi Vallis 50
3.13. Oblique view of northwest flank of Alba Patera 52
3.14. Braided lava channels, southwest rift of Mauna Loa 53
3.15. Cinder-spatter cones, Mauna Loa 53
3.16. Aa lava flows on southwest rift of Mauna Loa 53
3.17. Lava tube on Mauna Iki 54
4.1. Geomorphic map of a terrestrial drainage network 56
4.2. Methods of ordering the channels in a drainage network 57
4.3. Viking mosaic of heavily cratered terrain 58
4.4. Small valley networks in the heavily cratered terrain of Mars 59

4.5. Small valley networks in the heavily cratered terrain of Mars 59

4.6. Viking mosaic of heavily cratered terrain in the Margaritifer Sinus region 60

4.7. Valley systems along the boundary between hilly and cratered terrain and cratered plateau 62

4.8. Subparallel or "digitate" valley network development 62

4.9. Cratered plateau lowland serving as baselevel for valley system shown in Figure 4.8 62

4.10. Valley networks in the southeastern portion of the Margaritifer Sinus region 63

4.11. Regional setting of Nirgal Vallis 65

4.12. Viking mosaic of Nirgal Vallis 66

4.13. Morphological map of Nirgal Vallis 66

4.14. Upstream portion of Nirgal Vallis 68

4.15. "Meandering reach" of Nirgal Vallis 69

4.16. Downstream portion of Nirgal Vallis 69

4.17. Western member of the two large valleys comprising Nanedi Vallis 70

4.18. Viking mosaic of Bahram Vallis 71

4.19. Ma'adim Vallis 72

4.20. Portion of Ius Chasma showing tributary development 73

4.21. Gully development on the flanks of a tableland in Ophir Chasma 74

4.22. Upper reach of Reull Vallis 74

4.23. Tyrrhena Patera 75

4.24. Viking mosaic of Tyrrhena Patera 76

4.25. Sinuous valley northwest of Hecates Tholus 78

4.26. Sketch map of streamlined hills in an unnamed channel west of Elysium Mons 79

4.27. Regional setting of the Elysium channels 80

4.28. Chaotic terrain on the northern margin of the volcanic plains of Hecates Tholus 82

4.29. Landsat image of sandstone plateau terrains in eastern Utah 84

4.30. Headwardly eroding valleys in the southeastern Utah sandstone plateaus 85

4.31. U-shaped valleys cut in the Pololu volcanic series on the northeastern side of Kohala volcano, Hawaii 85

4.32. Amphitheater-headed valley on the southern flank of Haleakala volcano, island of Maui 85

4.33. Dissected slopes of the western Maui shield volcano 86

4.34. Summit caldera of Haleakala volcano 86

4.35. Kaupo Gap, Haleakala volcano 86

5.1. Lobate ridges on the western flank of Arsia Mons 89

5.2. Viking mosaic showing the western flank of Arsia Mons 90

5.3. Fractured plains in the Cydonia region 91

5.4. Possible thermocirque and alas development in the southern Chryse Planitia 93

5.5. Steep-sided, flat-floored depressions 93

5.6. Possible thermokarstic development in heavily cratered terrain 94

5.7. Talus slopes and debris flow in Kasei Vallis 95

5.8. Spur-and-gully topography in the Valles Marineris 95

5.9. Talus cone and debris fan development in northern Kasei Vallis 95

5.10. Geomorphic maps of debris aprons 96

5.11. Debris aprons surrounding fretted terrain uplands 97

5.12. Viking mosaic of the Nilosyrtis region 97

5.13. A portion of the mosaic in Figure 5.12 98

5.14. Upper part of Mamers Valles 99

5.15. Landslides in Ius Chasma 100

5.16. Viking mosaic of a large landslide, Gangis Chasma 101

5.17. Viking mosaic of eastern Valles Marineris showing landslides 102

5.18. Lobate landslide mass along the northern escarpment of Olympus Mons 103

5.19. Large landslide or debris flow lobe on the west basal scarp of Olympus Mons 104

6.1. Map of the Chryse Basin 109

6.2. Sketch map of Ravi Vallis 110

6.3. High-resolution images of Iani Chaos 110

6.4. Sketch map of Ares Vallis 111

6.5. Viking mosaic showing upper reaches of Ares Vallis 112

6.6. Viking mosaic of a constricted reach of Ares Vallis 113

6.7. Sketch map of outflow portion of Ares and Tiu Valles 113

6.8. Geomorphic map and Viking images of Ares Vallis 114

6.9. Scabland on the channel floor at the mouth of Tiu Vallis 116

6.10. Geomorphic map of the mouth of Tiu Vallis prepared from Viking images showing features that are also characteristic of catastrophic flood erosion 116

6.11. Geomorphic map showing streamlined uplands at the mouth of Ares Vallis 117

6.12. Eroded crater at the mouth of Tiu Vallis 117

6.13. Sketch map of the Maja Vallis system 118

6.14. Generalized geomorphic map of the Maja Vallis canyon region 119

6.15. Legend for Figures 6.14, 6.16, and 6.22 120

6.16. Geomorphic map (in three parts) of the canyon of Maja Vallis and adjacent channeling of the Lunae Planum 121

6.17. Viking mosaic of the canyon reach of Maja Vallis 124

6.18. Mare-like ridges and crater in Maja Vallis 127

6.19. Flow expansion of Maja Vallis onto the Chryse Planitia plains 128

6.20. Relatively small streamlined hills and small anastomosing channels 129

6.21. Probable mid-channel bars near the flow expansion of Maumee

Vallis onto the Chryse
Planitia   129

6.22. Geomorphic map of the
Maumee Vallis system   130

6.23. Large crater in the Maumee
Vallis system   132

6.24. Index map of the Lunae Planum
and western Chryse
Planitia   133

6.25. A portion of upper Kasei
Vallis   134

6.26. A portion of Kasei Vallis   134

6.27. Streamlined hills of the western
Chryse Planitia near the mouth
of Kasei Vallis   135

6.28. Geomorphic map of lower Kasei
Vallis   135

6.29. Geomorphic map of a portion of
lower Kasei Vallis immediately
east of Figure 6.28   136

6.30. Viking mosaic of a portion of
Kasei Vallis   137

6.31. Typical spur-and-gully wall
morphology in lower Kasei
Vallis canyon   138

6.32. Tributary canyon morphology
in the eastern canyon section of
Kasei Vallis   138

6.33. North Kasei Channel   139

7.1. Regional pattern of the
Channeled Scabland   141

7.2. Relationship of glacial Lakes
Missoula and Bonneville to
catastrophic Pleistocene
flooding   141

7.3. Exposure of the Mount St.
Helens ash "triplet"   143

7.4. Loess scarps, scabland, and a
major divide crossing in the
Cheney-Palouse Scabland
tract   143

7.5. Oblique photograph of minor
divide crossings   144

7.6. High-water surface profile
through a portion of the Cheney-
Palouse Scabland tract   145

7.7. Regional paleohydraulic
features of the Channeled
Scabland   146

7.8. Anastomosing channel pattern
in the Telford–Crab Creek
Scabland complex   147

7.9. Excellent association of
depositional macroform (bar)
with superimposed mesoforms
(giant current ripples)   148

7.10. Landsat photograph of the
southern part of the Cheney-
Palouse Scabland tract   149

7.11. Oblique aerial photograph of
the upstream end of a
quadrilateral loess residual on
the Palouse-Snake divide   149

7.12. Geomorphic map of the central
portion of the Cheney-Palouse
Scabland tract   150

7.13. Residual loess hill streamlined
by flood erosion   151

7.14. Vertical aerial photograph and
topographic map of a small
pendant bar in the Cheney-
Palouse Scabland   151

7.15. Oblique aerial photograph of
Bar 2 near Wilson Creek   152

7.16. Schematic development of a
hypothetical pendant bar   152

7.17. Oblique aerial photographs of
scabland cataracts and inner
channels   153

7.18. Oblique aerial photograph and
topographic map of Dry
Falls   154

7.19. Orbital image of the upper
Grand Coulee   154

7.20. Topographic map of the
Potholes Cataract   155

7.21. Representative cross sections of
scabland channels   155

7.22. Inner channel development in
Moses Coulee   156

7.23. Topographic map of lower
Moses Coulee   156

7.24. Butte-and-basin
topography   156

7.25. Longitudinal grooves developed
on the scabland surface near
Palm Lake   157

7.26. Hypothetical sequence of flood
erosion for a typical scabland
divide crossing   158

7.27. Typical sets of giant current
ripples in the Channeled
Scabland   159

7.28. Logarithmic relationship of
height as a function of chord for
forty sets of giant current
ripples   159

7.29. Mean ripple chord as a function
of stream power   160

8.1. Hydraulic data for various
scabland channels   162

8.2. Oblique aerial photograph of
the head of Lenore
Canyon   163

8.3. Erosion by scabland flooding at
the crest of High Hill
anticline   163

8.4. Joint-controlled rock basins and
channels on the Palouse-Snake
divide crossing   163

8.5. Cross section of an idealized
Yakima basalt flow   164

8.6. Boulder of basalt
entablature   165

8.7. Characteristics of a
macroturbulent kolk   165

8.8. Formation of a horseshoe-
vortex system at the front of a
vertical cylinder and scour hole
development near a
boulder   166

8.9. Longitudinal groove
development on a basalt
scabland surface   167

8.10. Morphometry of a streamlined
form   168

8.11. Somewhat generalized
illustration of drag coefficient
variation with Reynolds
number   168

8.12. Length of streamlined forms
versus planimetric form
area   169

8.13. Maximum width versus
planimetric form area for
streamlined forms   169

8.14. Maximum width versus length
for streamlined forms   171

8.15. Streamlined shape factor k
versus maximum flow
Reynolds number for peak
discharges   171

8.16. Ice drive on the Red Deer River,
Alberta   172

8.17. Critical conditions for
cavitation in flood flows on
Mars and Earth   173

9.1. Estimated variations in Martian
orbital eccentricity, orbital
inclination, and obliquity   181

Preface and Acknowledgments

Had this book been written only fifteen years earlier, its title would likely have relegated it to the realm of speculative philosophy, perhaps even to fictional accounts. That the channels of Mars are a matter of serious and intense scientific interest follows directly from the National Aeronautics and Space Administration program of planetary exploration. Prior to acquisition in 1965 of the first spacecraft image of Mars, only telescopic views were possible. The best terrestrial telescopes never revealed the surface of Mars any more clearly than that of the Moon as viewed through lower-power binoculars. The scientific study of the Martian surface was formerly less than objective, as astronomers openly discussed controversial interpretations made during periods of extraordinary "seeing."

After more than four years of acquiring orbital images of the planet Mars, the last of two orbiting Viking spacecraft was shut down on August 7, 1980. The total number of pictures produced by the two Viking orbiters is nearly 60,000. These pictures and those of the Mariner missions of 1965–1972 have made the landscapes of Mars as real to earth scientists as those of the Colorado Plateau, Hawaii, Iceland, and the Columbia Plateau. Prior to these missions scientific questions concerning the Martian landscape required considerable imagination. Given our assumptions concerning the Martian environment, what did we hypothesize concerning the Martian landscape? The present situation leads to new questions. Given the profound images that we now have of the Martian landscape, what do we hypothesize about the planetary environment that produced

that landscape? The new problems are as exciting as the old. Moreover, a lively imagination is still required to explain the terrain that has been imaged so well.

The channels and valleys of Mars serve as a central theme for a general exposition of Martian geomorphology, or perhaps more properly "aresmorphology." The scientific study of the Earth's landforms, geomorphology, did not begin in earnest until 200 years ago. The catalyst for the new science was James Hutton's realization that terrestrial valleys and the rivers flowing in them were not the products of supernatural causes. Rather, these fundamental elements of the landscape derive from processes of continental erosion and deposition that have characterized that geological history of our planet. Similarly the study of Martian valleys and channels has recently advanced to a phase of reasoned scientific inquiry. For a second time in the history of science, the origin of valleys and channels on a planetary surface is the critical question to be answered in the attempt to understand the processes that have operated on that surface. This book does not presume to fully answer the puzzle posed by the Martian channels. It does claim that the attempt to solve that puzzle will teach us much about Mars. Indeed, the puzzle is so interwoven with our thinking about Earth that it will provide new insights about our own planet as well.

Although this book was a personal project, it would not have been possible without the support of related Mars research by the National Aeronautics and Space Administration. My studies of Martian landforms were completed under the following grants and contracts: (1) NASA Grant NSG-7326, "Origin of Channels on Mars," administered by the Planetary Geology Program, Office of Space Science, (2) NASA Contract NAS 1-15164, "Studies of the Martian Channels from Viking Orbital Imagery," a part of the Viking Guest Investigator Program, (3) NASA Grant NSG-7557, "Morphogenetic Studies of Martian Channels," a part of the Mars Data Analysis Program. My field work in the Channeled Scabland

was supported by the National Science Foundation (Grant GA-21478), and work on the Hawaiian volcanoes was supported by NASA through the U.S. Geological Survey Hawaiian Volcano Observatory (Planetary Guest Investigator Program). Portions of this book were completed while I was a Fulbright-Hays Senior Research Scholar at the North Australia Research Unit, Australian National University. The Australian research was supported by the Australian-American Educational Foundation. Most of the Mars photographs in this book were supplied by the National Space Science Data Center.

Many individuals aided in the completion of this project. F. Earl Ingerson and Daniel J. Milton encouraged, facilitated, and supported my early efforts to study Martian geomorphology. Stephen E. Dwornik and Joseph M. Boyce provided the support at NASA Headquarters to fund my continuing work. My former students, now colleagues in research, R. Craig Kochel and Peter C. Patton, were sources of new ideas and cooperative projects. Many of the geomorphic maps in this book are largely the work of Dr. Kochel. I would also like to express my appreciation to Professors Arthur L. Bloom, James W. Head III, and Peter C. Thomas for their detailed review of earlier drafts of this book. Their comments have proved most helpful. More than anyone, however, my wife, Pauline M. Baker, has been the key individual in facilitating the completion of this book. She endured some of Earth's worst discomforts to help me learn of another planet's mysteries. Her dedication to quality has been my inspiration.

The Channels of Mars

Chapter 1
Telescopes, Canals, and Space Probes

"Planet" was the word used by the ancient Greeks for one of the wandering stars, those spots of light that moved against the background of the heavens. Of these, one displayed a peculiar orange-red color. The Romans named it Mars after their god of war. At closest approach Mars' elliptical orbit brings it to about 55×10^6 km from Earth. Because the Earth has a near-circular orbit, the Mars orbit may yield much larger separations. As a result, viewing conditions for Mars vary with time, adding difficulty to seeing its intriguing surface. Table 1.1 lists some important properties of Mars.

EARLY OBSERVATIONS

Telescopic observations of Mars probably began in 1610 when the planet was observed by Galileo. The first informative map of Mars was made by the Dutch physicist Christian Huygens in 1659. However, the most extensive early studies of Mars were made by the English astronomer Sir William Herschel. By 1784 Herschel had demonstrated an uncanny Earth-like nature for Mars. He showed that the Martian day lasts about 24½ hours and that Mars has a tilted axis of rotation. More exciting was his discovery that the Martian seasons correlated with an annual growth and recession of the planet's polar caps. By the early 1800's Mars was generally conceived to be a planet with oceans, polar snows, arid land masses, and clouds. This Earth-like landscape was further extended to implications concerning habitability. Herschel and many contemporary scientists speculated that Martians enjoyed an environment somewhat comparable to that on Earth.

The moons of Mars were not discovered until 1877, when Asaph Hall of the U.S. Naval Observatory sighted a star-like object circling the planet. The inner satellite, Phobos, is 27 km in diameter and only 6,050 km above the planet's surface, which it circles in 7 hours and 39 minutes. Viking pictures show that it is covered with a high density of impact craters (Figure 1.1). The smaller satellite, Demos, has an orbital radius of 23,500 km. It is about half the size of Phobos and has a surface littered with huge boulders.

Jonathan Swift confounded many readers by predicting in 1726 in *Gulliver's Travels* that Mars had two moons. Moreover, Swift even approximately predicted the orbital sizes and periods for the two satellites. Although such knowledge appears uncanny, it is almost certain that Swift obtained it from the astronomer Johannes Kepler. Kepler believed in a geometric progression of planetary satellites: Venus, zero; Earth, one; Mars, two; and Jupiter, four. Although wrong about Jupiter, this idea plus simple calculations according to Kepler's third law of planetary motion easily yields Swift's "prediction." Nevertheless, this affair has become another part of Mars lore. Carl Sagan (1973a: 104–105) concludes: "There is

TABLE 1.1. Mars Data

Mean diameter (km)	6,788
Mean distance from sun (km × 10⁶)	227.8
Distance at perihelion (km × 10⁶)	206.5
Distance at aphelion (km × 10⁶)	249.1
Sidereal period (length of year)	
(Earth days)	687
(Mars days = sols)	668.6
Axial rotation period (solar day)	24 hr: 37 min: 22 sec
Axial inclination (degrees)	25
Surface gravitation (% Earth)	39
Sunlight (% Earth-received light)	36–52
Mass (kg)	6.4×10^{23}
Mean density (g/cm³)	3.933

an entire genre of writing on how it was that Swift knew about the moons of Mars, including the suggestion that he was a Martian. Internal evidence suggests that Swift was no Martian, and the two moons can almost certainly be traced back to Kepler's speculation."

Giovanni Schiaparelli, the director of an observatory in Milan, made some of the most important telescopic observations of Mars. He was able to observe the planet during very favorable viewing conditions in 1877 and 1879, when Mars swung very close to Earth. Schiaparelli's mapping of the planet (Figure 1.2) introduced a terminology that became widely accepted by other astronomers. His map showed bright areas, which he believed to be land, and dark areas, which he designated as oceans ("mare"). He gave the Martian landscape a marvelous series of names, derived mainly from classical mythology.

During the late nineteenth and early twentieth centuries Mars was a focus of scientific scrutiny. Among the observations that captured both scientific and public imagination were the following:

White Clouds. White spots, distinct from the polar caps, were observed to occur repeatedly over some regions of Mars. Today we know that the favored localities are either high volcanoes, like Olympus Mons, or basins, like Hellas. One cloud, which forms repeatedly in the Tharsis region, assumes the shape of the letter W. This cloud forms because of the topographic barriers provided by the Tharsis shield volcanoes (to be discussed in Chapter 2). Most of the white clouds are probably water ice, but some very high altitude clouds may be condensed carbon dioxide (Snyder, 1979).

Yellow Clouds. These clouds were observed to form preferentially during perihelion, the time of closest approach of Mars to the sun. Clouds forming at this time enlarge from selected sites in the southern hemisphere. At times they can obscure the entire planet in a yellow shroud. The early suggestion that the yellow clouds were great dust storms has been confirmed by recent planetary orbiters.

Figure 1.1. Viking image of Phobos showing the gaping crater Stickney, which is 10 km in diameter. The impact responsible for Stickney may have shattered the satellite to produce the prominent pitted grooves. The crater is named for Angelina Stickney, wife of Asaph Hall, the discoverer of the moons of Mars. Hall's search for the tiny moons nearly ended in frustration. However, at the urging of his wife, Hall spent several more nights at the telescope and finally sighted the elusive objects.

Variable Dark Regions. Some of the bright and dark markings on the surface of Mars vary in size with Martian seasons and years. Some early observers even believed that the markings could be seen to change in color from grey to green with the advent of local spring. Many astronomers also described an apparently dark zone that advanced equatorward from the Martian poles as the polar ice caps retreated. This phenomenon, the "seasonal wave of darkening," is definitely not the result of growing vegetation responding to the onset of Martian spring. Modern explanations for the surface brightness variations involve eolian transport of very fine dust (aerosols) and mixtures with volatile ices of water and carbon dioxide (Wells, 1979).

Linear Markings. Father Pietro Angelo Secchi, who first noted the white clouds, introduced the Italian word *canale* ("channel") in 1869 to describe apparent lines on the planet's surface. Between 1877 and 1888 Schiaparelli mapped a profusion of *canali* (Figure 1.2), but he was initially guarded as to their origin. His map implies that the *canali* were water-filled channels dividing the land masses of Mars and connecting the various oceans. Although Schiaparelli recognized that the *canali* were very difficult to see, his prolific writings led to intense interest in them. For the next sixty years arguments raged between telescopic observers who saw the *canali* and observers who adamantly denied their existence.

Although the *canali* came to dominate public imagination concerning Mars, the large-scale, variable dark markings held considerable scientific interest. Even to the 1950's, the seasonal variations in the markings and their reputed greenish color were interpreted as evidence of vegetation on Mars. Such concepts led to the belief that, other than Earth, Mars was the most likely habitat for life in the solar system.

Amidst the various arguments over the role of vegetation on Mars one series of papers was prophetic. Dean B. McLaughlin, an astronomer at the University of Michigan, published eight papers between 1954 and 1956 in which he ascribed the variable dark markings on Mars to a combination of volcanic and eolian processes. McLaughlin's proposed wind circulation pattern is now known to be in error, and his identification of dark surface materials as volcanic ash is questionable. However, the general thesis of important volcanism and wind action on Mars is amazingly appropriate in view of the findings of Mariner 9 and other discoveries of the last decade (Veverka and Sagan, 1974). Unfortunately many of McLaughlin's contemporaries continued to favor plant life as an explanation for the dark markings (Kuiper, 1956).

Now that Mars has been observed from orbit and from the ground, we know that the light and dark markings have no fundamental significance for the planet. Seasonal variations in mi-

nor atmospheric constituents and surficial reflectivity are responsible for changes in the fuzzy telescopic images of Mars. Against the phenomenal stage of Martian geology (Chapter 2) this is but a modest drama.

The limitations of the telescope as a scientific probe of Mars, nevertheless, did not deter public fascination with the planet. Perhaps the most popular of the fictional accounts of Mars were the numerous novels by Edgar Rice Burroughs. Burroughs described a planet with retreating oceans, decayed civilizations, and monstrous green barbarians. Although more famous for his African novels featuring "Tarzan of the Apes," Burroughs was equally imaginative with his character John Carter, "Warlord of Mars." Around John Carter, Burroughs wove an entire lore and culture for "Barsoom" (the name given their planet by the Martians).

The War of the Worlds by H. G. Wells describes fictional Martians threatened by the deterioration of their planet, which is slowly losing its water and atmosphere. The Martians invade Earth in a last attempt to save their civilization. Eventually they are defeated, not by the ingenuity of the Earthlings, but by an unexpected enemy—lowly terrestrial bacteria to which the Martians have no immunity.

THE CANALS OF MARS

Schiaparelli's accounts of Martian *canali* led to the most famous of the controversies about Mars. In 1886 telescopic observers all over the world began to report their observations: Mars was covered with networks of fine streaks. The sightings were all considered to be at the limits of telescopic resolution and viewed under the optimum lighting conditions. Not all astronomers agreed, however. E. E. Barnard of Lick Observatory made detailed observations of many features not seen by Schiaparelli, but he could not see any straight line patterns.

Perhaps the debate over linear streaks on Mars would have remained an assessment of viewing conditions were it not for the flood of attention accorded the problem by Percival Lowell. Lowell, a wealthy Boston businessman, was so fascinated by Schiaparelli's accounts that he established a special astronomical observatory for planetary studies at Flagstaff, Arizona. Lowell mapped hundreds of linear markings. For him the translation of *canali* was literal. He considered the Martian canals to be the work of intelligent creatures living on a dying, arid planet. The canals were their means of transporting water from the polar caps to the equatorial deserts of Mars.

Lowell's semipopular accounts of Martian geography were not universally acclaimed. The English scientist E. W. Maunder performed some simple psychological experiments that demonstrated the eye's tendency to produce imaginary linear connections for random arrangements of discontinuous blotches and streaks. The experiments can be easily replicated by drawing a random arrangement of marks on a large sheet of paper. Dots, circles, ovals, straight lines, wavy lines, and irregular smudges should be thus displayed to a group of young children in a classroom. Children sitting at different distances from the display will produce quite different sketches of what they see. Those far from the pattern will often draw an arrangement of regular straight lines. Maunder actually performed this experiment in 1913 with 200 English

schoolboys. Yet confidence in visual observation remained strong until telescopic views of Mars were supplanted by high-resolution space probe images.

Even those scientists who accepted the reality of linear markings on Mars were not unanimous in their explanations of those markings. The proposals for their origin included giant crustal fracture systems (Fielder, 1963), upward water seepage and plant growth along fractures (Jamison, 1965), linear sand dunes (Gifford, 1964), igneous dikes (Fairbridge, 1972), and rift zones (Sagan and Pollack, 1966). However, the Mariner and Viking pictures of Mars have failed to confirm any hypothesis. Mutch and others (1976) concluded that regular, linear elements of the appropriate size are not observable on orbital images of the planet. Maunder's "English schoolboy" experiment provided the most likely explanation for what is perhaps the most famous astronomical telescope observation.

This conclusion was anticipated by the French observer E. M. Antoniadi in 1930. Antoniadi had the advantage of the largest refracting telescope in Europe, and he drew his maps during periods of the very best observing conditions. What he found were irregular, patchy arrays of spots and splotches that gave an illusion of lin-

earity when viewed under slightly poorer conditions (Figure 1.3).

It is possible that, more by chance than design, some of the great channels that really do exist on Mars were mapped as *canali*. Without adequate resolution of the feature seen, Schiaparelli and others probably mapped Ares Vallis as "Indus" (Hartmann, 1974a). In general, however, there is little correspondence between the Mars maps made by telescopic observers and the subsequent detailed imagery of the planet obtained by the Mariner and Viking spacecraft.

The first controversy over Martian channels (*canali*) therefore centered on the reality of the features. As noted by Sagan (1973a), there was never any doubt that the Canals of Mars were a product of intelligence. The only question was on which side of the telescope the intelligence was located. This book is concerned with the second controversy over Martian channels, initiated nearly a century after Schiaparelli's work. When Mariner 9 images showed fluvial-like channels on Mars, the questions that emerged no longer concerned the existence of the depicted features. The questions now clearly focus on the fascinating scientific problem of genesis. How were these features produced? What does their nature and distribution imply about the com-

Figure 1.2. Schiaparelli's map of Mars, based on his observations in 1877–1888. Reproduced from NASA Spec. Publ. 337. The map shows the nomenclature introduced by Schiaparelli and the system of linear "canali" which he recognized. Note that the astronomical convention has north at the bottom of the map. Comparison to the modern geologic map (Figure 2.2) will show little correspondence.

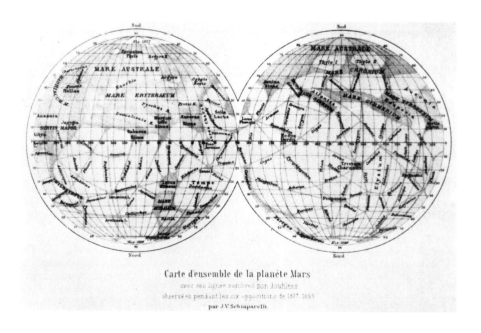

Carte d'ensemble de la planète Mars
avec ses lignes sombres non doublées
observées pendant les six oppositions de 1877–1888
par J.V. Schiaparelli

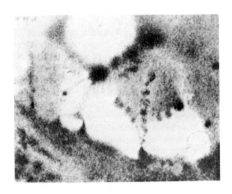

0   1000   2000

Figure 1.3. Comparison of E. M. Anto-niadi's mapping of Mars markings (top) with a "canal" interpretation drawn by earlier observers under poorer viewing conditions when dark spots and pattern boundaries are seen as linear streaks (bottom). The region mapped is south of Elysium (bright oval, top). From Hartmann and Raper (1974).

position and history of the planet? These are no longer questions of observational astronomy. They are problems in modern geology and comparative planetology.

THE MARINER MISSIONS

The Mariner 4 spacecraft ended the era when Mars studies were limited to Earth-based observations. Eight months after its launch on November 28, 1964, Mariner 4 returned twenty-two pictures of Mars as it sped past the planet. The best images (Figure 1.4) were fuzzy depictions of lunar-like craters. In retrospect, we realize that the Mariner 4 fly-by probably could not have picked a more bland region of the planet to survey.

Prior to Mariner 4 most scientists felt Mars would resemble the Earth more closely than the Moon. The pre-Mariner consensus was that Mars would have an atmosphere of nitrogen and carbon dioxide with small amounts of argon, water vapor, and oxygen. Estimates of the atmospheric surface pressure were in the range 20–125 mb. Mariner 4 showed that Mars was far more harsh an environment. The new estimate of atmospheric surface pressure was 5–10 mb, and the atmosphere appeared to be entirely carbon dioxide.

In 1969 two spacecraft with more powerful cameras added more data to the Martian file. Mariner 6 returned thirty-three images and Mariner 7 returned ninety-one. The new images mainly showed cratered terrain, and Mars retained its reputation for blandness. Mutch and others (1976) summarized the prevailing scientific wisdom immediately after the Mariner 7 mission, as follows: (1) Mars has few if any Earth-like landforms, and therefore it never experienced any of the internal processes that affect Earth. (2) The dominant terrain type on Mars is a primordial cratered surface that was formed in approximately the same way and at the same time as the lunar highland surface. (3) Mars

Figure 1.4. Picture returned by the first imaging mission to Mars, Mariner 4. This crater occurs in the Atlantis region. Note that image quality is far less advanced than achieved by Mariner 9 and Viking.

never had an Earth-like atmosphere. (4) Water never played an important role on the surface of Mars. Because of conclusions (3) and (4) life was considered to be nearly as unlikely on Mars as on the Moon. Because of conclusion (2) some scientists even wondered if future Mars exploration would produce any new knowledge that was not being generated in more exciting fashion by the then-current manned exploration of the Moon.

How fortunate that Mars exploration did not end at this point! On November 13, 1971, Mariner 9 went into orbit around Mars. The spacecraft carried an advanced camera system, similar to that used in commercial television. Signals received in an image tube, or vidicon, were converted to digital form and returned to Earth. Each image produced by this process consists of 582,400 picture elements (pixels) each of which records 1 of 512 levels of brightness. Two cameras were used: the wide-angle (A-frame) camera with an approximate 1-km ground resolution from periapsis altitude and the narrow-angle (B-frame) camera with an approximate 100-m ground resolution from periapsis altitude. Periapsis is the point of closest approach to a planet's surface by a satellite's elliptical orbit. The wide-angle A camera had a 50-mm focal length lens and the narrow-angle B camera had a 500-mm focal length lens. Since the resolution of objects on images is directly proportional to the focal length of the lens, this explains the two different resolutions. Ground resolution is simply a measure of a distance on the planet's surface corresponding to the smallest distance resolvable on the image.

The Mariner 9 cameras provided virtually complete coverage of Mars with A-frame images (1–3 km ground resolution) and 1 percent coverage with B-frame images (100 m resolution). On October 27, 1972, Mariner 9 had to be shut down because of depletion of the supply of gas used to keep the spacecraft positioned with its radio transmitter directed toward Earth. In nearly 700 orbits the spacecraft had returned more than 7,000 pictures of the Martian surface.

The features imaged by Mariner 9 all required new names. All place names on Mars are now established by the International Astronomical Union. Table 1.2 lists some of the commonly used terms for features on Mars.

Mariner 9 arrived on Mars when the planet was engulfed in a great dust storm. To the dismay of the mission geologists, the early pictures failed to reveal any of the planet's secrets. This dismay was later allayed in one of the most dramatic revelations of planetary exploration. Against the background of atmospheric dust four black spots were consistently visible. Narrow-angle camera views of each spot showed a crater-like depression in its center. It was as though the great dust storm was one last attempt by the old war-god Mars to fool the inquisitive of Earth. Successions of previous scientists had been led to believe in Earth-like conditions, canals, plant life, and finally a lunar clone. The first Mariner fly-bys imaged only the least interesting features. Now a vehicle capable of seeing all had finally arrived, but Mars cloaked itself in its cover of dust. What Mars could not conceal was the summits of its highest mountains, each one capped with the tell-tale caldera of a shield volcano. As the dust storm slowly settled, the Mariner 9 cameras revealed the secrets that Mars had hidden for so long. Mars was a world with both Earth-like and Moon-like features, but all were integrated into an order that was uniquely Martian. No one among practitioners of pre–Mariner 9 science or fantasy had quite conceived of the actual Martian landscape.

Although the Martian volcanoes had been anticipated by Dean McLaughlin, the Martian channels had not been seriously expected by any scientists. Sinuous valleys such as Mangala Vallis (Figures 1.5 and 1.6) created an immediate impression of formation by running water. This posed a paradox because water in liquid form is unstable at the present surface temperatures and pressures of Mars. Either the channels had formed by some nonaqueous process or Mars had to have had a very different atmospheric environment in its past. Some Mariner 9 scientists favored the former view because the mission failed to reveal other clear indicators of aqueous processes besides the channels. Despite the anomalous channels, Mars displayed an abundance of lunar-like tectonism. A common post–Mariner 9 view was that Mars was in a developmental stage analogous to that of Earth perhaps 3 billion years ago as it was slowly accumulating an atmosphere. Somehow Mars

TABLE 1.2. International Astronomical Union Names for Features on Mars

Name	Feature	Probable Origin
Catena	Crater chain	Volcano-tectonic
Chasma	Large canyon	Structural
Dorsum	Elongate prominence	Lobate fault scarp, or anticline
Fossa	Linear depression	Structural graben
Labyrinthus	Set of intersecting linear depressions	Structural
Mensa	Flat-topped mesa	Erosional remnant
Mons	Large isolated mountain	Volcano
Patera	Irregular or complex crater with scalloped edges	Ancient volcano
Planitia	Plain that is low relative to surrounding terrain	Impact basin covered with volcanic flows
Planum	Relatively smooth, less heavily cratered plateau	Elevated volcanic plains
Tholus	Isolated domical hill	Steep-sided volcano
Vallis	Sinuous valley	Fluvial and/or sapping
Vastitas	Extensive plain	Mantled terrain

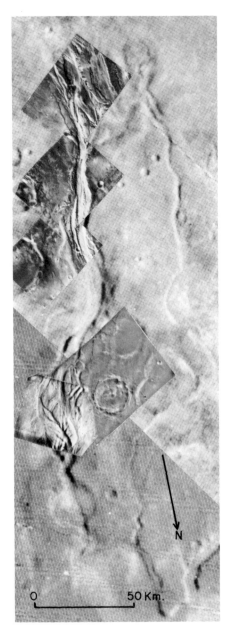

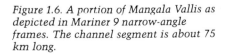

Figure 1.5. Composite of Mariner 9 wide- and narrow-angle frames showing the downstream portions of Mangala Vallis. The three top narrow-angle frames are shown in Figure 1.6. The "braided reach" (left center) is shown in Figure 3.11.

Figure 1.6. A portion of Mangala Vallis as depicted in Mariner 9 narrow-angle frames. The channel segment is about 75 km long.

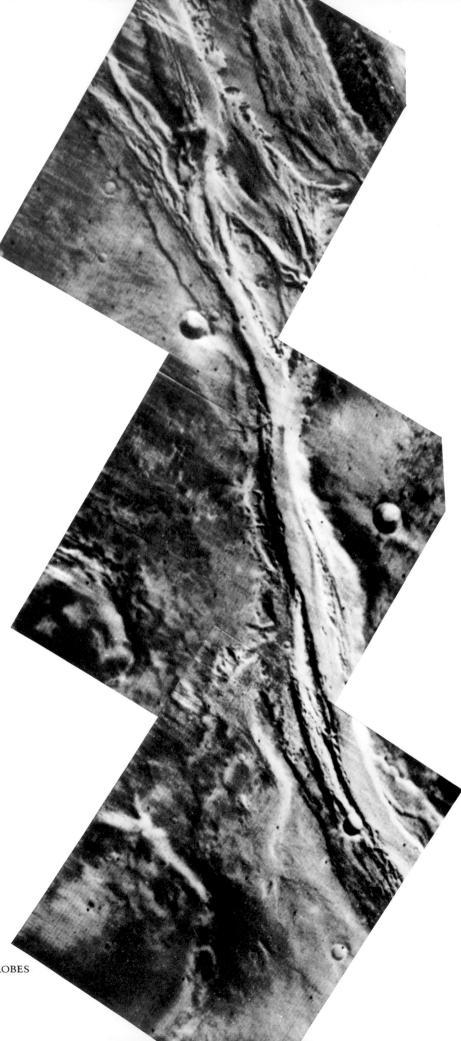

was trapped in this incomplete evolutionary state. Perhaps its small size never allowed it to retain a large atmosphere or to accumulate sufficient volatile substances to generate them. But, while this theory might explain craters, volcanoes, and the present thin atmosphere, how could it explain fluvial-like channels? The second Mars-channel controversy was launched.

On December 2, 1971, the Soviet Union soft-landed a capsule "Mars-3" on Mars, but the mission achieved little other than an engineering success. The lander returned only a twenty-second signal. In March 1974 two Soviet spacecraft orbited Mars, returning about 70 pictures of the planet, and two Soviet landers attempted soft landings. Neither landing was successful, but one lander returned some data that indicated a possible atmospheric argon content of 20 percent or more. Actually this estimate depended on a very tenuous analysis of a mode of instrument failure. The estimate implied that Mars may have had a primordial atmosphere much more dense than at present. Now the Viking mission has provided firm data on the argon content of the Martian atmosphere. Mars does have considerable argon, but 1–2 percent, not 20 percent. The anomalous Soviet result preserved some optimism among Viking scientists that proof might be found for the controversial atmospheric implications of the channel systems discovered by Mariner 9. Chapter 9 of this book outlines current thoughts about the history of Mars' atmosphere.

THE VIKING MISSION

Viking is the latest, most complex, and most successful scientific mission launched to Mars. The mission involved four spacecraft: two landers and two orbiters. The primary goal was to land and make observations on the surface of Mars. The orbiters were mainly conceived as platforms for remote sensing instruments that would aid site certification and acquire other data independent of the landers. In retrospect, considering the supplementary role accorded the orbiters, it is ironic that some of the most impor-

tant mission results, including most of the data to be discussed in this book, were derived from them.

The Viking orbiters each had science scan platforms that could be aimed at the surface of Mars. These platforms were motor-driven mountings for instruments that could be aimed at the Martian surface. Each scan platform had a visual imaging subsystem (VIS), an infrared thermal mapper (IRTM), and a Mars atmospheric water detector (MAWD). The VIS consisted of two television cameras and their associated electronics. Each camera recorded overlapping fields of view measuring 1.5° × 1.7°. The maximum picture resolution occurred at periapsis, when the satellite made its closest pass to the planet's surface. For both Viking orbiters periapsis was approximately 1,500 km during the early phases of the mission. By October 1978, the main lander science was complete, and periapsis was lowered to only 300 km. Thus, the late phases of the Viking mission generated some orbital pictures of remarkably high resolution.

As the orbiters passed across the planet their cameras recorded two-frame-wide strips. The cameras operated alternately, each repeating its cycle every 4.5 seconds. The result was continuous coverage of slightly overlapping frames along the ground tract. At the 1,500-km periapsis altitude this produced continuous coverage 80 km wide and about 1,000 km long. Each image in this swath resolved a picture element (pixel) of 38 m on individual frames measuring 35 × 40 km. The resolution was thus slightly better than that of the Mariner 9 B-frame cameras, and the long Viking mission also encountered excellent atmospheric clarity. Moreover, the Viking cameras were able to preserve high-frequency information that was lost by the Mariner 9 system. Nearly 60,000 Viking orbital pictures are now available.

The IRTM was an infrared radiometer that measured thermal emission from the Martian surface. The MAWD was an infrared spectrometer that measured the proportion of incident solar radiation passing through the Martian atmosphere at wavelengths that tend to be absorbed by water.

Thus, it could estimate the water content of the atmosphere. Both these instruments yielded spectacular results. As the IRTM scanned the summertime north polar cap of Mars in 1976, temperatures were found too warm for frozen carbon dioxide. Here was solid proof for water ice comprising the residual summer cap. The result was confirmed by the MAWD, which showed that the highest atmospheric water abundance during the northern summer occurs over the perennial north polar ice cap. Although the very thin Martian atmosphere can hold little water, it is often completely saturated, proving that substantial reservoirs of frozen water must occur on the planet.

Viking 1 was inserted into orbit on June 19, 1976, and its cameras were immediately trained on the planet to certify a landing site. The landing was originally scheduled for July 4, 1976, the bicentennial of the United States. The original proposed site was at the mouth of a large channel system that had been mapped from Mariner 9 data. Site certification was complicated by the large surface area within which the landing had to be expected after commands were issued to the spacecraft (an ellipse measuring approximately 100 × 200 km). Moreover, the Viking lander had only 22 cm of ground clearance. The VIS pictures showed that the proposed site was a scoured channel zone, rather than a depositional fan at a channel mouth. Irregular "scabland" topography indicated the possibility of eroded bedrock blocks of large size (NASA, 1976). The Viking 1 Orbiter was forced to search other areas to identify a smooth landing zone.

Eventually the site was selected in the central basin of Chryse Planitia at latitude 22.38°N, longitude 47.49°W (Figure 1.7). The Viking 1 lander touched down at 5:12:07 A.M. (local time, Jet Propulsion Laboratory, Pasadena, California), July 20, 1976. The Viking 2 landing was accomplished on September 3, 1976, at Utopia Planitia, latitude 47.97°N, longitude 225.71°W. Each lander was a fully automated station (Figure 1.8) capable of acquiring data in the following sciences: meteorology, seismology, dual-camera (stereo) imagery, physical and

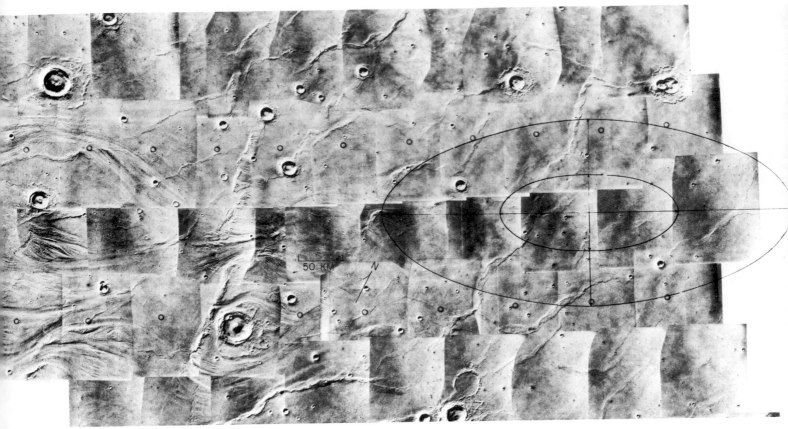

Figure 1.7. The Viking 1 landing ellipse in the Chryse Planitia region of Mars. This mosaic of Viking orbiter frames shows pronounced channeling from Maja Vallis at left. (National Space Science Data Center [NSSDC] picture 17026.)

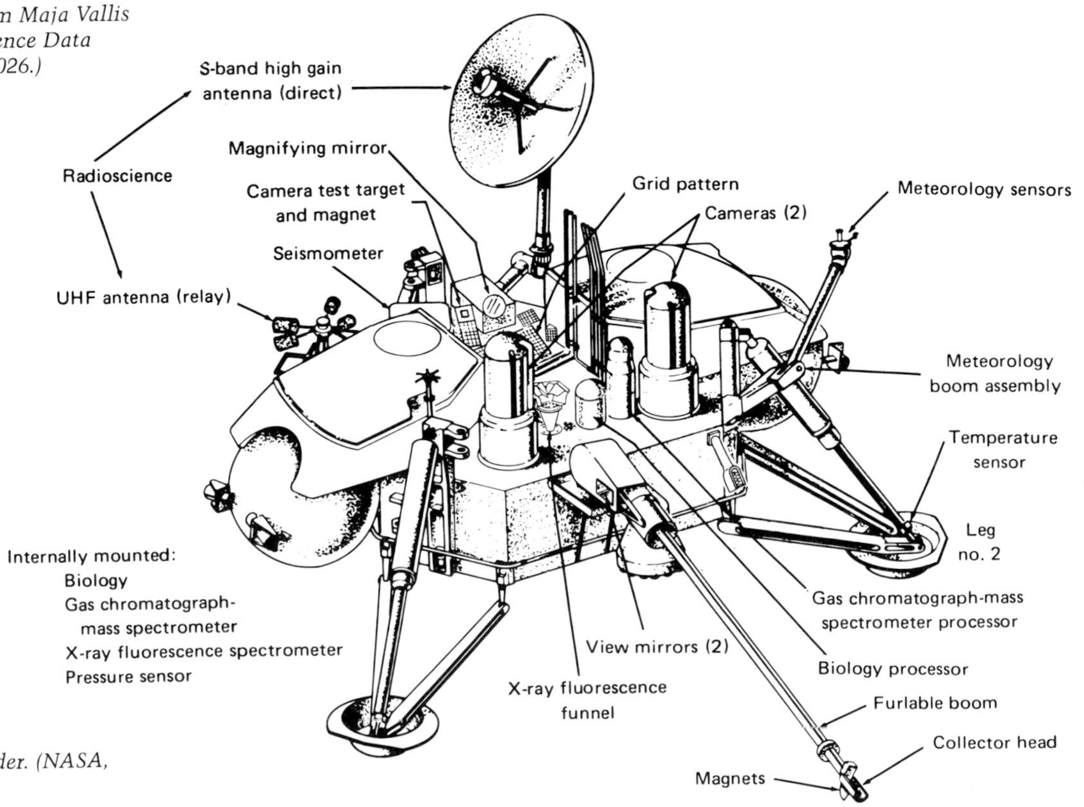

50 KM

N

S-band high gain antenna (direct)

Magnifying mirror

Camera test target and magnet

Seismometer

Radioscience

UHF antenna (relay)

Grid pattern

Cameras (2)

Meteorology sensors

Meteorology boom assembly

Temperature sensor

Leg no. 2

Gas chromatograph-mass spectrometer processor

Biology processor

Furlable boom

Collector head

Magnets

Internally mounted:
Biology
Gas chromatograph-
    mass spectrometer
X-ray fluorescence spectrometer
Pressure sensor

View mirrors (2)

X-ray fluorescence funnel

Figure 1.8. The Viking lander. (NASA, 1976.)

Figure 1.9. Viking lander picture of the Viking 1 landing site. The sediment accumulations are drifts of very fine-grained particles stabilized by the large rocks. The largest rock at the far left is about 2 m across and is located about 8 m from the camera. The meteorology boom of the lander is visible in the center foreground.

Figure 1.10. Viking lander picture of the Viking 2 landing site. The larger rocks are about 0.3–1.0 m in diameter. Several small troughs are also visible, as well as small drifts of very fine sediment.

magnetic properties of the soil, and direct soil-chemistry analyses.

The lander pictures show a block-littered surface at the Viking 1 site (Figure 1.9). Blocks as large as 2 m across were probably ejected from nearby craters; many of the rocks appear to have been faceted by wind erosion processes. Fine-grained material occurs as drift complexes, which do not appear to be active at present. The Viking 2 site presents a flat plain of fine-grained sediment overlain by rocks with an amazingly broad and uniform distribution (Figure 1.10). The rocks are uniformly pitted and fluted. Somewhat similar rock shapes are produced by eolian erosion in the extremely arid deserts of southwestern Egypt (McCauley and others, 1979). An unusual topographic feature at the Viking 2 site is a network of shallow, linear troughs. These may be a type of patterned ground, similar to features developing in terrestrial environments of intense freezing and thawing of ground ice (see Chapter 5). One enigmatic revelation of the Vik-

ing landers was an apparent absence of sand- and pebble-sized particles of solid rock. Between the large rocks prominent on the lander photographs practically all the fine soil consists of micron-sized particles. Some of this soil is apparently cemented to form larger particles of "duricrust" by the precipitation of minerals in the top few centimeters of the surface (Baird and others, 1977).

The Viking landers observed the wind, temperature, and pressure regimen on Mars. These show diurnal cycles, transient weather patterns, seasonal effects, and responses to global dust storms. Martian weather is much more controlled by topography, surface radiation, and dustiness than is Earth's, but the same basic dynamical processes occur on both planets (Ingersoll and others, 1979).

Prior to the Viking landings, eolian activity was given prominence as a major geological process presently active on Mars. There had even been concern for the safety of the landers in the high winds of a Martian dust

storm (Snyder, 1979). However, the only observable changes at the landing sites were slight variations in the brightness or color of portions of the surface, two small slumps near Lander 1, and the deposition of a thin frost layer near Lander 2 (Figure 1.11). The brightness changes are explained by the deposition or removal of a dust layer a few microns thick associated with the two observed dust storms (Jones and others, 1979).

Certainly the most eagerly awaited results from the Viking mission concerned the biology experiments. Three experiments constituted the much-heralded search for life (Figure 1.12). The pyrolytic release experiment (PR) tested for atmospheric cycling of carbon into organic compounds. The gas exchange experiment (GEX) tested for uptake and release of various gases that occur in organic decomposition. The labeled release experiment (LR) added radioactive nutrients to the soil to see if radioactive gas was released by decomposition. Unfortunately the results of all experiments are more puzzling than conclusive. Results from the PR and GEX experiments seem best explained nonbiologically (Klein, 1979). The GEX experiment shows that reactive oxidents occur in the Martian soil. No organic compounds at all have been detected in the Martian soil. This is surprising, considering that many nonbiologic processes introduce organic compounds. The surficial Martian soil must possess intensely oxidizing agents capable of destroying all organic material. However, the experiments do not exclude the possibility of life inside or just beneath the surface of porous rocks or at some depth in the ground.

The Viking labeled release experiment yielded data that satisfied the pre-mission criteria for metabolic activity. However, this is a conclusion possible only in isolation. The current consensus among Viking investigators is to ascribe the few Viking biological "signals" to nonbiological causation (Klein, 1979). The question of life on Mars is certainly not resolved by two very isolated sampling stations of limited analytical capability. However, obvious signs of life on Mars have not been recognized in any experiment, orbiter or lander. On the other hand, Viking has confirmed that ingredients for organic synthesis are present on Mars. The atmosphere has nitrogen. The channels tell of considerable water in a past epoch. Although life has not been proven on Mars, neither has it been disproven.

The differing views from the Viking landers and orbiters should not be surprising. Consider two selected landing points on Earth. If both points were selected from orbit for landing safely, they would probably be relatively low-altitude, featureless plains, as were selected on Mars. The views from these landing stations would therefore be unlikely to convey the richness of the terrestrial landscape. Rather than great canyons, majestic peaks, or winding rivers, the imaginary Earth-landers would probably reveal monotonous scenes such as the arid plainslands of central Australia or the wind-swept steppes of northern Asia. While the data from these sites would reveal much to an extraterrestrial intelligence about our planet, the data would nevertheless be limited. Orbital images, by contrast,

Figure 1.11. Patches of white frost as viewed at the Viking 2 landing site on September 25, 1977. The view is partly obscured by the lander's camera reference charts and its Earth-communication antenna in the foreground. The frost is concentrated in the shadow zones of rocks 10–75 cm in diameter.

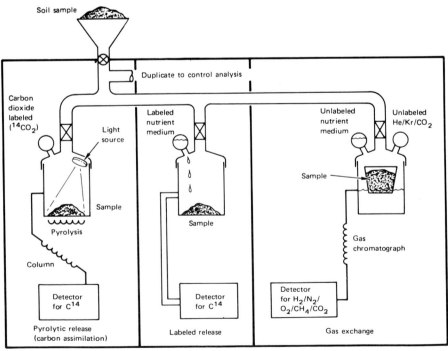

Figure 1.12. Simplified schematic diagram of the Viking biology experiments. (NASA, 1976.)

would reveal the great mosaic of Earth environments that had not been sampled.

Despite the phenomenal technical achievements of Viking, the mission's lasting value is in the scientific questions, both those answered and those raised. Mars always has been and will continue to be a treasure trove of secrets and mysteries. The channels of Mars are merely the latest mystery in a long succession.

THE MARS CHANNEL MYSTERY

Before the advent of modern space probes, the early telescopic observers of Mars struggled with the task of exploring a fascinating object with less-than-adequate tools. It is a wonder that so much was really discovered concerning such a difficult object to observe. As noted by Moore (1980), the best telescopes never showed Mars any more clearly than the Moon viewed through lower-power binoculars.

The most famous of the Mars observations of the pre–space probe period was that of the so-called "canals," which had first been termed "channels" (canali) by the Italian astronomers Secchi and Schiaparelli. Percival Lowell's accounts of intelligent design for the canals created a scientific furor over what might be considered the first Mars channel mystery. One of Lowell's disciples, C. F. Housden, even analyzed the pumping hydraulics required to transfer water from the melting Martian polar caps to the equator. Housden proved, to his complete satisfaction, that the pipes had to be about six feet in diameter.

After over fifteen years of detailed Mars survey by space probes, we can now easily dismiss many earlier hypotheses. However, we must also view these earlier ideas in the context of unsolved problems in our present understanding. An in-depth analysis of Lowell's work (Hoyt, 1976) reveals that he made many useful, though less publicized, contributions to astronomy. D. S. Evans (1976) writes, "Divested of their nutty overtones his observations of Mars were fundamental and comprehensive." This moderation of criticism derives from the realization that even the acquisition of images beyond the dreams of earlier astronomers has not yet solved many fundamental problems concerning Mars.

Lowell and his contemporaries were pioneers in the modern science of comparative planetology. It is true that Lowell's geology was naïve, his theories too superficial, and his writings too sensational. However, he drew attention to the study of the planets as a specialized field, one involving fascinating interdisciplinary work in geology, atmospheric science, geophysics, and even biology.

This book will emphasize the new Mars channel mystery. Whereas the sleuths of the canal mystery were frustrated by poor viewing conditions and inadequate resolution, the current Mars channel investigators are nearly embarrassed by a mountain of data. Both mysteries have water as a leading suspect for activity on a planet that otherwise appears to be remarkably dry.

Lowell envisioned Mars as an aging planet, afflicted with the fatal planetary disease of "desertism." The canals were the adjustment by the Martians to a fate which would eventually befall the younger Earth. We now know that Mars is indeed a desert planet, a very cold one. The evolution of Martian landforms and the Martian atmosphere can tell us much about Earth, its past and future. The Martian polar caps do contain water ice, but it is probably a last vestige, not a source, of the water responsible for the channels. Lowell's canals required an alien intelligence for their design, but the channels that do occur on Mars are natural landforms. Their regular shapes and patterns require our intelligence for their explanation. Lowell's canals supposedly irrigated the crops of a Martian civilization. The biological implications of the Martian channels are more obscure. If the channels did form during an ancient aqueous epoch, did that epoch allow life to evolve on Mars? Despite the fact that these questions have not yet been fully answered, one conclusion does emerge from the modern research. The process of trying to solve the Mars channel mystery, as in all good mysteries, leads to discovery, stimulation, and satisfaction.

Chapter 2
The Geomorphology of Mars

Nearly 200 years ago a Scottish physician named James Hutton read a paper before the Royal Society of Edinburgh. His thesis, subsequently elaborated in print (Hutton, 1788, 1795), was that the relatively slow-acting geological processes which we observe today are responsible for features that developed during the geological past. Hutton formulated this principle of uniformitarianism through his observations of terrestrial rivers and their valleys. Instead of valley formation by violent cataclysms, Hutton emphasized the prolonged action of soil erosion and river transport. By the mid-nineteenth century Hutton's uniformitarian principle had become the paradigm of geology, thanks mainly to the writings of Charles Lyell (1830–1833).

Geomorphology is the study of landforms and the processes that shape them. Given its tradition of uniformitarianism, this science has largely concerned itself with studies of slow-acting processes that could be related to landforms. One exception to this trend appeared during the 1920's in a series of studies of proglacial stream channels in the Columbia Plateau region of eastern Washington. J Harlen Bretz (1923, 1925, 1928a, 1928b) proposed that this region, which he named the Channeled Scabland, was the product of a cataclysmic flood. Considering the nature and vehemence of the opposition to Bretz' flood hypothesis, its eventual scientific verification constitutes one of the most fascinating episodes in the history of modern science (Baker, 1978a, 1981).

It is a long jump from the valleys of Scotland to the channels of Mars—a leap of at least 50 million kilometers

and 200 years of scientific tradition. As the Viking spacecrafts were orbiting Mars in the summer of 1976, their cameras revealed landforms that had also been described on Earth—in the Channeled Scabland of eastern Washington (see Chapter 7). Fifty years before, Bretz' hypothesis had been dismissed as either outrageous or merely applicable to a unique region. Now Viking scientists were using Bretz' scabland studies as the major Earth-analog to the largest of the spectacular Martian channels. This is but one example of the renewed interest in Earth-surface processes generated by the space program (Sharp, 1980).

The study of landforms on other planets stretches the science of geomorphology to its limits, both in method and in philosophical basis. Concepts as basic as uniformitarianism must be seriously questioned. The entire range of Earth-surface features must be considered as possible analogs to discoveries on the surfaces of other planets. The relatively recent discovery of alien landscapes poses many disturbing questions for a science grown complacent with the study of the familiar.

INTERPLANETARY COMPARISONS

Many geomorphologists would probably agree with Thornbury (1969), who referred to the last quarter of the nineteenth century as "the heroic age in American geomorphology." Perhaps the premier geomorphologist of that era was Grove Karl Gilbert. Gilbert's monographs on the Henry Mountains (Gilbert, 1877) and Lake Bonneville (Gilbert, 1890) remain standards by which to judge scientific excellence. One reason for the profound development of geomorphology between 1875 and 1900 was the intense scientific study of newly discovered terrains in the American West (Baker and Pyne, 1978). Most of the twentieth century has not offered many new landscapes to geomorphologists as stimulation for new ideas, unless one includes the sea floor in the realm of geomorphology. The exception to this trend has only recently emerged, and the newly discovered landscapes are on the surfaces of

other planets. The new frontier of geomorphology lies in the comparative study of planetary surfaces.

The comparison of planetary surfaces is mainly accomplished with orbital images or photographs. The interpreter of the landforms on those images relies on analogic reasoning to reconstruct the complex interaction of processes responsible for the observed features. Mutch (1979) has summarized the difficulties of this approach: (1) The method often assumes a unique correlation between the observed landforms and the responsible processes. Actually geomorphologists recognize that some landforms may be generated by different combinations of processes converging on the same result. This problem of "equifinality" is a continuing limitation on geomorphic analysis. (2) Photointerpreters are artificially constrained in their analyses by their range of familiarity with natural landscapes. For this reason the proposed analogs must be exhaustively pressed for their limitations as explanations for the phenomena under study.

Mutch (1979) observed that the origins of landforms on other planets are established not so much by the individual study of analogs as by a consensus among the active investigators. After the photographs and images have been studied for many years, one explanation remains that explains the majority of terrain features and is not incompatible with the remaining ones. The decade that has elapsed since Mariner 9 has allowed consensus explanations to emerge for Mars, and these will be discussed in this chapter. However, controversial issues remain. These will be identified wherever appropriate.

Mars has long been compared to other, better-known bodies in the solar system. The nineteenth-century astronomers believed the planet to be Earth-like. The early Mariner missions produced the impression of a Moon-like body. The mosaic of Viking images in Figure 2.1 illustrates the dilemma. At first glance one sees a lunar-like terrain of impact craters and one huge impact basin. This is all that might be revealed by a low resolution picture. However, close inspection of Figure 2.1 shows distinctly nonlunar features. In the foreground the craters are seen to be severely degraded in comparison to lunar craters. More intriguing are the numerous filament-like furrows that appear to dissect the intercrater areas. Near the center of the picture a large valley transects the rim of mountains around the impact basin. Finally, the background shows that this planet has an atmosphere. The parallel white streaks above the horizon are clouds, probably of frozen carbon dioxide. Mars clearly has experienced geomorphic processes that never operated on the Moon.

In their comparison of the geologic evolution of the terrestrial planets (Mercury, Venus, Earth, Moon, and Mars), Head, Wood, and Mutch (1977) reached the following conclusion: "In terms of areal coverage, volume, and time duration, impact cratering and volcanism are the two processes dominating the surface histories of the terrestrial planets." Only on Earth and Mars are atmospheric and hydrospheric processes known to be important agents of terrain modification. The Moon and Mercury have surfaces that had essentially evolved to their present appearance 2.5 billion years ago or earlier. Both these bodies experienced an early history of crater and impact basin formation that fractured and mixed their primordial crusts. The Moon and Mercury then experienced extensive volcanic activity, producing the dark mare plains of the Moon and extensive intercrater smooth plains on Mercury. Radiometric dates from returned Apollo rock samples suggest that the lunar outpourings occurred between 3.9 and 3.2 billion years ago (Taylor, 1975).

The Voyager mission in 1979 showed that Io, one of the satellites of Jupiter, probably has the most active surficial geology of any planetary body in our solar system. However, the Earth is more active than any planetary body closer to the sun than the Jovian system. In contrast to the Moon and Mercury, 98 percent of the terrestrial surface is less than 2.5 billion years old, and 90 percent is less than 600 million years old. This youthful surface has resulted from the continual destruction of old terrain and the creation of new terrain by plate tectonic motion. Most of this activity occurs in the Earth's ocean basins, which are floored largely by oceanic basalt. When continental volcanic rocks are considered, about 70 percent of the Earth's surface is volcanic (Head, Wood, and Mutch, 1977). This contrasts with the 17 percent of the lunar surface that is covered by mare basalt units (Head, 1976).

The Martian surface may represent an intermediate stage in an evolutionary sequence of terrestrial planet development. Approximately one-third of its surface consists of volcanic plains, and perhaps 50 percent of its surface is less than 2.5 billion years old (Head, Wood, and Mutch, 1977). Like the Earth, Mars has a planet-wide contrast in terrain elevations. For the Earth this contrast occurs along the continental-oceanic boundaries; for Mars the contrast occurs between the heavily cratered terrains of the southern hemisphere and the volcanic plains of the northern hemisphere (see following section). Both planets have significant polar deposits and both show features of probable fluvial origin. Before considering the important implications of this latter phenomenon, it is necessary to establish the background of Martian geology.

THE GEOLOGICAL MAP OF MARS

With the exception of the remote analyses by the two Viking landers, our knowledge of Martian geology derives from the study of images transmitted by the various orbiters. The images display landforms and landscapes. The landscapes may be constructive, such as those formed by lava flows, or they may be destructive, formed largely by erosion. On the Earth geologists can study geologic history through the depositional record of rocks in canyons, wells, or other exposures. Both constructive events (deposits) and destructive events (unconformities) are analyzed in the science of stratigraphy. On the Moon stratigraphic analysis can also be rigorously applied (Mutch, 1972). The Moon does not display active surficial activity, so the record of overlapping crater ejecta and lava flows can be discerned from remote photographic sensors.

Mars poses an interesting problem for the photogeologist. Some terrains, like those on the Moon, are very ancient and perhaps not appreciably modified by erosional processes. Other regions display pronounced modification by a variety of surficial processes. Thus, the planet consists of many exposed surfaces of ages extending through the whole history of the planet. In the study of Mars, geomorphology (the study of landforms)

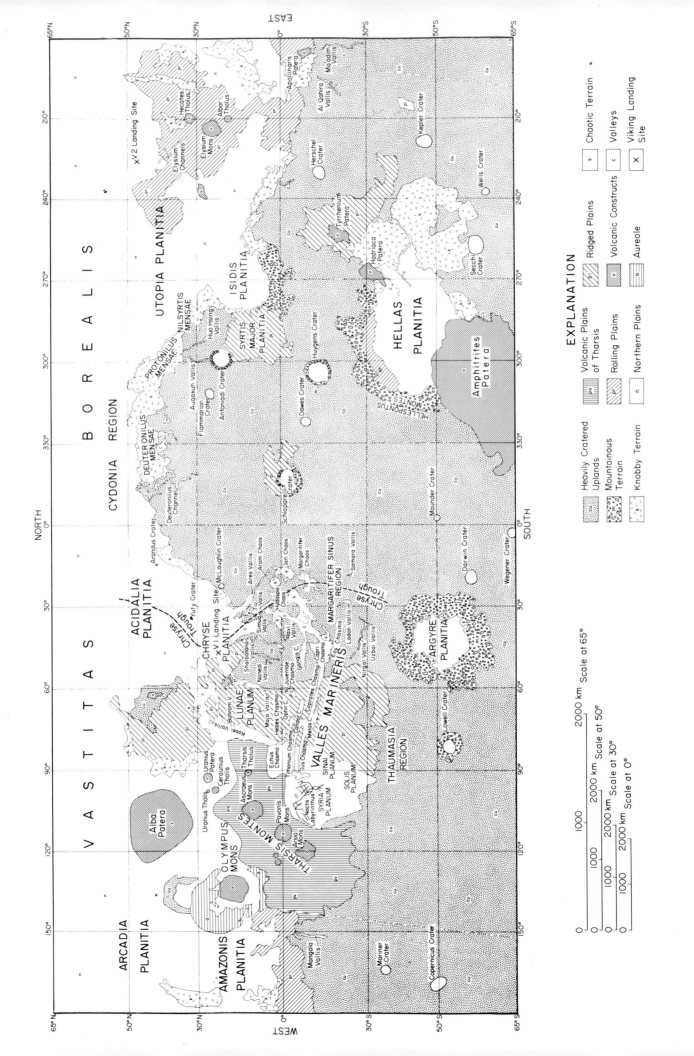

and stratigraphy are inextricably entangled (Milton, 1975).

Age relations of geologic units can be established either by relative or by absolute means. Relative ages can be discerned from remote photographs or images when certain relationships exist between adjacent surfaces. A younger constructive surface may have deposits that clearly blanket an adjacent, older terrain. Sometimes the younger deposits will occupy depressions in an irregular older terrain. Likewise, destructive surfaces may cut older terrains. Erosion surfaces formed by stream incision or scarp retreat are examples. Geologic structures, such as faults and folds, may also intersect older terrains, indicating age relationships.

On the Earth and the Moon, absolute ages have been established by radiometric determinations. Martian surfaces must be assigned absolute ages on the basis of the density of impact craters on those surfaces. The method assumes that the density of craters is proportional to surface age. Crater densities can be compared planet-wide and used as a tool for stratigraphic correlation. However, the validity of absolute ages derived from crater densities has severe limitations. The rate of impacting may not have been constant; surfaces may become saturated; and craters may be eroded or buried shortly after formation.

The surface of the Earth consists of two major terrains: the relatively high-standing continents and the more extensive ocean basins. Similarly, Mars presents two very different terrains of first-order significance. A highly cratered plateau of hilly terrain occupies equatorial regions and much of the southern hemisphere. This relatively high-standing region contrasts with more sparsely cratered plains that lie mostly in the northern hemisphere. These two ter-

TABLE 2.1. Geological Mapping Units for Mars

Heavily cratered uplands	The extensive hilly and cratered terrain of the southern highlands. In many areas the intercratered regions are relatively smooth and dissected by small dry valleys.
Mountainous terrain	Ejecta and uplifted blocks of ancient terrain, caused by large impacts.
Knobby terrain	Fretted and knobby zones, probably produced by scarp retreat that isolated erosional remnants of heavily cratered upland regions.
Volcanic plains of Tharsis	Relatively young lava flows that emanated from the Tharsis volcanoes.
Rolling plains	Intermediate-age lava flows that emanated from the Elysium volcanoes or other sources.
Northern plains	A complex of plains units showing extensive evidence of volcanism, permafrost processes, and eolian modification.
Ridged plains	Relatively old lava plains showing prominent wrinkle ridges and other features similar to the lunar maria.
Volcanic constructs	Mostly shield volcanoes varying in age from relatively young (Tharsis Shields) to intermediate (Elysium) to ancient (various patera).
Aureole	Elongate hills or ridges around the bases of certain shield volcanoes.
Chaotic terrain	Blocky, fractured terrain apparently developed by collapse of heavily cratered uplands.
Valleys	Highly modified terrains on the floor of the Valles Marineris canyon system or valleys resulting from channel development.

Figure 2.2. Geologic map of Mars between latitudes 65°N and 65°S showing location of major channels and other important surficial features of the planet. This map is slightly modified from that of Scott and Carr (1978).

rains constitute a great dichotomy of planetary scale. The boundary between the two terrains is a great girdle circling the planet and inclined approximately 35° to the equator (Mutch and others, 1976).

The above considerations are the basic principles for mapping the geology of Mars. Figure 2.2 shows the geologic map of Mars. The major mapping units are modified from the compilation by Scott and Carr (1978) that summarized the geologic mapping of the planet. These units are briefly described in Table 2.1. The heavily cratered uplands unit includes both hilly and cratered terrain and cratered plateau areas that exhibit plains development between impact craters. The various mapped plains units include volcanic plains of differing ages, from youngest to oldest: (1) the volcanic plains of Tharsis, (2) the rolling plains, and (3) the ridged

plains. The northern plains unit is of more complex origin, but includes numerous probable volcanic and permafrost features. The various mapping units will become more meaningful in the discussion that follows. More extensive treatments of Martian geology are provided by Mutch and others (1976); Gornitz (1979); Arvidson, Goettel, and Hohenberg (1980); Carr (1980); and Murray, Malin, and Greeley (1981).

CRATERS

When Galileo turned his telescope on the sky in 1609, he discovered the craters of the Moon. Craters on Mars were not discovered until pictures were returned by the Mariner 4 spacecraft in 1965. Although cratering may at first appear to be a monotonous quality of the inner planets in our solar system, closer study of craters can

yield a wealth of information concerning planetary histories.

The morphology of craters is studied for several purposes (Mutch and others, 1976): (1) different morphologies for craters of similar size and age may indicate different origins (endogenic or exogenic) or differences in the impacting body (comet or meteoroid); (2) different morphologies for craters of similar origin and age may show properties of the planetary crust, since larger crater diameters imply deeper penetration (or higher impacting energy); (3) morphologies for craters of similar size and origin may vary with crater age, thereby indicating various degradational processes; and (4) the material ejected from craters (ejecta) may overlap or degrade surrounding terrain. Overlapping ejecta blankets constitute one of the principal means of establishing stratigraphic relationships on the Moon.

Lunar craters display a well-studied progression of morphologic change with increasing size. Craters less than about 20 km in diameter generally display a simple bowl shape and an upraised circular rim. The crater walls are relatively smooth unless severely degraded. Craters between about 20 and 200 km in diameter display interior terraces on the crater walls, flat crater floors, and central peaks in the interior. The crater rims have a hummocky, subcircular shape. A patchy radial ray system extends out from the hummocky rim zone. This radial facies further changes outward to discontinuous clusters and chains of secondary craters (Shoemaker, 1962). A transitional range of sizes occurs to about 300 km. Craters larger than 300 km constitute basins and are frequently surrounded by multiple concentric mountain ranges with intervening troughs.

Martian craters show a progression somewhat similar to that of the lunar craters, but the size ranges are different. Simple bowl-shaped craters are replaced by complex forms at about 5–15 km diameter rather than about 20 km as on the moon (Mutch and others, 1976). This effect probably results from the greater surface gravity of Mars in comparison to the Moon. Large ejecta blankets and secondary crater fields occur on Mars (Figure 2.3), but they are far less common than on the Moon. Mars also has large impact basins, but these are much more degraded than those on the Moon. The multiple rings are generally deeply eroded. Plains deposits (probably volcanic flows) fill the basins and extend through the eroded rims. The largest Martian impact basin is Hellas, centered at 293°W, 42°S, and over 1,600 km in diameter.

Lunar crater degradation is thought to occur by (1) superposition of younger craters on older ones and (2) gradual wearing by the numerous impacts of very small ejecta particles. Martian craters appear to have experienced more extreme degradation over the planet's history. Ejecta deposits have probably been removed, and terraced walls are modified to hummocks and radial gullies (Mutch and others, 1976). Martian craters are probably degraded by lunar-style impact erosion, eolian modification, water-related processes, volcanic mantling, and periglacial processes.

The excellent Viking pictures revealed that many Martian craters have a unique morphology, different from that observed elsewhere in the solar system. The ejecta surrounding the crater is layered, and each layer has an outer edge terminating in a low ridge or escarpment (Figure 2.4). The surfaces of well-preserved ejecta blankets typically display radial striae, ramparts, and concentric features. Following McCauley (1973), who recognized this crater morphology on Mariner 9 frames, Carr and others (1977a) named these craters "rampart craters." Whereas Mariner 9 investigators attributed rampart crater morphologic features to eolian modification of ejecta, Carr and others (1977a) used the high-resolution Viking imagery to demonstrate a primary origin. An important discovery was that the ejecta flowed around

Figure 2.4. Mosaic of Viking orbital images of rampart craters in the Chryse Planitia (approximately 23°N, 43°W). The depicted scene is about 350 km wide. Note the lobate margins of the ejecta. Some ejecta layers terminate distally at low ridges or escarpments. Some craters have multiple overlapping ejecta lobes. (NSSDC picture 16911.)

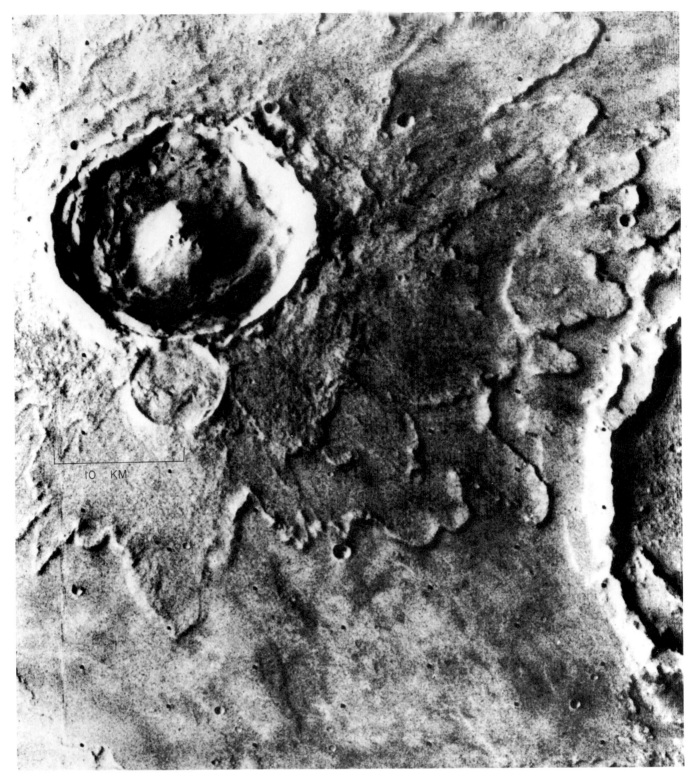

Figure 2.5. The crater Yuty, measuring 18
km in diameter and located at 22°N,
34°W. Note the pre-Yuty crater, just south
of the rim of Yuty itself. This crater is
buried by a thin ejecta blanket, one of
several complex ejecta lobes around
Yuty. (Viking frame 3A07.)

obstacles to the radial movement, such as older craters and hills.

The crater Yuty (Figure 2.5) shows a smaller crater just south of its rim. The ejecta thins over this crater, but it continued to move on the other side. Such evidence rules out the ballistic emplacement of ejecta, since the latter process would uniformly mantle the topography. Carr and others (1977a) noted that the deflection of ejecta layers by low obstacles suggests that, after ballistic ejection from the crater, the ejecta continued its motion outward from the crater as a debris flow.

Carr and others (1977a) attributed the peculiar form of Martian craters to either of two factors: (1) the entrainment of atmospheric gases in the ejecta, or (2) the incorporation into the ejecta of volatiles such as water that were formerly in the ground. Additional evidence for the extensive presence of ground ice in the Martian crust (Carr and Schaber, 1977) is compatible with factor 2. The water vapor or liquid probably acted as a lubricant to permit the observed flow features around the craters.

Rampart craters are widely distributed on Mars, at all latitudes and on nearly every major geologic unit on the planet (C. C. Allen, 1979a; Mouginis-Mark, 1979). The maximum development of ejecta fluidization occurs at low elevation and high latitude, as in the northern plains. Craters developed at higher elevations near the equator show indications of less mobile ejecta (Mouginis-Mark, 1979). These variations may reflect properties of the target material, perhaps the nature of ground ice in the permafrost.

The prolonged bombardment of a planetary surface by impacting bodies produces a layer of fragmented or brecciated material, known as regolith. The lunar regolith consists almost entirely of impact-generated fragments of rock and ejecta. This brecciated zone probably extends to depths of many kilometers. The whole zone is therefore known as "megaregolith" to distinguish it from the near-surface layers of regolith. The Martian regolith has been observed only at the two Viking landing sites (Figures 1.9 and 1.10). It consists of an-

gular blocks and fine-grained soil that derive both from primary impact and from subsequent weathering and transport processes. It is probable that the Martian megaregolith is also kilometers in thickness and that it is locally frozen to comprise permafrost.

CRATERING AND SURFACE AGES

The technique of crater counting is especially useful in age comparisons of various Martian surfaces (McGill, 1977). The standard technique is to select surface areas that appear by visual inspection to have a homogeneous crater population. The investigator may then count all craters larger than 1 km in diameter within this sample area. This count is then converted to a density of craters with diameters greater than 1 km in a standard surface area of 10^6 km². A problem with correlations made by this technique is that some crater sizes may be preferentially masked by various "resurfacing" processes. For ex-

ample, old smaller craters may be covered by material ejected by the formation of younger craters or they may be buried by younger lava flows. Wise, Golombek, and McGill (1979) attempted to overcome this difficulty by counting only the craters in what they considered to be the most reliable size range. They then extrapolated along a standard curve developed by Neukum and Wise (1976) to determine the 1-km crater density. This value is named the "crater density number" or simply the "crater number."

Crater numbers are useful in the relative dating of different surfaces on Mars. Their conversion to absolute ages, however, requires assumptions that remain a matter of controversy among various investigators. The most important assumption concerns the rate of crater production during Martian history. Unfortunately the only well-known crater production curve for the solar system is that derived for the Moon, where radiometric dates have essentially calibrated

TABLE 2.2. Ages for Some Geomorphic Surfaces on Mars

Crater Number[a]	Region	Geologic Mapping Unit	Approximate Age of Surface (Assuming Lunar Curve)
20–200	Olympus Mons	Volcanic construct (young)	<0.5 B.Y.
100–2,000	Shields and plains	Volcanic plains and constructs (Tharsis Montes)	<0.5–2.5 B.Y.
1,500–5,000	Alba Patera	Volcanic construct (intermediate age)	2.0–3.6 B.Y.
2,000	Viking 1 landing site (Maja Vallis)	Valley	2.5 B.Y.
3,500	Older parts of Chryse Planitia	Northern plains	3.4 B.Y.
5,000	Formation of Kasei Vallis	Valley	3.6 B.Y.
6,500	Tyrrhena Patera	Volcanic construct (old)	3.7 B.Y.
7,500	Hadriaca Patera	Volcanic construct (old)	3.7 B.Y.
9,000–20,000	Northern plains	Northern plains	3.8–3.9 B.Y.
20,000	Lunae Planum	Ridged plains	3.9 B.Y.
100,000	Cratered highlands	Heavily cratered uplands	4.0 B.Y.

[a]Neukum and Wise (1976); Wise, Golombek, and McGill (1979).

the curve. Whether and how Mars crater production rates differ from those on the Moon remain open questions. In recent years studies of asteroids and comets that pass into the inner solar system have yielded some limits to just how different production rates may be. The consensus seems to favor roughly similar crater production rates for all planets and satellites of the inner solar system, perhaps within a factor of 3 (Hartmann, 1977).

Another conclusion of comparative planetary cratering studies is that cratering rates for the inner planets have probably varied with time. Very high impact rates characterized the first billion years of planetary history, the so-called "accretion" phase of planetary development. This phase was then followed by 3.5 billion years of a lowered cratering rate that has remained essentially constant to the present.

In comparing Mars cratering to lunar cratering, Soderblom and others (1974) assumed that the cratering rate on Mars would be higher than that on the Moon by a factor of 1.5. This results in younger ages on Mars than on the Moon for a given cumulative crater frequency. Neukum and Wise (1976) also concluded that the impact fluxes for the Moon and Mars were very similar. However, they further assumed differences in the impact velocities of bodies hitting the two planets. The result is slightly older ages on Mars than on the Moon for a given cumulative crater frequency. Recently most investigators have agreed that curves for Martian crater density versus age tend to converge toward the lunar curve within a factor of ± 2 (Wise, Golombek, and McGill, 1979). Table 2.2 presents some approximate ages for various surfaces on Mars, assuming that the lunar impact flux applies directly to Mars.

Various studies of Martian cratering have also yielded an interesting indication of a period of erosion and deposition in Martian history (Chap-

man, 1974; Jones, 1974; Hartmann, 1973; Soderblom and others, 1974). The arguments for this are complex, but center on the absence of relatively small craters in the ancient Martian highlands. Small craters there are highly degraded and masked by deposition (Arvidson, 1974). The age of this erosive event is uncertain, but current thinking that the Martian and lunar flux rates were similar would place the event early in Martian history, perhaps over 3.5 billion years ago (Mutch and others, 1976). A likely explanation for the period of increased crater obliteration is the presence of liquid water and a very dense atmosphere early in the planet's history (Jones, 1974; Mutch and others, 1976).

VOLCANISM

Volcanoes were the first geologic features to be revealed by the 1971 Mariner 9 mission to Mars. This was certainly appropriate considering the revolution in geologic knowledge about Mars that was occasioned by the Mariner 9 mission. The four volcanoes that first appeared on the Mariner 9 images are the four largest and probably the youngest on the planet. They occur on an immense topographic rise called the "Tharsis bulge," and their elevations reach to about 27 km above the Mars reference level (Carr and others, 1977b). They are the highest topographic features on the planet.

Three of the four great volcanoes, Arsia Mons, Pavonis Mons, and Ascraeus Mons, form a long line running from northeast to southwest. The fourth and largest volcano, Olympus Mons, occurs 1,000 km to the northwest of this line. All four volca-

noes are morphologically similar to, though much larger than, terrestrial shield volcanoes (Figure 2.6). Such volcanoes are composed of numerous thin sheets of basalt that was erupted as very fluid, low-viscosity lava.

The Tharsis shields are among the youngest geologic features on Mars. Carr and others (1977b) found crater numbers on Arsia Mons to average about 200. For Olympus Mons the range was 20–50. These values indicate extreme youth on the various planetary surface dating scales. Considering the various potential errors involved, the volcanoes certainly were active within the last few hundred million years, and it is possible that Olympus Mons is still active today. Crater counts on the volcano flanks and margins give much higher densities, up to 2×10^3 craters >1 km diameter per 10^6 km^2. This range of values indicates that the volcanoes were active over an extremely long time span, certainly for hundreds of millions of years.

Olympus Mons

Olympus Mons (Figure 2.7) measures approximately 700 km across and stands about 25 km above the surrounding plains. It has a prominent summit caldera that measures approximately 80 km across (Figure 2.8). By analogy to terrestrial shield volcanoes, this caldera probably resulted from collapse along circular fractures after withdrawal of the supporting magma. The upper flanks of Olympus Mons consist of relatively thin, short lava flows that form concentric terraces around the summit. On the lower flanks the terraces disappear and the lava flows show prominent

Figure 2.6. Photograph of Mauna Loa shield volcano on the island of Hawaii. (Photograph by V. R. Baker, August 1979.)

Figure 2.7. Viking mosaic of Olympus Mons. (Viking mosaic 211-5360.)

Figure 2.8. Mosaic of high-resolution Viking orbiter images showing detailed features in the immense summit caldera of Olympus Mons. The collapse crater at the top center measures about 25 km across and is nearly 3 km deep. Much of the caldera floor at lower right is characterized by "wrinkle" ridges similar in form to ridges that form on basalt flows in the lunar mare regions. (Viking mosaic 211-5601; 18°N, 133°W.)

levees and chain craters. The flow structures are directly analogous to features on terrestrial shield volcanoes (Greeley, 1973).

An unusual feature of Olympus Mons is the great basal escarpment, up to 6 km high, that surrounds its margin. This scarp displays huge mass movement features that are discussed further in Chapter 5. Lava flows have poured over this scarp and formed prominent lava fans (Carr and others, 1977b). The origin of the scarp, whether erosional or structural, is unclear.

Beyond the scarp is a ring of plains formed by broad lava flows. Outside this ring, about 100–200 km beyond, is an aureole of grooved terrain, 200–700 km wide. The origin of this unusual topography also remains unresolved.

Why is Olympus Mons so large in comparison to terrestrial shield volcanoes? Carr (1975) noted that plate movements in the Earth's crust do not allow shield volcanoes to have a very long life span. The Hawaiian volcanoes remain active for only a few million years (Dalrymple, Silver, and Jackson, 1973). Chains of new volcanoes are produced as old ones are rafted away by the thin, mobile crust of the Earth. This apparently does not happen on Mars. The Martian crust seems to be much thicker and more rigid than that of the Earth. It is capable of supporting great upwarps such as the Tharsis bulge. On Earth such a bulge would be short-lived because of flow and movement in the thin crust. On Mars a volcano can remain over its magma source for an extremely long period of time and thereby build itself to magnificent proportions.

The Tharsis Volcanic Province

The Hawaiian shield volcanoes generate most of their lavas along linear rift zones that flank their central conduit. The linear arrangement of Arsia, Pavonis, and Ascraeus Montes clearly suggests a similar rift arrangement, but on an immense scale. Carr and others (1977b) demonstrated that various pit craters and secondary vents also line up on this northeast-trending alignment.

The Tharsis region is one of the most fascinating of the geologic provinces on Mars. It has experienced a long sequence of events in which the great shield volcanoes are but the most recent. Tharsis development began with bulging and the development of a vast radial fault system (Carr, 1974a). Wise, Golombek, and McGill (1979) used crater counting techniques to show that the bulk of this faulting took place during a time equivalent to crater number 10,000. The bulging and faulting was followed by an extremely long-lived phase of volcanic eruption.

The Tharsis lavas form extensive plains throughout the region. The volcanic nature of the plains is clearly shown by numerous low lobate escarpments that resemble flow features on the lunar maria. These lavas mostly derived from the four large volcanic centers, Arsia, Pavonis, Ascraeus, and Olympus Montes. Schaber, Horstman, and Dial (1978) used crater counts to show that the oldest flows have crater densities of 3,000 craters larger than 1 km per 10^6 km^2 and the youngest flows have 90 craters of this size range per 10^6 km^2. Depending on the various cratering rate assumptions, the eruptive history may have extended over 2×10^9 years (Schaber, Horstman, and Dial, 1978).

Other Martian Shield Volcanoes and Patera

The second most prominent volcanic province on Mars is the Elysium region. The three major volcanoes of this area are Albor Tholus, Elysium Mons, and Hecates Tholus. As in Tharsis, the volcanoes are perched on a broad bulge in the planetary surface. Elysium Mons is the largest shield, approximately 15 km high and 200–300 km across (Carr, 1975). This volcanic province is somewhat older than Tharsis. The crater number for Elysium Mons is approximately 1,000 and that of the surrounding volcanic plains is about 2,000 (Neukum and Wise, 1976). As with the Tharsis province, Elysium is characterized by extensive volcanic plains that were produced by prolonged volcanic flooding.

Malin (1977) suggested that Elysium Mons was a composite volcano, and thus significantly different from the Tharsis shield volcanoes. Composite volcanoes consist of intercalated lavas and pyroclastic materials

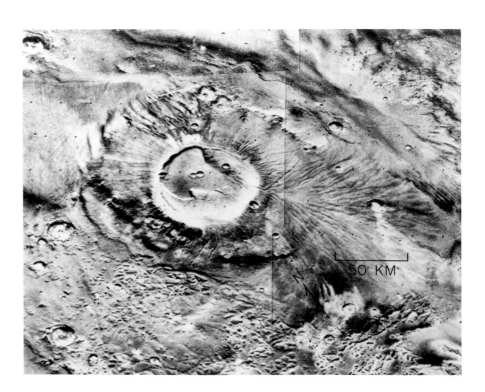

Figure 2.9. Appollinarsis Patera, located at 8°S, 186°W. (NSSDC picture 18143.)

which indicate a history of alternating effusive and explosive volcanism. Malin noted that the flanks of Elysium Mons are covered with blanketing deposits that are interpreted as volcanic ash. Studying especially Hecates Tholus, northernmost of the Elysium volcanoes, Reimers and Komar (1979) also presented evidence for considerable explosive volcanism. They cited the following morphological features: steep slope angles, caldera morphology similar to that of explosive volcanoes like Krakatoa in Indonesia, numerous radial channels, and blanketed volcanic flanks (presumably dune- or fan-like deposits of ash). Reimers and Komar attributed the explosive volcanism to interaction with ice-rich permafrost. Water seeping into the volcanic edifice periodically produced a phreatic eruption rich in pyroclastic debris.

Mars possesses a type of low, broad central-vent volcanic structure known as a "patera" (meaning "saucer"). This feature has no known equivalent on any other planet. The largest example is Alba Patera, which has a maximum diameter of 1,600 km but rises only about 6 km from the Martian surface (Carr, 1975). The volcanic origin is inferred from a central

collapse caldera complex from which lava flows radiate. Alba also possesses a concentric ring of fractures with a diameter of about 600 km. The Viking imagery of Alba reveals well-preserved lava flow features extending from the central caldera complex through the fracture ring and to a total distance of 1,000 km. Carr and others (1977b) showed that some lava flows were fed by lava tubes, others formed broad sheets, and still others comprise an anastomosing complex of channel-fed and tube-fed flows. The presence of lava tubes is believed to indicate that the lavas were basaltic in composition, or at least that they possessed the rheology of basaltic flows.

Alba Patera is interpreted as an older volcanic feature than the Tharsis volcanoes. Wise, Golombek, and McGill (1979) obtained crater numbers of 1,500–5,000 for the lava plains around this volcano. As in Tharsis, this range indicates a long eruptive time sequence, but much further back in Martian history. Perhaps Alba was once an enormous shield volcano that collapsed to its present low profile.

Even older patera features occur in the heavily cratered southern hemisphere of Mars. Appollinarsis Patera (Figure 2.9) is about 200 km wide, and it possesses a great rift zone from which flows emanate in similar fashion to flows on the rift zones of Mauna Loa in Hawaii (Figure 2.10). Tyrrhena Patera measures 500 km across. It is severely degraded and embayed by long valleys. Crater numbers for this volcano average 6,500 (Neukum and Wise, 1976) to 38,000 (Masursky, Dial, and Strobell, 1980).

Lava Plains

Like the Moon with its vast mare regions, Mars possesses extensive plains of probable volcanic origin without obvious source vents. Terrestrial lava plains are also common with lavas of basaltic composition. The Deccan lavas of India cover an area of 256,000 km². In the northwestern United States the Columbia River Basalt Group covers an area of 200,000 km² with an estimated volume of 200,000 km³ of basalt (Swanson and Wright, 1978). It is interesting that the latter

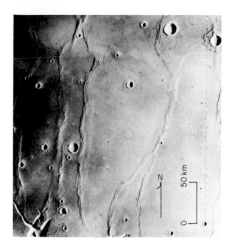

Figure 2.11. The Lunae Planum region of Mars, showing prominent "wrinkle ridges." The depicted area is at approximately 20°N, 66°W. (Viking frame 519A26.)

region also displays catastrophic flood channels that are similar to channels on Mars (see Chapter 7).

An extensive province of lava plains on Mars occurs in the Lunae Planum region (Figure 2.11). These plains are considerably older than the Tharsis and Elysium lavas. Neukum and Wise (1976) demonstrated crater numbers of about 20,000 for Lunae Planum, indicating an age of about 3.9 billion years on the lunar crater curve. This age also generally corresponds to the onset of lunar volcanism (Taylor, 1975). Moreover, the Lunae Planum also possesses long sinuous ridges, similar to the "wrinkle ridges" of the lunar maria.

The origin of lunar wrinkle ridges, also called "mare ridges," is controversial. The cited origins generally fall into two categories, volcanic and tectonic. Advocates of volcanic origin ascribe the features to igneous intrusions, pressure ridges in flows, or lava-lake processes. Most of the ridges show clear mare associations, often forming concentric rings around the circular mare basins. Some ridges grew between successive lava flows, perhaps as squeeze-ups of lava liberated in a settling lava lake (Hodges, 1973). Advocates of tectonic origin cite post-mare deformational processes. Howard and Muehlberger (1973) interpreted some wrinkle ridges

Figure 2.10. The southwest rift zone of Mauna Loa volcano on the island of Hawaii. (Photograph by V. R. Baker, August 1979.)

as thrust faults. The fact that some ridges extend from the mare ridges into the adjacent highlands seems to favor the tectonic origin.

In their survey of wrinkle ridges on Mars, Lucchitta and Klockenbrink (1979) recognized diverse origins. The evidence for tectonic origin includes (1) dominance of certain ridge trends, (2) preferential erosion of ridges and indications of upturned beds, and (3) indications of draping from subjacent topography. The evidence for volcanic origin includes (1) the restriction of the best ridge development to regions of deep lava flooding (although a few ridges locally transect nonvolcanic terrains) and (2) ridge development inside calderas. Lucchitta and Klockenbrink (1979) noted that both the lunar and the Martian ridges resemble the anticlinal ridges of the Columbia Plateau basalts near Yakima, Washington.

The most extensive lava plains region on Mars occurs north of the great planetary dichotomy. This area also shows some zones of wrinkle ridges, as in the Chryse Planitia. In other portions of the northern plains a volcanic origin is indicated by lava flow fronts or by small volcanic constructs. Viking images show that plains-forming lava flows locally embay the highlands-lowlands boundary of Mars (Scott, 1978). This indicates a complex history of the northern plains in which crustal lowering and highland scarp retreat preceded lava extrusion and embayment of the heavily cratered uplands.

Volcano-Ice Interactions

The Mariner 9 mission provided mostly low resolution images of the northern plains of Mars. Thus, when the Viking orbiters returned crisp, detailed views of this region, they illustrated many unforeseen topographic features. One of these is the occurrence of small craters on what appear to be raised platforms or pedestals. In other areas the pedestals and platforms occur without craters. When similar features were observed during the Mariner 9 mission a common hypothesis was that they resulted from erosive stripping, probably by wind, of the terrain around resistant impact craters. However, the Viking pictures

indicate that these features are primary structures little modified by erosion.

Hodges and Moore (1978) suggested that the platform and pedestal structures form in a manner similar to the table mountains of Iceland. The Icelandic table mountains are small shield volcanoes formed by eruptions beneath thick ice sheets (C. C. Allen, 1979b). The magma-meltwater interactions produce a friable rock known as "moberg." As the volcano builds through the ice, the eruption may spread tephra over the surrounding glacier and then cap the mountain with flat-lying basalt flows. The result is a mesa-like platform with a resistant caprock. Cones and craters may be superimposed on this platform by later eruptions.

Hodges and Moore (1978) hypothesized that Olympus Mons may have begun its development as a table mountain. Certainly their idea explains the enigmatic basal scarp that surrounds the mountain. Moreover, the aureole of grooved terrain may be supraglacial tephra deposits that were deformed by the underlying melting of ice. The mass movements along the basal scarp may have developed as the caprock of hard, basaltic lava flows was undermined by erosion of the underlying "moberg." In isolation this idea might seem extremely speculative. Further chapters, however, will show that Mars has evidence for the presence of abundant ice in its ancient lithosphere.

Many of the large Martian volcanoes may have exhibited phreatic (explosive) phases in their early eruptive history. This would occur as magma was erupted through water-saturated (or ice-rich) megaregolith materials (Greeley and Spudis, 1981). These early eruptions produced extensive ash deposits, but the style of volcanism changed as the megaregolith was depleted in water near the eruptive sites. Subsequent volcanism consisted of effusive lava production, constructing prominent shields and domes. The resulting stratigraphy of relatively hard lava flows overlying more easily eroded ash may explain the extensive erosion around the margins of some Martian volcanoes.

MODIFIED TERRAINS

Canyons

The Mariner 9 mission discovered the immense troughs of the Valles Marineris. McCauley and others (1972) referred to this region as a great equatorial canyon system. About 5,000 km long and 150–700 km wide, it stretches across the planet from approximately 20°W to 100°W longitude, between the equator and 15°S latitude. Some local segments of the canyon are as much as 9 km deep. Sharp's (1973a) study of Mariner 9 pictures revealed an important role for massive landslides and dry avalanches in creating the troughs. Canyon walls showed abundant evidence of hillslope processes that were active relatively recently in Martian history.

Viking imagery of the Valles Marineris has revealed fault scarps, downdropped blocks, landslide morphology, eolian dunes, and areas of peculiar bedded sediments (Blasius and others, 1977). The canyons show a progressive change in morphology from west to east. The westernmost region is an intersecting assemblage of grabenlike troughs known as Noctis Labyrinthus (Figure 2.12). This zone of isolated troughs shows evidence of faulting as the mechanism of downdropping the trough floors.

East of Noctis Labyrinthus are the long, narrow canyons Tithonium Chasma and Ius Chasma. The floor of Ius is choked with landslide debris (Figure 2.13). Ius also displays tributary canyons with elongate light and dark markings on the canyon floors, suggesting the down-valley flow of debris. The pronounced development of these tributaries on the south wall of Ius Chasma may result because the regional slope enhanced sapping processes along this wall (Sharp and Malin, 1975).

The central troughs of the Valles Marineris are the broad Melas and Coprates Chasmas (Figure 2.14). These canyons display some remarkably straight, low scarps along the bases of their walls. The scarps locally cut across spurs and tributary gullies, leaving hanging valleys. Some landslide lobes overlie these scarps, which are almost certainly the traces of normal faults. The scarp traces often par-

Figure 2.12. Viking mosaic of the western Valles Marineris showing Noctis Labyrinthus. Area is centered at approximately 10°S, 100°W. (Viking mosaic 211-5271.)

Figure 2.13. Viking imagery of Tithonium Chasma (northern canyon) and Ius Chasma. (Viking mosaic 211-5469.)

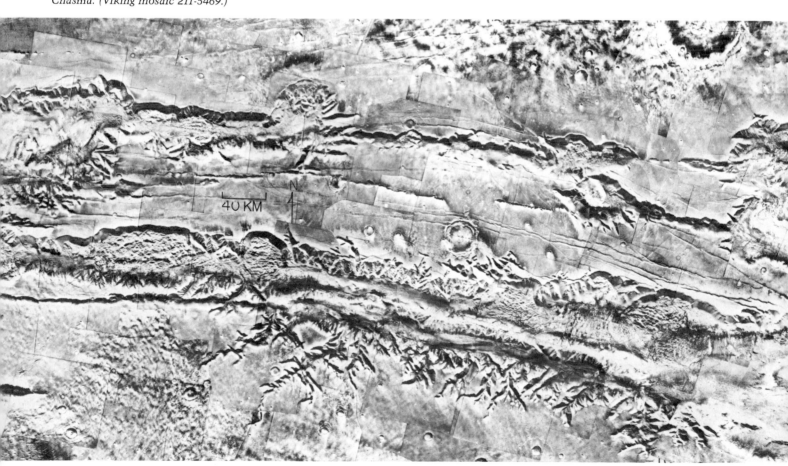

Figure 2.14. Viking mosaic of the central portion of the Valles Marineris, including Melas Chasma (left) and Coprates Chasma (right). Area is centered at approximately 12°S, 67°W. (Viking mosaic 211-5751.)

allel fracture faces on the cratered upland. The conclusion of Blasius and others (1977) was that the canyons formed largely by extensional tectonics. The western Valles Marineris probably are underlain by discrete elongate blocks of Martian crust that shifted vertically and tilted in relation to one another. Portions of the canyon floor have similar crater densities to the adjacent uplands, supporting the hypothesis of vertical movement along great faults.

The tectonic activity of the Valles Marineris acted over a considerable time span. The vertical faulting was in competition with erosion and deposition that tended to induce filling and broadening of the troughs. The canyon walls reveal that the upper one-third to one-fourth of the exposed rock (perhaps 1–2 km thick) is layered, probably a layered sequence of basalt flows. The material underlying this is more homogeneous. Its incorporation into huge lobate flows indicates that it can exhibit very low strength. Layered materials, including some very regularly bedded sediments, are widespread on the canyon floors. Blasius and others (1977) attributed these floor deposits to cyclic variations in sedimentation conditions, perhaps driven by some planetary cycle of climatic change.

The eastern Valles Marineris include Capri Chasma and present a morphology that is transitional to chaotic terrain (Figure 2.15). These areas have been highly modified by erosional processes and are discussed more fully in the next section. To the north of the Valles Marineris proper are two large troughs, Echus and Juventae Chasmas, which also contain considerable zones of chaotic terrain. Echus and Juventae are the respective "source" regions for the two large channel systems Kasei and Maja Valles (Figure 2.2).

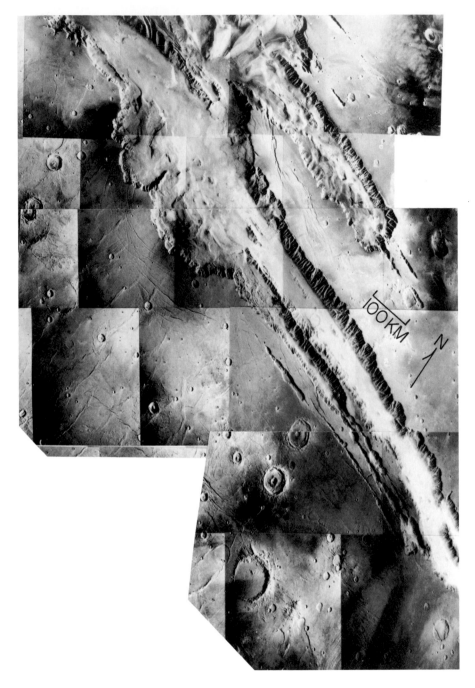

Figure 2.15. Viking mosaic of the eastern portion of the Valles Marineris, showing Capri Chasma and the complexes of chaotic terrain at the head of Simud Vallis. Area is centered at approximately 15°S, 55°W. (Viking mosaic 211-5750.)

Chaotic Terrain

Sharp (1973b) defined chaotic terrain as follows: "a jumbled chaos of slump and collapse blocks in lowland depressions bounded by steep walls with arcuate fractures." The individual blocks may be up to 10 km long, and the chaos zones may be several hundred kilometers wide. The bounding arcuate fractures appear to form as arc-shaped slump blocks moved from adjacent escarpments (Figure 2.16). Arc-shaped fractures may extend into adjoining upland terrains, perhaps indicating incipient stages of chaos development. Chaotic terrain was recognized as a unique Martian feature as early as the 1969 Mariner 6 mission (Sharp and others, 1971).

Many chaos zones display a progressive change in block shape with distance from the bounding scarps or fractures. Large equant blocks occur near the scarp, but smaller blocks of pyramidal shape occur further away, and finally isolated blocks may occur on a floor that is otherwise relatively smooth. The transition implies an erosional process and removal of the eroded debris.

Some chaos zones show abrupt bounding escarpments or irregular planimetric configuration. Sharp (1973b) estimated the relief on these escarpments as 1–3 km. The escarpments are locally bounded by crescent-shaped slump blocks bearing remnants of the upland surface and now rotated backward into the escarpment face.

Well-developed chaotic terrain on Mars is restricted to areas of the heavily cratered uplands around the southern margins of the Chryse Planitia. Zones of chaotic terrain occur near the heads of all the outflow channels of this region, including Kasei, Maja, Shalbatana, Simud, Ravi, Tiu, and Ares Valles. Chaotic terrain also occurs around the margins of major troughs, especially at the eastern margin of the Valles Marineris system. The chaotic terrain not connected either to channels or troughs, such as Margaritifer Chaos, lacks bounding escarpments and appears to be in an incipient stage of development.

Fretted Terrain

Sharp (1973b) defined Martian fretted terrain as a complex of smooth, flat, lowland areas separated by abrupt escarpments from relatively heavily cratered uplands. The planimetric pattern is strikingly irregular (Figure 2.17). Outliers of the heavily cratered uplands are often separated from the main escarpments, and sinuous flat-floored chasms often are developed for hundreds of kilometers back into the main zones of cratered uplands. These chasms were named "fretted channels" by Sharp and Malin (1975).

The Martian fretted terrain is best developed in a 500-km-wide band along the cratered upland/northern plains boundary from about 220°W to 30°W longitude (Mutch and others, 1976). Along this band, which extends halfway around the planet, the fretted terrain has clearly developed at the expense of the old cratered uplands (Sharp, 1973b).

The escarpments along the fretted terrain margins seem to have a remarkably uniform height of about 1–2 km. Sharp (1973b) suggested that this may result because the otherwise relatively homogeneous near-surface Martian material has a sharp, planar, physical discontinuity at a depth of 1–2 km. This discontinuity may have formed because of the development of ice-rich permafrost to that depth. The planet-wide evidence for such a layer has been summarized by Soderblom

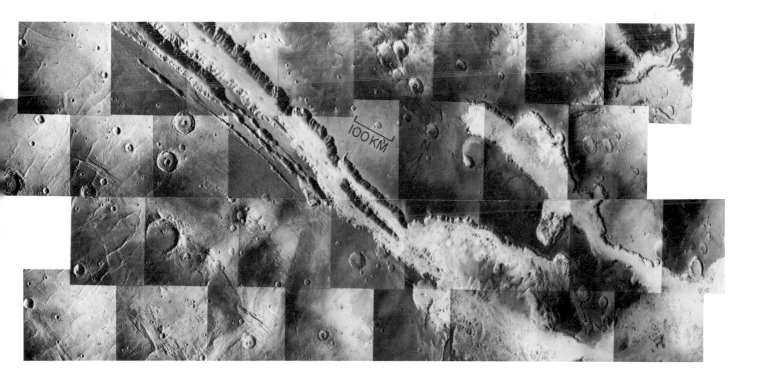

100 KM

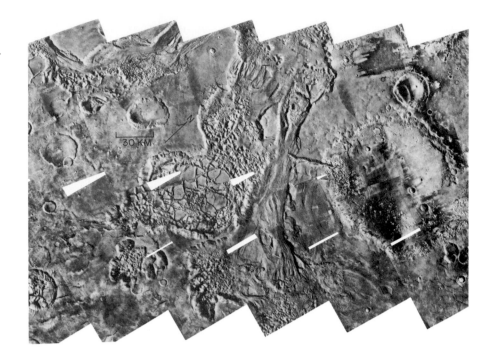

Figure 2.16. Chaotic terrain at Hydaspis Chaos (H). The large channel Tiu Vallis (T) emanates from this chaos zone. (Viking mosaic 211-5556.)

Figure 2.17. Fretted terrain in the Protonilus Mensae region of Mars (approximately 44°N, 313°W). (Viking mosaic 211-5582.)

and Wenner (1978). Chapter 5 presents a discussion of the Martian permafrost problem.

Some fretted terrain on Mars displays a linear network of uplands and troughs (Figure 2.18). These areas probably resulted from the localization of escarpment recession and trough formation along linear zones of structural weakness, such as faults and joints.

The lowland floors of fretted zones on Mars appeared smooth and featureless at the resolution of Mariner 9 imagery. The higher-resolution Viking imagery provided a surprise. Spectacular debris aprons were found to surround many of the residual upland areas. Flow features in the debris imply that ice may play a role in the mobilization of this debris (Carr and Schaber, 1977; Squyres, 1978).

POLAR TERRAINS

The annual expansion and shrinkage of white cappings at both Martian poles has been studied for at least 200 years by telescopic observers. Mariner 7 confirmed that relatively thin frost blankets of carbon dioxide advanced and retreated from the polar regions. During its first Martian winter (1977), the Viking 2 lander, located at 47.97°N latitude, 225.71°W longitude, observed that a condensate formed on the ground as temperatures fell to the condensation margin of carbon dioxide. Although this condensate was originally thought to be pure carbon dioxide, strong arguments have been made that much of the frost observed was transported water ice coatings on micron-sized dust particles. These were blown by wind from equatorial latitudes and settled in cold latitudes because atmospheric carbon dioxide condensed on them, increasing their weight (Jones and others, 1979).

The frost caps occupy as much as 30 percent of the hemispheric surface area of Mars in winter. These areas shrink to less than 1 percent in summer. Residual ice caps occupy the Martian poles during summer in the respective hemispheres, after recession of the frost blankets which surround them in other seasons. The northern cap must be water ice. This was confirmed by the infrared radiometers of the Viking orbiters (Kieffer and others, 1976). The observed temperature over the summertime north polar cap ranged from 200 K to 215 K (−73°C to −58°C). These temperatures are much too warm for a perennial ice cap of frozen carbon dioxide, since carbon dioxide begins to precipitate from the Martian atmosphere at 148 K (−125°C) (Miller and Smythe, 1970). However, the south polar residual cap was found to be much colder, with temperatures just under the

condensation point of carbon dioxide. The south polar cap may be a mixture of both water ice and dry ice that retains its cover of carbon dioxide frost all year round.

Results from the infrared spectrometers on the Viking orbiters correlate well with the thermal mapping data on the perennial north polar ice cap. These instruments essentially map atmospheric water vapor. They showed the greatest water abundance over the residual north polar cap, where the atmosphere was saturated with 100 precipitable microns of water (Farmer, Davies, and LaPorte, 1976). During the Martian year atmospheric water vapor was observed to redistribute itself away from the north pole. Peak values similar to those over the northern cap were not repeated, even in the vicinity of the southern summer polar cap. After a full year of observation, Farmer and Doms (1979)

Figure 2.18. Fretted terrain developed from a moderately cratered upland surface at approximately 30°N, 72°W. The upland surface is cut by intersecting linear troughs that probably represent differential erosion of a fracture pattern. Scattered mesa-like outliers occur in the lowland plains to the lower right of the center. (Viking frame 519A12.)

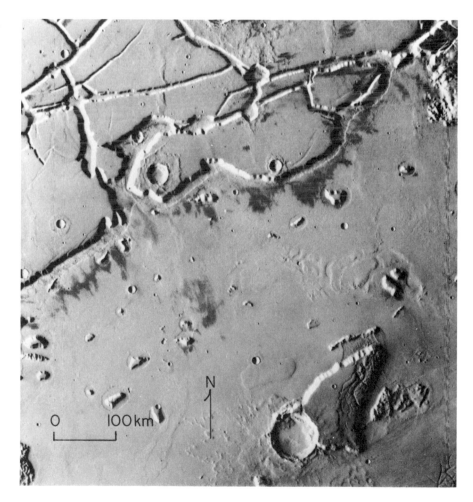

Figure 2.19. Canyons eroded in the Martian north polar cap. The layered deposits are exposed in the canyon walls. Dark areas that are frost-free and perennial occur at the top of the mosaics. At bottom left is a large field of dark-colored dunes in Borealis Chasma. The frames in this mosaic are about 100 km wide. (NSSDC picture, Viking 18234.)

concluded that Martian atmospheric water vapor is probably in equilibrium with the regolith at polar and mid-latitudes. They suggest that a permanent reservoir of water ice is probably buried beneath the Martian surface at depths of 0.1–1 m at all latitudes poleward of 40°.

The obvious form for a subsurface reservoir of water ice on Mars is permafrost in the regolith. Unfortunately the atmospheric data do not indicate the thickness of the Martian permafrost, merely that some of it occurs quite near the planet's surface at the polar and mid-latitudes. However, extensive geological evidence (Chapter 5) shows that the ancient Martian

permafrost was extremely thick, perhaps a few kilometers.

The morphology of the residual polar caps is well known from both Mariner 9 and Viking imagery. The northern cap, with a diameter of approximately 1,000 km, is the largest. This is because the present axis of rotation with respect to perihelion favors development of the northern cap. Both caps display unusual spiral patterns of dark bands (Figure 2.19). These intracap dark bands are really canyons that expose layered and terraced deposits. Erosion of the layered deposits along the canyon walls has produced a cliff-and-bench topography. The cliffs are 100–1,000 m high

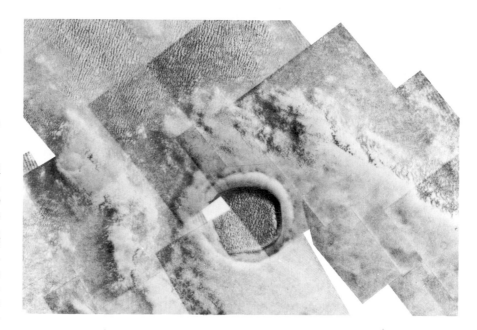

Figure 2.20. Large dunefields associated with a crater at 71°N, 51°W. The crater measures approximately 10 km in diameter, and it has acted as a trap for sand moving from left to right (west to east). Northwest of the crater (upper left) is an immense field of crescentic dune ridges which grades eastward into scattered barchan dunes. (Viking mosaic 211-5641.)

and they expose a remarkable layered sequence of deposits (Cutts, Blasius, and Roberts, 1979). The layers, up to 30 m thick, probably represent some sort of cyclic depositional process, perhaps with alternating periods of dust-rich and dust-free ice accumulation (Cutts, 1973). Descending masses of cold air at the poles are probably responsible for the dust transport. It is intriguing to speculate that the depositional alternation might correspond to cycles of planetary climatic change (Murray and others, 1972).

The great canyon-like cliffs of the north polar cap are of uncertain origin. A. D. Howard (1978) proposed an insolation-controlled ice ablation and accumulation process to explain the cliffs. His model suggests that any arbitrary random perturbation on a regional slope will develop by this process into nearly parallel curvilinear steps that face roughly downslope. Scarp retreat eventually straightens out any initial irregularities. The spiral structure of the north pole results from slope orientation; the southwest facing scarps achieve the highest daily temperatures and therefore the maximum ablation.

Cutts, Blasius, and Roberts (1979) offered evidence that the layered terrain is exposed in troughs, rather than in scarps facing away from the pole. They noted that the undulating upper surface of the residual cap has a wave-like relief that parallels the troughs. They proposed that the undulating surface is created by advancement and recession of the perennial ice cap during geologic time. The troughs evolved from the undulating plains by an insolation-controlled process of preferential accumulation of dust and ice.

The layered polar deposits overlie a more massive terrain that is generally smooth and undulating, but locally marked by extraordinary mesas and hollows. Sharp (1973c) referred to the latter topography as pitted and etched terrain. These regions appear to have undergone erosion by selective deflation to reveal a floor of ancient cratered terrain beneath a massive sedimentary blanket.

The northern polar cap of Mars is surrounded by an enormous field of sand dunes that are very dark in color (Figure 2.20). The dunes have advanced into some craters, in some cases completely filling them with sand. Other craters acted as obstacles, causing the dunes to back up on the windward crater rims and flow around their margins (Breed, Grolier, and McCauley, 1979; Tsoar, Greeley, and Peterfreund, 1979).

SUMMARY

The surface of Mars consists of heavily cratered uplands, mainly at equatorial and southern latitudes, and more sparsely cratered plains, mainly in the northern hemisphere. Many Martian craters are surrounded by ejecta lobes that may indicate impact into watery or icy target materials. Crater densities on various Martian landscapes indicate that many of the planet's surface features formed prior to 1–3 billion years ago.

Volcanism on Mars produced huge shield volcanoes, vast lava plains, and a variety of more complex phenomena. Some volcanoes may be composite, with their explosive activity facilitated by interaction with an ice-rich permafrost. Volcano-ice interactions explain a variety of unusual Martian landforms.

The immense trough system called Valles Marineris is a terrain of downfaulting and landsliding. At its eastern end it is transitional to chaotic terrain, consisting of slump and collapse blocks in lowland depressions surrounded by steep scarps. Large scarps also characterize the boundary between the heavily cratered uplands and the northern plains. Here a zone of outliers from the main upland comprises the fretted terrain.

The Martian polar caps contain both water ice and frozen carbon dioxide. Layered deposits beneath the seasonal frost layers probably represent cyclic deposition of dust-rich and dust-free ice, perhaps a function of long-term climatic change on Mars.

Thus, the general geomorphic features of Mars are dominated by the same processes as on Mercury and the Moon: impact cratering and volcanism. However, secondary features associated with water and ice occur only on Earth and Mars. The most spectacular aqueous features are the channels of Mars.

Chapter 3
Channel Types, Distribution, Ages, and Proposed Origins

The abundant variety of channels and channel-like features on Mars was revealed by the Mariner 9 pictures returned in 1971–1972. Early reports of the mission results recognized that running fluids were necessary to carve the channels and speculated that the most likely candidate fluid was water (McCauley and others, 1972). However, the abundance and the awesome size of the channels were not consistent with the prevailing view of a water-impoverished atmosphere (Leighton and Murray, 1966). McCauley and others (1972) noted that some of the largest channels originated from chaotic terrain regions, and they proposed that melting of extensive ground ice on Mars had resulted in both the chaotic terrain and the channels.

Figure 2.2 shows the location of various named channels that will be discussed in this book. Only the largest Martian channels have been plotted on Figure 2.2 to show their relationship to the planet's major geological features. Table 3.1 lists these channels as an aid to subsequent discussion.

CLASSIFICATION

The tremendous variety of channels and channel-like features on Mars has defied a rigorous classification. Indeed the name "channel" is probably not appropriate in many cases. As pointed out by Schumm (1974) and by Sharp and Malin (1975), "valley" or "trough" would be more accurate. Terrestrial channels are open conduits through which a fluid has passed. They are much smaller in scale than valleys and troughs, which often have been considerably enlarged by tectonic influences or hillslope processes. Of course, fluvial channels and river valleys are intimately associated on Earth, and it is tempting to translate that experience to Mars. Despite these difficulties, the term "channel" has been so widely utilized for various elongate narrow depressions on Mars that its abandonment would only produce further confusion. In this book the term will be generally applied to all Martian valleys and troughs which display at least some evidence that fluid flow processes contributed to their genesis.

The first suggested classification of Martian channels was posed by Masursky (1973) and subsequently applied more rigorously by Masursky and others (1977). The largest features are the "broad channels" that are associated with source areas in the chaotic terrain. These channels occur around the Chryse Planitia and extend from the heavily cratered uplands out onto the more sparsely cratered plains. Included in this category are Ares, Tiu, Simud, Shalbatana, Maja, and Kasei Valles.

The second category of Masursky (1973) was that of narrow, sinuous channels of intermediate size that were not associated with chaotic terrain. This category displays considerable morphological variation, including channels with and without tributaries, braided patterns, and well-developed meanders. Examples are Ma'adim, Mangala, and Nirgal Valles.

The third category is the "small and closely spaced channels" of the ancient cratered terrain. These channels form complex networks that run down the sides of large craters and through the intercrater plains. These small channels and some of the intermediate-scale channels were previously thought to have required rainfall for their formation because of their network patterns of tributaries (Masursky, 1973).

The final category in Masursky's classification is that of volcanic channels. These are features with attributes best explained by the channeling action of molten lava flows. They are restricted to volcanic terrains, especially the flanks of the large shield volcanoes. Some volcanic channels on Mars are morphologically similar to the lunar sinuous rilles, which are generally believed to be volcanic in origin.

Sharp and Malin (1975) attempted to provide a rigorous classification of Martian channels based on the available Mariner 9 imagery. Although a purely descriptive classification would be desirable, the great variety of channels would clearly make it unwieldy. Instead, Sharp and Malin presented a scheme with strong genetic implications. Channels and channel-like features were first categorized as the principal results of either external (exogenic) or internal (endogenic) processes. Endogenic channels might result from such processes as subsidence, volcanism, or faulting. Exogenic channels seemed to require some surficial process of origin. The exogenic channels were further divided into the following categories:

Outflow Channels. These generally correspond to Masursky's "broad channels." They are large features that start full-born from localized sources (Figure 3.1). Sharp and Malin (1975) concurred with the suggestion (Milton, 1973; Baker and Milton, 1974) that these channels display numerous features suggestive of catastrophic flooding.

Runoff Channels. These generally correspond to Masursky's "small and closely spaced channels." They display tributary branches in their headwaters, presumably as collectors for the fluids that excavated them (Figure 3.2). Varieties were classed as "integrated channels," "dendritic tributaries," and "slope gullies."

Fretted Channels. These features were not considered by Masursky (1973). They generally occur in the regions of fretted terrain, as defined by Sharp (1973b). The channels are steep-walled and have wide, smooth, concordant floors (Figure 3.3). Their planimetric configuration is often quite complex, displaying irregularly in-

Figure 3.1. Outflow channels northeast of the Valles Marineris system. The channels emanate from zones of chaotic terrain (C) and flow northward (arrows) toward the Chryse Planitia. Figure 3.7 is a sketch map of approximately the same region. (Viking mosaic 211-5821.)

300 KM

N

Figure 3.2. Small-scale "runoff" channels (valleys) in the heavily cratered terrain of Mars. Note the large craters that are superimposed on the high-density networks at the center of the picture. These networks must be quite ancient, perhaps 3.5+ billion years old. (Viking frame 097A25.)

TABLE 3.1. Large Channels on Mars

Channel Name	Approximate Location		Channel Type	Figure Number in This Book	Reference
	Latitude	Longitude			
Al Qahira*	18°S	197°W	Runoff	2.2 (location only)	Mutch and others (1976)
Ares*	7°N	20°W	Outflow	3.1, 3.7, 3.8, 6.1, 6.3–6.8	Sharp and Malin (1975)
Auqukuh*	30°N	300°W	Fretted	2.2 (location only)	Malin (1976)
Bahram*	21°N	57°W	Outflow?	4.18, 6.1	Malin (1976)
Elysium	37°N	220°W	Outflow	4.25–4.27	Malin (1976)
Hamarkhis* (Hellas)	37°S	270°W	Outflow	2.2 (location only)	Malin (1976)
Huo Hsing*	30°N	293°W	Fretted	2.2 (location only)	Malin (1976)
Kasei*	23°N	67°W	Outflow	6.1, 6.24–6.33	Baker and Milton (1974)
Ladon*	23°S	28°W	Outflow?	4.3, 4.11, 6.1	Malin (1976)
Ma'adim*	21°S	182°W	Runoff	4.19	Sharp and Malin (1975)
Maja*	17°N	55°W	Outflow	6.6, 6.13–6.20, 6.22	Baker and Kochel (1979a)
Mamers* (Deuteronilus)	36°N	343°W	Fretted	3.3, 5.14	Sharp and Malin (1975)
Mangala*	7°S	151°W	Outflow	1.5, 1.6, 3.11	Milton (1973)
Maumee*	19°N	55°W	Outflow	6.13, 6.17, 6.21–6.23	Baker and Kochel (1979a)
Nanedi*	6°N	49°W	Runoff	4.17, 6.1	Sharp and Malin (1975)
Nirgal*	28°S	40°W	Runoff	4.11–4.16, 6.1	Sharp and Malin (1975)
Parana*	24°S	10°W	Runoff	4.8–4.10	This report
Ravi*	0°	42°W	Outflow	3.12, 6.2	This report
Reull*	42°S	253°W	Outflow	4.22	Malin (1976)
Samara*	22°S	20°W	Runoff	4.23, 6.1	This report
Shalbatana*	5°N	44°W	Outflow	3.1, 3.7, 6.1	Sharp and Malin (1975)
Simud*	5°N	36°W	Outflow	2.15, 3.1, 3.7, 6.1	Malin (1976)
Tiu*	10°N	32°W	Outflow	2.16, 3.1, 3.7, 6.1, 6.7, 6.9, 6.12	Malin (1976)
Uzboi*	28°S	37°W	Outflow?	4.11, 4.12, 6.1	This report
Vedra*	19°N	55°W	Outflow	6.1, 6.13, 6.17	Baker and Kochel (1979a)

Note: Names followed by asterisks are official names of the International Astronomical Union. Names in parentheses were used unofficially in the cited references.

Figure 3.3. Upper reaches of Mamers Valles, a fretted valley system in the Deuteronilus region described by Sharp and Malin (1975). This frame is centered at approximately 32°N, 341°W, and it has a width of approximately 150 km. North of the imaged reach, the valley widens to about 35–40 km and eventually debouches into Ismenius Lacus at about 40°N, 339°W. The wide flat floors and abrupt steep walls with irregular scallops and alcoves are typical of fretted channels. The short deep tributaries and the integration of craters into the channel system suggest an origin by headward sapping processes (Sharp and Malin, 1975). See p. 39 for explanation of A1, A2, etc. (Viking frame 529A22; spacecraft range 6068.3 km; resolution 151.7 m per picture element.)

dented walls, fracture control, integrated craters, and isolated butte- and mesa-like outliers.

Excavated Channels. This category was provided to accommodate channel-like features that are closed at both ends. Presumably exogenic processes such as deflation or evaporation might produce these landforms, although endogenic mechanisms are probably more likely.

The general category of "endogenic channel-like features" was applied to forms resulting from nonerosive processes, including fracturing, tectonic movements, lava tube collapse, and subsidence resulting from the removal of underlying support. Sharp and Malin (1975) further suggested that composite channels, initiated by endogenic processes and subsequently exogenically modified, were probably abundant on Mars. For example, a channel-like feature formed by collapse might be subsequently modified by fluid erosion, such as by water or wind.

The acquisition of excellent imagery of the Martian channels by the Viking orbiters has not led to an improved channel classification. Instead, each detailed view has shown the complexity of individual channels, probably reflecting a whole spectrum of processes that operate to initiate, enlarge, and modify primary landforms. Given the immense variety of valleys on the Earth, this realization should not be surprising. A hard and fast classification of the

Martian channels can probably neither be given nor make sense.

For purposes of informal discussion, the divisions of outflow, runoff, fretted, and volcanic channels are useful. Subsequent elaboration of the details of individual channels will serve to focus on the responsible processes on channel origins.

DISTRIBUTION OF CHANNELS ON MARS

The Mariner 9 pictures revealed an apparent dichotomy in the distribution of channels on Mars. The large channels displayed an erratic and widely spaced occurrence, implying a random distribution (Sharp and Malin, 1975). The Viking data have not modified this view, other than to note a high concentration of outflow-type channels around the southern and western margins of the Chryse Planitia region (see Figure 2.2).

The small Martian channels seemed to display a definite association with equatorial latitudes. The pertinent data were summarized in an analysis by Sagan, Toon, and Gierasch (1973). Their work was based on observations from the low-resolution (A-frame) imagery of Mariner 9. A striking peak occurred a few degrees north of the equator, and abundances declined markedly toward high latitudes. An attractive explanation for this distribution was that liquid water was stable at the warmer equatorial latitudes, and perhaps an earlier, "nonglacial" Mars had an equatorial rainfall maximum such as that on contemporary Earth.

Systematic examination of Viking data by Pieri (1979) and by Carr (1979a) showed that the small runoff networks do not concentrate near the equator. Instead of a latitudinal control they demonstrate a geologic one: they occur on almost all the ancient cratered terrain units but are absent on plains units younger than Lunae Planum (Pieri, 1980a). Pieri (1979) suggested that the misleading Mariner 9 results derived from the incomplete coverage of the planet by Mariner 9 imagery and from the partial burial of channels by younger intercrater plains lavas and eolian debris mantles.

CHANNEL AGES

The methods for determining the ages of various surfaces on Mars have already been discussed. When the Martian channels were first observed on Mariner 9 imagery, their implication of an aqueous epoch in the planet's history immediately led to attempts at estimating the age of that epoch. The first suggestions were that the channels were quite young because of their fresh appearance and a paucity of superimposed impact craters (Masursky, 1973; Hartmann, 1973). Subsequent studies revealed, however, that many craters were indeed superimposed on the channels, and it was concluded that they must be old, at least 10^8 years (Hartmann, 1974a).

The most complete channel age analysis attempted from Mariner 9 imagery is that of Malin (1976). He counted all visible craters 0.25–16 km in diameter on Mariner 9 pictures of twenty-five major channels on Mars. He found that the large Martian channels exhibited a crater density of about 3,000 craters ≥ 1 km diameter per 10^6 km^2. Of course, this figure represents an average age of the twenty-five large channels, mostly outflow and fretted varieties. Malin's sample included none of the small-scale dry valley networks of the heavily cratered terrains. Malin concluded, from the available impact rate assumptions, that the channels must be very old, certainly 10^9 years or older. He further argued that stratigraphic relationships supported this antiquity. Some channels, such as Mangala Vallis, are partially overlain by volcanic plains materials, presumably about

10^9 years old. Moreover, because the channels clearly eroded very large craters and impact basins, they must be younger than those very ancient features.

The Viking pictures have provided a much more extensive data base for channel age determinations. Table 3.2 presents some of the published crater counts on individual channels. In general, the new data confirm the great antiquity of the Martian channels. Assuming the lunar impact flux rate, only portions of Mangala Vallis and a surface in upper Tiu Vallis would be as young as 10^9 years B.P. Most of the large outflow channels have crater densities in the range of 1,500–2,500 craters larger than 1 km per 10^6 km^2 (Masursky and others, 1977; Carr, 1979b). The lunar curve places that age range at 2–3 billion years B.P. Assuming the Neukum and Wise (1976) curve, the age would be about 3.5 billion years.

Masursky, Dial, and Strobell (1980) recently presented additional data on channel and terrain ages obtained from study of high-resolution images obtained late in the Viking mission. These new data reveal relatively low crater numbers for some outflow channels. These revised values are indicated in parentheses after the channel name: Vedra Vallis (360 ± 120), youngest portion of Kasei Vallis (600 ± 250), Tiu Vallis (650 ± 300), youngest portion of Ares Vallis (750 ± 250), and Maja Vallis (750 ± 250).

The small runoff channel systems of the ancient cratered terrains are generally older than the large outflow channels. Carr (1979b) estimated that crater counts for relatively fresh appearing small runoff channels exceed those of the outflow channels by about a factor of 4. The more degraded runoff channels are presumably even older, perhaps comparable in the age to the heavily cratered terrain in which they occur. If the lunar impact flux curve is assumed, the small channel networks are at least 3.5 billion years old.

The crater-count age difference between large outflow channels and

TABLE 3.2. Ages for Some Martian Channel Surfaces

Crater Number[a]	Approximate Age (Lunar Curve)	Channel	Reference
280	0.5 B.Y.	Upper Tiu Vallis	Masursky and others (1977)
850–1,150	1 B.Y.	Mangala Vallis	Scott and others (1979)
1,500	2 B.Y.	Younger portion, Kasei Vallis	Baker and Kochel (1979a)
1,500	2 B.Y.	Ares Vallis	Masursky and others (1977)
1,500–2,500	2–3 B.Y.	Ares, Kasei, Maja, and Vedra Valles	Carr (1979b)
2,000	3 B.Y.	Maja Vallis at Chryse Planitia	Masursky and others (1977)
2,200	3 B.Y.	Kasei Vallis near mouth	Masursky and others (1977)
2,900	3.3 B.Y.	Bahram Vallis	Masursky and others (1977)
3,400	3.4 B.Y.	Ma'adim Vallis	Masursky, Dial, and Strobell (1980)
3,600	3.4 B.Y.	Nanedi Vallis	Masursky, Dial, and Strobell (1980)
5,000	3.6 B.Y.	Oldest portion, Kasei Vallis	Wise, Golombek, and McGill (1979)
6,000–9,000	3.7 B.Y.	"Fresh-appearing" valley networks	Carr (1979b)

[a]Number of Craters ≥1 km Diameter per 10^6 km^2

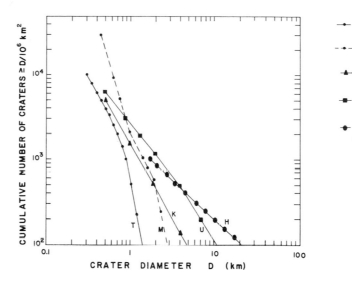

Figure 3.4. Crater counts for several Martian outflow channels compared to nearby upland and heavily cratered surfaces. (From Baker and Kochel, 1979a, Jour. Geophys. Res., v. 84, p. 7961–7983, copyrighted by American Geophysical Union.)

small runoff channels is consistent with the geological distribution of the channels. The small runoff channels are restricted to the old densely cratered terrain. None dissect the younger plains of Chryse Planitia and the Lunae Planum, although these plains are extensively scoured by several large outflow channels. Carr (1979b) concluded that the consistently greater ages of small runoff channels may imply an antecedent relationship to the outflow channels. He presented a model of planetary channel evolution in which the small runoff channels formed early in the planet's history, presumably when a warmer, denser atmosphere permitted their development. Subsequent planetary cooling then formed a thick permafrost cap over entrapped ground water beneath the heavily cratered terrain. Periodic outbreaks of the confined ground water then produced the outflow channels during an epoch when dendritic drainage basin development was no longer possible.

It should be remembered that crater age determinations are subject to both practical and theoretical limitations (Malin, 1976). Figure 3.4 shows some of the crater curves that are used for channel age studies. Note that the portions of the curves at large crater diameters are clearly separate, but some curves cross as lower crater sizes are considered. This is especially true of curves from the upland (Lunae Planum) and heavily cratered regions. These older areas have probably experienced resurfacing processes that preferentially obliterated the smaller craters, while the larger craters were more likely to be preserved. Unfortunately, the small runoff channels occur in terrains that have been extensively resurfaced. Large outflow channels like Tiu, Maja, and Kasei Valles generally show more reliable crater curves.

Figure 3.3 (Mamers Valles) illus-

trates another problem encountered in assigning crater ages to some Martian channels. Craters A1 and A2 clearly predate the channel, since they are cut by a continuous flat-floored valley. Several tributary canyons have even developed in the walls of these craters, which appear quite ancient because of their lack of upraised rims and ejecta materials. Crater P appears to postdate the channel. Its upraised rim has not been cut by the channel. The problem posed by this relatively fresh-appearing crater is the distribution of its hummocky ejecta. The ejecta has been removed at both points of contact with the channel. Moreover, a small channel tributary at T has dissected part of the ejecta material. The channel floor near this point shows longitudinal striations, suggesting that it may contain debris that has been slowly moving downvalley (see Chapter 5). Channel development here seems to have extended through a considerable period of time, and it has included secondary erosion of a crater that was imposed during a primary phase of channel history. The lack of any small craters on the channel floor, similar to those on the adjacent upland, is consistent with this implication of prolonged development. Another tributary at C has worked its way headward and intersected an older branch (B) of Mamers Valles. This older branch joins the main valley north of the imaged reach. The mor-

phology of the tributaries suggests that a progressive sapping process may explain these relationships. The sapping probably operated over a considerable span of time.

The large outflow channels of Mars appear to occupy a pivotal position in the channeling chronology of the planet. They present large surfaces that have allowed the necessary statistics for the best crater counts. Although there is evidence for various secondary modifications of these channels (Baker and Kochel, 1979a), the outflow channels nevertheless have extensive evidence of primary fluid-flow generated topography. Their crater ages also show a fairly narrow clustering. For this reason it will be useful to define a working terminology of Mars channeling history. Events prior to the episode of major outflow channeling will be termed "antediluvian." This epoch encompasses the considerable span of time necessary to form the heavily cratered terrain, the small valley systems, and older plains areas such as Lunae Planum. Events following the outflow channeling in any local area will be termed "postdiluvian." In many cases these will be secondary modifications of a primary channel or valley. The outflow channeling itself will be considered the "diluvian" episode of channeling history on Mars. It should be emphasized that this is an informal terminology for discussing outflow channels, the terrains they have incised, and the modifications of the channeled terrains. It is not a formal time-stratigraphic division of Martian geology.

CHANNEL ORIGINS

Fretted Channels

Fretted channels are a peculiar feature of Mars associated with the planet's great escarpments. The most extensive development of these channels occurs along the planet-wide boundary that separates the somewhat sparsely cratered plains of the northern hemisphere from various older elevated terrains. These ancient terrains include the heavily cratered uplands and cratered plateaus, such as the Lunae Planum. The escarpment along that boundary is generally about 1–2 km in height (Sharp, 1973b) and often consists of a complex of erosional outliers. Occasional flat-floored valleys extend back into the escarpment proper, appearing to have developed at the expense of the upland surface. These valleys comprise the "fretted channels." Two very large fretted channels have received International Astronomical Union names: Auqukuh and Huo Hsing Valles. However, most fretted channels remain unnamed.

Sharp and Malin (1975) described the fretted channels of the Deuteronilus-Protonilus region between latitudes 30° and 45°N and longitudes 300° and 350°W. The channels appear to be transitional into the outlier topography of the fretted terrain proper. Steep-walled "islands" and residual knobs often form plexus groupings of channels. Crater rims and crustal frac-

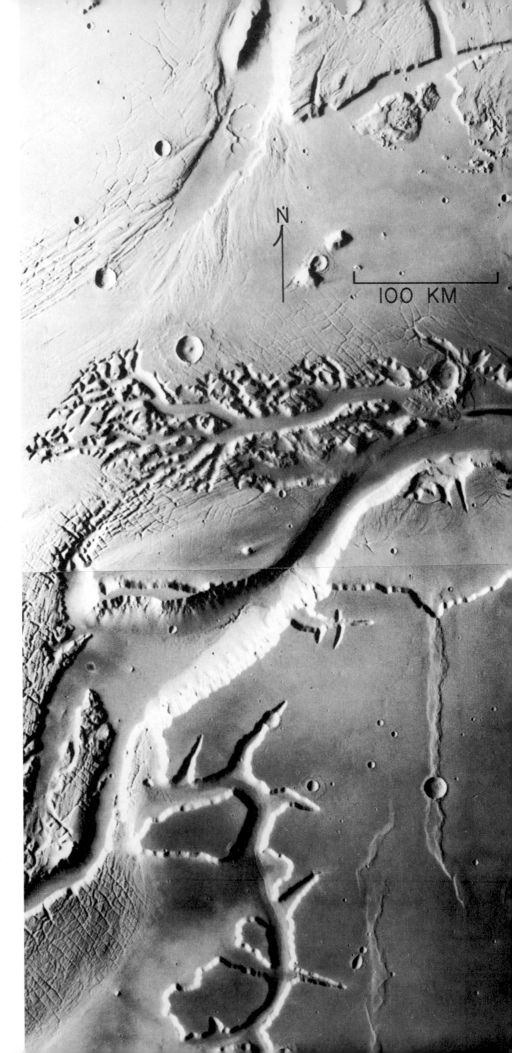

Figure 3.5. Viking imagery of upper Kasei Vallis (centered at approximately 21°N, 72°W). Note the labyrinthine inner channel complex at the upper center. A very large fretted plateau margin occurs in the Sacra Fossa area (bottom center). Steep scarps separate the Lunae Planum plateau (bottom right) from the broad, grooved floor of Kasei Vallis (bottom left). The prominent reentrants into the plateau upland probably formed by a sapping mechanism in which erosion was concentrated along zones of structural weakness in the plateau-forming rocks. (Viking mosaic 211-5770.)

tures appear to have controlled channel development. Probably the same erosional process is responsible for producing both the fretted terrain and the associated channels. The channels merely represent erosion along zones of less resistance than is common at the escarpment. Probably once the channel begins to form it exerts a self-enhancing influence on its own development, concentrating erosive processes to contribute to its own headward growth.

Viking pictures of the Protonilus and the adjacent Nilosyrtis regions of Mars show that the fretted terrain and associated channels are floored with debris that has flowed off adjacent slopes (Carr and Schaber, 1977; Squyres, 1978). These debris blankets appear to be produced by the slow movement of hillslope detritus mobilized by interstitial ice. The process may help explain how mass wastage debris is carried away from active slopes and scarps on Mars. Although the process may not be primarily responsible for the fretted channels, it certainly is an important process of secondary valley modification (see Chapter 5).

Figure 3.5 illustrates a fretted channel complex that occurs within the much larger outflow channel Kasei Vallis. This labyrinthine inner channel complex is located at 73.5°W, 22°N. It provides an excellent example of fracture control, probably by the polygonal network of cracks visible to the south of the complex. The valleys all display steep walls and head abruptly in "box canyons." The valley floors comprise a flat level that seems to be developing at the expense of the upland. An origin by sapping seems to be the most likely explanation of these relationships. However, morphology alone does not distinguish possible aqueous sapping (Milton, 1973), as would be expected on Earth, from the dry sapping of ground ice, as proposed for Mars by Sharp (1973b).

Probably sapping along cliffs and escarpments plus various debris removal processes constitute the primary erosional mechanism for the fretted channels (Sharp and Malin, 1975). Certainly the lack of obvious fluid-flow bed forms on the valley floors precludes any appeal to rapidly flowing fluids. There is, however, considerable evidence of secondary valley modification, probably by periglacial processes as discussed in Chapter 5.

Runoff Channels

Sharp and Malin (1975) originally defined this class of Martian channels to include fluvial-like valleys with tributary branches. They included in this category the numerous small channel networks in the heavily cratered terrain. In addition, they included several quite large channels of complex morphology, but including some tributary development. Nirgal Vallis and Ma'adim Vallis are examples of this latter group. Although many runoff channels do increase in size and depth distally from their proximal tributary heads, they do not show bed forms within the channelways that would be expected for flowing water. For this reason, it is useful to redesignate these features as "valley networks." The term allows that a fluvial component may formerly have contributed to the genesis of the features, as indicated by the tributary development. However, it leaves open the questions of how the possible fluvial component was derived, how much it contributed to valley development, and whether nonfluidal processes might mimic a fluvial valley. The term "channel" implies a major conduit of fluid conveyance, which can clearly be demonstrated for the outflow channels but not for runoff channels. The term "runoff" implies similarities to the terrestrial hydrologic cycle in which the surface water component is largely derived from precipitation.

It is useful to distinguish the relatively small valley networks of the heavily cratered terrain from several large Martian valleys with some tributary development. The latter valleys, including Ma'adim and Nirgal, are best described individually, since they appear to represent a more complex history of development. These large valleys have lengths of several hundred kilometers, and widths that may exceed 10 km. Their size generally exceeds that of the small valley networks by an order of magnitude.

The small valley networks were first described as "furrowed terrain" by McCauley and others (1972). The Mariner 9 imagery showed that small valleys commonly occurred in parallel patterns, imparting a corrugated texture to the heavily cratered terrain. Some systems radiated from the larger craters, while others seemed confined to the intercrater plains. Because of their planimetric form and because of their high density on the cratered terrains, Milton (1973) suggested that the valley networks provided the best evidence for rainfall during an ancient Martian epoch of aqueous activity. Typical development of small valley systems is illustrated in Figure 3.6.

Sagan, Toon, and Gierasch (1973) cited the small Martian valleys as evidence of a past epoch of higher pressures and abundant liquid water on the planet. They presented a model for climatic change on Mars which assumed that enough carbon dioxide was frozen at the Martian polar caps to yield, when thawed, an atmospheric pressure of 1 bar. Variations in Mars' orbital parameters, especially its obliquity, variations in the luminosity of the sun, and albedo changes at the polar caps were all thought capable of establishing an advective instability in the Martian atmosphere. This instability would permit two stable climates—one similar to the present (atmospheric pressure of 7 mb) and one with a pressure of as great as 1 bar. The ancient dense atmosphere could have permitted flowing water.

The Viking project results have shown that the Martian polar caps cannot possibly contain enough carbon dioxide to allow this mechanism to function. However, the model does illustrate the importance of the Martian channels and valleys in describing atmospheric evolution on the planet. Obviously it is necessary to make the most complete analysis of the channels and valleys possible before using that analysis to constrain atmospheric models. Some current atmospheric models are discussed in Chapter 9. Chapter 4 presents recent results on the analysis of Martian valley networks.

Outflow Channels

The outflow channels of Mars pose a complete contrast to the small valley networks (Table 3.3). The outflow channels are much larger features. Their widths are one to two orders of magnitude greater, and some outflow channels (Maja, Ares, and Kasei) can be traced phenomenal distances, 2,000 km or more. The outflow channels are true channels because a broad variety of fluid-flow landforms characterizes their beds. These fluid-flow bed forms were first recognized on Mariner 9 imagery (Baker and Milton, 1974; Sharp and Malin, 1975; Nummedal, Gonsiewski, and Boothroyd, 1976), but their variety and abundance has been better revealed by Viking pictures (Baker, 1978b; Baker and Kochel, .1979a). The outflow channels also appear full-born on the landscape, involving a relatively small number of localized sources, generally collapse zones or chaotic terrain regions. They transect terrains of varying ages. Most emanate from sources within the heavily cratered terrain. However, unlike the small valley networks, most outflow channels traverse the boundary between the ancient cratered uplands and the younger plains units of the northern hemisphere. The channels surrounding the Chryse Planitia can be traced from the cratered uplands to considerable distances out on the Chryse plains (Greeley and others, 1977).

The outflow channels are best described at three different scales, each relating to different scales of fluid flow phenomena. At the macroscale the outflow channels display a regional anastomosis across large areas of planetary surface (Figure 3.7). They trend downgradient from discrete source areas, most often zones of chaotic terrain. The individual channels are separated by residual uplands, many of which have been partly shaped or streamlined by the fluid flows (Figure 3.8). The channel walls are not uniformly spaced, but show abrupt expansions and constrictions.

Figure 3.6. Small valley networks on a heavily cratered terrain surface. Several varieties are illustrated, including subparallel slope gullies (G) draining outward from a crater rim, digitate networks (D) on the intercrater plains, and sinuous stem valleys (S) without major tributaries. (Viking frames 097A20 and 097A22.)

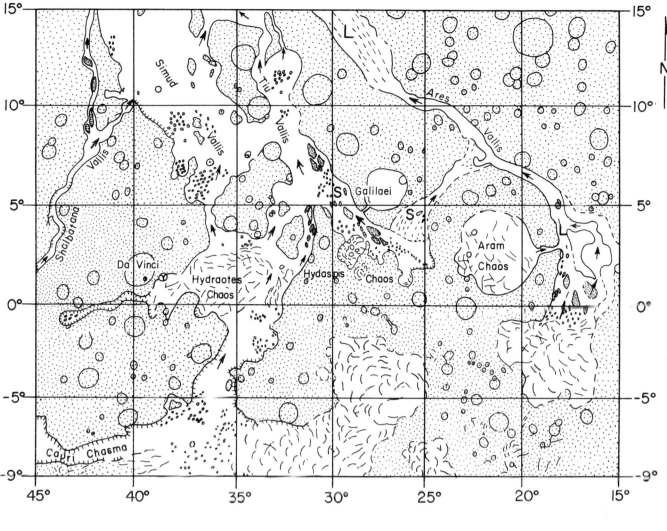

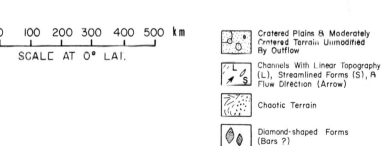

Figure 3.7. Sketch map of outflow channels south of the Chryse Planitia. The channels conveyed fluids in the directions indicated by the arrows. The heavily cratered uplands (dot pattern) have been locally modified to form chaotic terrain from which several channels emanate. Streamlined uplands (S) and longitudinal lineations (L) occur on the channel floors and indicate flow directions. Figure 3.1 is a Viking mosaic of approximately the same region.

Figure 3.8. Streamlined uplands at the outflow of Ares Vallis into the southern Chryse Planitia. The raised crater rims acted as barriers to fluid flows coming from the lower left. Ejecta deposits were eroded from the upstream sides of these crater obstacles, but they are preserved on the downstream sides. The long "tail" of the lower upland shows layering, indicating that a layered bedrock was probably preserved from fluid erosion on the downstream side of the crater. Figure 6.11 is a morphological map of this region. (Viking frames 4A50—4A57.)

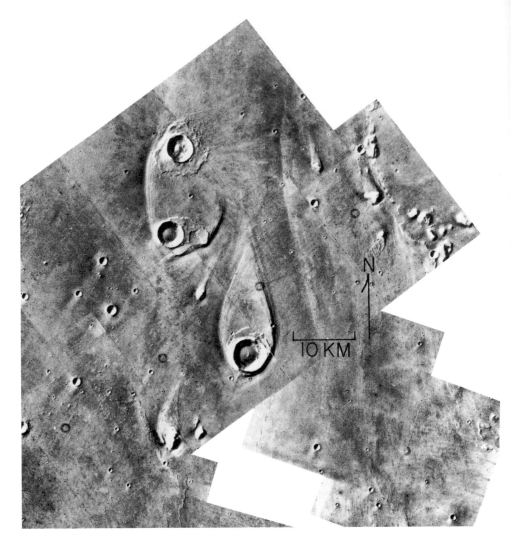

Often the channel margins show a distinct upper elevational limit to terrain that was eroded by the channelized fluid. Transected divides between channels and hanging valleys occur below this elevational limit. The channels may extend across diverse rock types for hundreds, sometimes thousands of kilometers from the probable fluid source areas. Low sinuosity and high ratios of channel width to channel depth characterize this route. Many details of the pattern are controlled by rock structure and by the differential resistance of layered rocks. Finally the eventual downstream channel terminations may display indistinct relationships, including the lack of obvious depositional fans and deltas.

The mesoscale of outflow channel characteristics applies to the larger features on the floors of the outflow channels. These features are presumably scaled to the depth of the fluid responsible for their genesis. They constitute bed forms on the channel floors, rather than major elements of the regional channel pattern. Baker and Milton (1974) first brought attention to the suite of bed forms in the Martian outflow channels. A common association is that of longitudinal grooves and inner channels that terminate upchannel at headcuts (Figure 3.9). Some sections of channel floor show irregular zones of erosive stripping (Figure 3.10). This topography is similar to the "scabland" topography that results from the plucking and scouring of bedrock by catastrophic flood flows. Another erosional bed form is the scour mark that develops adjacent to a flow obstacle on the channel bed. In the Martian

TABLE 3.3. Martian Outflow Channels Compared to Small Valley Networks

	Small Valley Networks	*Outflow Channels*
Macroform	Hierarchical tributary systems	Regional anastomosis extending from discrete sources; few tributaries
Source	Numerous first-order tributaries	Zones of chaotic terrain
Bed forms	Not visible	"Scabland" complexes
Downstream changes	Increase in width downstream	Constant size or size fluctuation downstream
Location	Restricted to heavily cratered terrain	Cut terrains of varying age
Width	1 km	10–100 km
Length	10–100 km	400–2,000+ km
Craters ($\geqslant$1 km)/10^6 km^2	10^4	2×10^3
Age (Lunar Curve)	>3.5 B.Y.	<3.0 B.Y.

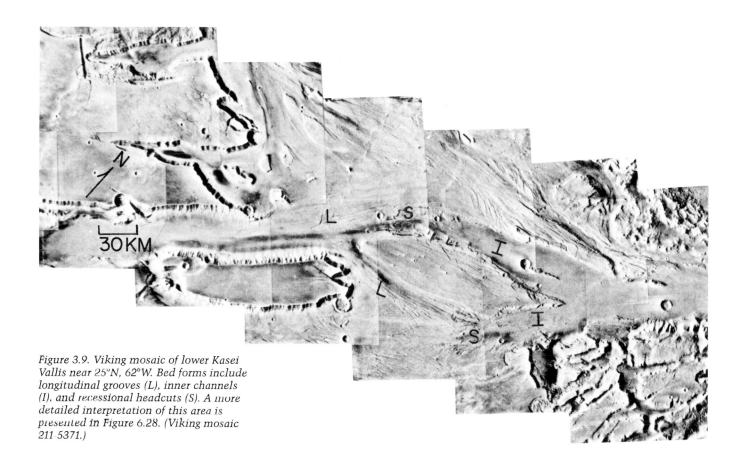

Figure 3.9. Viking mosaic of lower Kasei Vallis near 25°N, 62°W. Bed forms include longitudinal grooves (L), inner channels (I), and recessional headcuts (S). A more detailed interpretation of this area is presented in Figure 6.28. (Viking mosaic 211-5371.)

Figure 3.10. "Scabland" topography in the southern Chryse Planitia at the outflow of Ares Vallis. Some process appears to have stripped off irregular zones of the dark surface material to expose an underlying light toned material. (NSSDC picture 16829.)

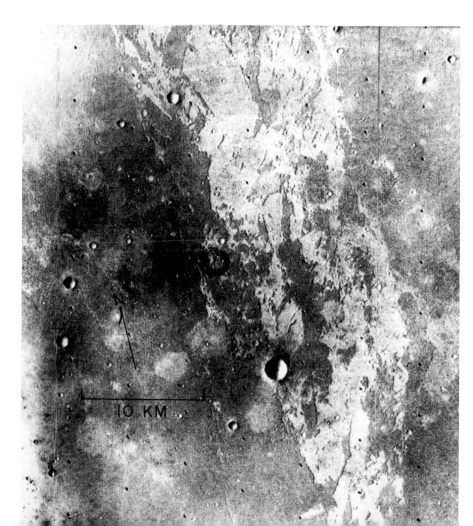

outflow channels, flow obstacles include bedrock projections, remnants of chaotic terrain blocks, and resistant crater rims. Crescent-shaped scour depressions occur immediately upchannel of these obstacles (Baker, 1978b). Pendant-like forms are often developed downchannel from the obstacles. These small-scale streamlined hills were completely overtopped by the responsible fluids. They may be remnants of antediluvian rock, preserved downchannel from the obstacle, or they may be depositional bars. Larger complexes of possible depositional bars occur immediately downchannel from flow constrictions. These were named "expansion bars" by Baker and Milton (1974).

The microscale of outflow channel description relates to smaller features on the bed, such as the actual particles eroded and transported by the fluid. Unfortunately this scale has not yet been evaluated on Mars.

Table 3.4 summarizes the major morphological attributes of the outflow channels. Not included are the various postdiluvial modification features, such as talus slopes and debris cones (Baker and Kochel, 1979a).

Several fluids have been proposed as responsible for creating the outflow channels: water, wind, viscous earthflows, glaciers, and lava.

Water. When Mariner 9 began returning pictures of Mars in 1971, the prevailing wisdom was that Mars was a dry planet. Its atmosphere contained only a negligible trace of water, and its poles were thought to have ice caps of carbon dioxide. The channel-like landforms revealed by Mariner 9 seemed the sole anomaly in this general picture of a water-impoverished planet. The interpretation of the channels led in either of two directions. On the one hand, Mars must have had one or several aqueous epochs in its past, implying surficial conditions vastly different from those of today (Milton, 1973). Or, on the other hand, the channels, while displaying some general similarities to terrestrial fluvial features, must have originated by some combination of nonfluvial processes.

Milton (1973) argued that certain features of the large Martian channels

TABLE 3.4. Characteristic Morphological Features of Martian Outflow Channels

A. MACROSCALE

Channel Pattern	Shown in These Figures
1. Anastomosing pattern of channels	3.1, 3.7, 6.4, 6.8, 6.16, 6.28, 6.29
2. Discrete source areas, such as chaotic terrain	2.16, 3.1, 3.12, 6.3, 6.24
3. Residual uplands separating channels	3.1, 6.8, 6.12, 6.14, 6.16, 6.17, 6.28, 6.29
4. Many uplands partly smoothed and streamlined by fluid flows, especially on their upstream ends	3.8, 6.8, 6.11, 6.12, 6.16, 6.17, 6.18, 6.25, 6.26, 6.27, 6.28, 6.29
5. Pronounced flow expansions and constrictions	3.1, 6.14, 6.17, 6.18, 6.23, 6.28, 6.29
6. Distinct upper elevational limit to eroded terrain	3.8, 6.5, 6.8, 6.12, 6.17, 6.18, 6.23
7. Transected divides and hanging valleys	6.6, 6.12, 6.17, 6.18, 6.23
8. Erosion of diverse rock types thousands of kilometers from probable fluid source areas	6.13, 6.14, 6.17, 6.24
9. Low sinuosity	1.5, 1.6, 6.6, 6.14, 6.17, 6.18
10. High width-depth ratio	1.5, 1.6, 6.17
11. Differential erosion of terrain controlled by structure and lithology	3.8, 6.12, 6.17, 6.18
12. Indistinct channel terminus, including the lack of obvious large-scale fans and deltas	6.17, 6.19, 6.21

B. MESOSCALE

Bed Forms	Shown in These Figures
1. Longitudinal grooves	3.9, 6.17, 6.18, 6.19, 6.25, 6.26, 6.28, 6.30
2. Inner channels	3.9, 6.17, 6.26, 6.28
3. Recessional headcuts (cataracts)	3.9, 6.17, 6.26, 6.28
4. Scabland	3.10, 6.10, 6.12
5. Scour marks near flow obstacles	6.28
6. Pendant forms (small-scale streamlined hills, which may be either residual or depositional)	6.16, 6.17, 6.20
7. Expansion bar complexes, or fan deltas	3.11, 6.17, 6.20

could have been produced only by running water. The implication was that an aqueous origin was therefore likely for a broad range of other Martian channel-like features for which the responsible fluid was not so obvious. Milton concluded that Mars had experienced a fluvial stage at some time during its history.

Milton (1973) pointed to the "braided reach" of Mangala Vallis as the strongest argument for running water on Mars. The Mangala Vallis channel

flows northward along the 151°W meridian from approximately 17°S to 4°S, where it debouches from a heavily cratered upland on to a volcanic lowland plain. Its braided reach (Figure 3.11) occurs at about 5°S, 151°W, where the channel broadens before constricting again further downstream. Milton (1973) argued that this braided pattern could be produced only by a high-density, low-viscosity liquid moving rapidly over a particulate bed. He noted that this requirement ex-

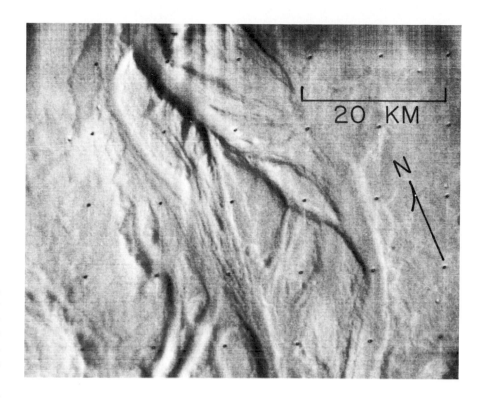

Figure 3.11. The "braided" section of Mangala Vallis located at about 5°S, 151°W. Note the cross-cutting relationships of secondary scour channels. Similar patterns develop on braid bars because of the fluctuating discharge regimen of terrestrial braided streams. Because of the immense scale of this feature and its location in an expanding channel reach immediately downstream from a constriction, Baker and Milton (1974) suggested that it was more likely an expansion bar or fan that developed during catastrophic flood discharges and that the depositional surface was subsequently scoured by waning flows or by subsequent floods.

cluded wind action, lava flows, and ash flows as likely agents. The remaining possibilities were water, ammonia, hydrocarbons, and liquid carbon dioxide. Of these, water required the least extreme deviation from the present Martian regimen. Moreover, it was the only fluid which produced adequate terrestrial equivalents to the Martian features.

Milton (1973) and McCauley and others (1972) recognized that the Martian outflow channels resembled catastrophic flood channels rather than the more common, evolved river systems of Earth. The most appropriate analogy for the Martian outflow channels was to the great flood-carved channelways produced when ice-dammed Lake Missoula released several catastrophic floods during the Pleistocene (Baker and Milton, 1974). The Lake Missoula floods were the most violent water flows in the terrestrial geological record, and they produced the Channeled Scabland region of eastern Washington (Bretz, Smith, and Neff, 1956; Baker, 1973a).

Schumm (1974) concluded that many of the largest Martian channels, including Mangala, Kasei, and Nirgal Valles, originated by tensional fracturing of the planetary surface locally modified by volcanic activity, wind action, and mass wasting. Schumm rejected a fluvial genesis for these channels because of their great size, the lack of clearly defined source areas, and the superficial resemblance of these large channels to tensional fractures in a variety of materials. Schumm did note that catastrophic

flooding could produce the scale of features noted on Mars, but he was dubious because of the problem of identifying a source for the prodigious quantities of erosive liquid. A major limitation at the time of his study, as he recognized (Schumm, 1974, p. 374), was that the available Mariner 9 imagery did not easily resolve unequivocal details of fluvial scour, sculpture, and deposition in the channels. However, other investigators (Baker and Milton, 1974; Sharp and Malin, 1975) came to a different conclusion from Mariner 9 images of Mangala, Ares, and Kasei Valles. They concluded that those details of fluvial scour were present, but that the scour was of the rare type achieved in catastrophic flooding, rather than the common type of terrestrial rivers like the Mississippi. The structural pattern of these channels was easily explained by the tendency of catastrophic floods to scour rock along structurally controlled zones of weakness (Baker, 1973b).

The identification of streamlined hills and other obvious fluid-flow features on Viking pictures of the Martian outflow channels has shown that a purely structural origin is untena-

ble. Nevertheless, structural influences on channeling processes have been profound. The erosive fluids certainly flowed preferentially along pre-existing troughs, many of which probably had tectonic origins. Fluid erosion also appears to have been localized along zones of structural weakness.

Because the current Martian environment poses severe problems for the persistence of water on the planet's surface, several investigators have continued to seek alternatives to the proposed aqueous origin of the Martian channels. Perhaps the bluntest justification of this work was provided by Lambert and Chamberlain (1978). They concluded that water cannot explain the channels on Mars for the following reasons: "the temperature range . . . is wrong; the atmospheric composition is wrong; there are no other fluvial erosion features; and the rest of Mars shows no evidence of the glaciation which should surely accompany fluvial erosion in the warmer equatorial climate." To this condemnation they added the following: "Any satisfactory H_2O-based theory is necessarily nonuniformitarian or even catastrophic in nature and unreasonable to that extent." Science would certainly proceed more swiftly if we could discard hypotheses so easily.

More serious objections to the aqueous hypothesis were summarized by Schonfeld (1977) and by Blasius, Cutts, and Roberts (1978). Many of these objections center on the source and sink relationships for the water flows presumed responsible for the channels. The major source problem involves a comparison between the assumed volume of water released to form the existing chaotic terrain versus the volume of the channels downstream from the chaos zones. Blasius, Cutts, and Roberts (1978) suggested that the source volumes are much too small to account for the indicated erosion. Their analysis assumes that (1) the present chaotic terrain area is the only zone from which fluid was derived and (2) the total downstream channel (really *valley*) volume was achieved by fluid erosion. Neither assumption is likely for the Martian outflow channels (Baker, 1978b).

Outflow channels such as Tiu, Simud, and Ares Valles appear to have developed by a headward growth through the heavily cratered terrain (Baker, 1977, 1978b). As fluids were released, producing collapse zones at the channel heads, the downstream zones of previous collapse were scoured by the subsequent flooding. Thus, the preserved zones of collapse (chaotic terrain) represent only the last phases of the process. The total channel volume was largely produced by collapse over earlier fluid-release zones and subsequently modified by the flow of fluids released at the younger collapse zones upstream. There channels, such as Maja and Kasei Valles, show ample evidence that the fluids followed antediluvian low points or troughs on the topography (Baker and Kochel, 1979a). The fluids merely generated a suite of bed forms and scour features in pre-existing valleys. Moreover, various postdiluvial processes, especially hillslope retreat and sapping, appear to have enlarged the channels after the fluid release episodes (Baker and Kochel, 1979a).

The sink problems for water as a channel-forming fluid center on features that presumably should be present at the channel mouths (Blasius, Cutts, and Roberts, 1978; Schonfeld, 1977). These include deltas and large accumulations of sediment at such locations as the Chryse Basin Viking 1 landing site. Such large-scale alluvial deposition is very difficult to recognize on orbital imagery of Mars (Sharp, 1978). However, a similar situation occurs on Earth in the Channeled Scabland, where delta-like sediment accumulations are also very difficult to recognize on orbital photographs. This occurs because the sediments consist largely of coarse basalt blocks eroded from nearby outcrops and spread over large depositional basins. Even the mounded gravel bars are difficult to identify because of their similarity to streamlined hills eroded from relatively soft preflood sediments (Baker, 1978c). Scabland erosion and deposition do not follow the well-known patterns of most terrestrial rivers. The assumption that certain sink relationships should be present if the channels are fluvial appears to derive from comparisons of Martian channels to fully evolved terrestrial river systems (Baker, 1978b). The catastrophic flood channels on Earth do not display these relationships, so the source-sink arguments are largely irrelevant in attempting to disprove one fluid or another. A more appropriate approach is to consider the various possible fluid-erosional systems and establish their compatibility with Martian channel morphology and bed forms.

Wind. An eolian hypothesis for the Martian channels has been developed by James Cutts and coworkers (Cutts, Blasius, and Farrell, 1976; Cutts, Blasius, and Roberts, 1978; Cutts and Blasius, 1981). Their model involved the erosive effects of saltating grains moving poleward up the outflow channel systems, most of which trend from equatorial areas northward toward the northern plains. Cutts and Blasius (1979) calculated that, provided the wind can be loaded to near its sediment transport capacity, the large outflow channels could be carved by bedload eolian transport alone in less than a million years. Blasius and Cutts (1979) stated that the major limitation of this model is the need to obtain immense quan-

tities of particles in the saltation size range (about 100 mm to 1 mm on Mars). They suggested that the saltating particles are derived from the chaotic terrains, perhaps through weathering of the country rock or because the eroded rocks are weakly bound aggregates of grains in the appropriate size range.

Eolian processes on earth are capable of creating a number of landforms that may be similar to features in the Martian outflow channels, including streamlined hills (yardangs) and longitudinal grooves (McCauley, Grolier, and Breed, 1977). Moreover, there are theoretical arguments that eolian processes should be more efficient on Mars than on earth (Arvidson, 1972), provided that strong winds are present and that suitable particulate matter exists on the surface. Unfortunately it has been difficult to verify the existence of saltation-sized particles on Mars. The Viking landers reveal either very fine particles that would move in suspension or very large particles, above the competency of the Martian winds. No large-scale saltation movement of sediment has been recorded by either of the Viking landers during the mission (Jones and others, 1979). The main evidence for saltation on Mars is the existence of sand dunes.

The large dunefields on Mars, especially in the polar latitudes, suggest that the saltation particles on Mars may have been concentrated in great ergs, similar to the "sand seas" of terrestrial deserts. Breed, Grolier, and McCauley (1979) speculated that Mars has had a long eolian history during which much of the saltation-sized material has already been stripped from the plains and transported to the great north polar erg and to various smaller ergs on crater floors.

Cutts, Blasius, and Farrell (1976) suggested that streamlined hills at the mouth of Ares Vallis (Figure 3.8) might be giant yardangs formed by wind erosion. Yardangs are most easily developed in relatively soft sediments, such as the lacustrine clay and silt of a dry playa (Blackwelder, 1934). The features begin as broad ridges and intervening troughs, perhaps aligned along former gullies, and eventually are transformed to beautifully streamlined hills (A. W. Ward, McCauley, and Grolier, 1977). Perhaps the largest terrestrial example occurs in the Kerman Basin of Iran. Here soft clayey sandstone has been fashioned to a landscape of parallel steep-sided ridges and corridors to a local relief of 60 m (Dresch, 1968). The ridges are spaced about 1–2 km apart and cover an area measuring 110 × 60 km. Wind is not the only agent of their formation, since streamfloods, solution processes, and the cracking of clay crusts also contribute (Dresch, 1968). Probable yardangs on Mars occur in relatively young, easily erodible materials (A. W. Ward, 1979).

The eolian hypothesis has a number of attractions. Material carried in eolian suspension can be widely dispersed on the planet, and the air itself is also easily disposed of. Thus, source and sink relationships pose fewer problems than with other erosive fluids. While the hypothesis has theoretical attractions, however, it suffers badly in tests of practical geology. Sharp and Malin (1975) pointed out that terrestrial regions of eolian erosion (deflation) generally have nearby accumulations of the eolian bedload material. These have not been recognized on Mars. Moreover, while terrestrial wind erosion can produce magnificent streamlined landforms (A. W. Ward, McCauley, and Grolier, 1977), it has not been known to produce long, sinuous channels eroded deeply into such rock as layered basalt flows. Because eolian erosion does not require a downhill slope to topography, the universal conformity of channel direction with regional slopes (Carr, 1979b) becomes an unbelievable coincidence. Milton (1973) questioned why wind on Mars would have become decoupled from larger air masses and confined to sinuous channels. Finally, many outflow channels have a distinct trim line on their margins, showing that the fluid did not erode above a distinct level. Such a sharp demarcation would only occur in a fluid with an upper boundary, such as water, lava, or a mudflow. These relationships cannot be reconciled with eolian erosion as the primary fluid for shaping the outflow channels. However, they do not preclude the action of wind as a secondary agent of modification. By its prolonged action wind may have reshaped some landforms that were generated by other processes.

Prior to the Viking mission eolian erosion was considered to be a dominant process on Mars (McCauley, 1973). Sagan (1973b) estimated that the rate of abrasion on Mars could be as high as 3×10^{-3} m/yr. However, high-resolution Viking images show that ancient cratered surfaces on Mars have crater-size frequencies similar to the lunar maria for crater diameters as small as 100 m. The 2 to 3 billion years that have elapsed since the emplacement of these craters were apparently not sufficient for their removal by eolian erosion. Arvidson, Guiness, and Lee (1979) therefore estimated an average deflation rate of only 10^{-6} m/yr.

Viscous Earthflows. Nummedal (1978) proposed that the large Martian outflow channels were generated by mudflows resulting from the liquefaction of large areas of the Martian crust. Liquefaction is a loss of cohesion in a material that occurs when sudden strain induces an excess pore pressure that approaches the effective overburden pressure. A closely related phenomenon is retrogressive flow sliding, also called failure by lateral spreading. In spreading failures there is a flowage of subjacent material that causes the overlying firmer rock or soil to break into units and spread apart. The most spectacular such failures occur when a fine-grained sediment actually changes state from a fairly hard, strong, brittle solid into a liquid of negligible strength. Materials exhibiting this remarkable property include the so-called "quick clays," which occur in the St. Lawrence Lowland of Canada and in Norway and Sweden. Individual "quick clay" slides of Norway may involve millions of cubic meters of material. The slide scar takes on the appearance of a bowl-shaped depression, usually floored by large flakes of the original dry surface soil layer that was not involved in the liquefaction process (Nummedal, 1978).

Several theories have been advanced to account for the distribution and properties of terrestrial quick clays. One hypothesis holds that the clays originated during glacial periods as fine-grained sediment deposited in marine coastal areas. The clays were coagulated by the electrolytes in sea water and settled quickly. Because clay minerals carry a negative charge, they coagulated by bonding of the flat, platy particles edge-to-edge. A "house of cards" structure was thus created, and sodium (Na^+) and other cations were important in providing the bond strength that made this initial soil system fairly strong. However, this structure is only stable so long as the electrolyte "glue" remains. In postglacial times, land that had been depressed below sea level by the weight of glacial ice was unloaded. Its rebound locally elevated glacial-marine clay deposits as much as 200 m above sea level. Subsequently, fresh rainwater leached the original salty water (and the sodium "glue") from the clays, leaving them in a very unstable arrangement. Upon shaking, the metastable structure collapses and expels the interstitial water, liquefying the clay and allowing it to flow on very gentle slopes.

Another hypothesis was derived to explain the occurrence of fresh-water quick soils. Again glaciers probably supply very fine clay-sized rock particles, but these are not clay minerals. Because these clay-sized particles (mostly quartz) are so small, the normally weak attraction between two particles caused by mutual distortion of electronic structures (van der Waals forces) become very effective. When particles are small enough, the ratio of the van der Waals bond strength to the weight of the particle is high, and the substance will be cohesive, strong but brittle, and very sensitive. These van der Waals forces are effective only when very small particles are very close together. When the clay-sized particles are distributed, these short-ranged forces are ineffective, and a total loss of strength occurs when the bonds are broken. If there is sufficient pore water, the material will liquefy and flow on very gentle slopes.

Nummedal (1978) suggested that the morphology of some outflow channels is consistent with an origin by liquefaction. Ravi Vallis (Figure 3.12) illustrates the pertinent features. Chaotic terrain near its headwaters

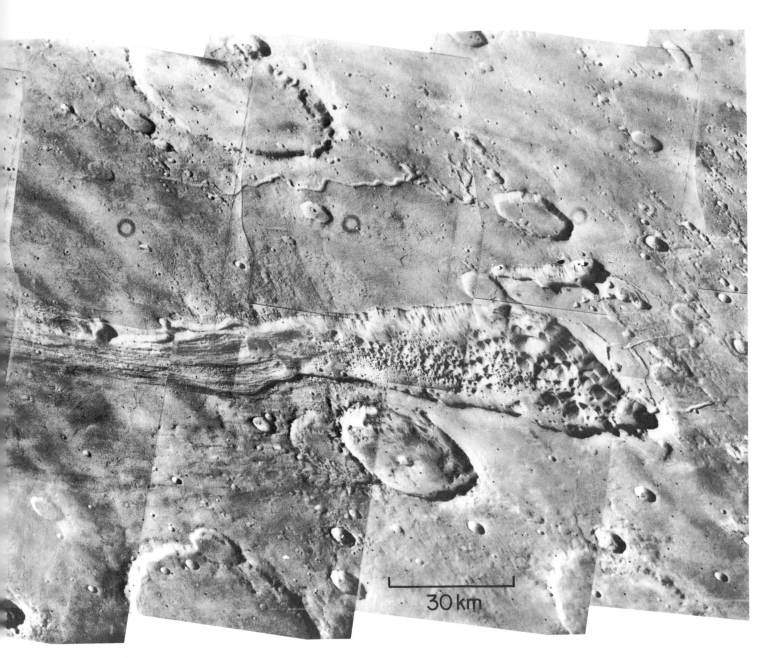

30 km

Figure 3.12. Oblique Viking image mosaic of Ravi Vallis, 2°S, 43°W. The channel is 120 km long. The grooved terrain occurs in an outlet zone that is 20 km wide. The chaotic collapse zone (right) is Aromatum Chaos. This view is toward the south; the channel fluid flowed east (toward the left). For a scaled map of Ravi Vallis, see Figure 6.2. (NSSDC picture 16983.)

(Figure 3.12, right) shows collapsed blocks of a competent material, presumably the non-liquefied crust of the spreading failure. Downchannel from the constricted portion of the head depression (Figure 3.12, center) there is ample evidence of extensive flowage of the liquefied sediment (grooved terrain in Figure 3.12, left).

Some varieties of subaqueous mass-movement processes also produce a morphology that is broadly similar to that of some outflow channels. The morphology has only recently become known through the use of improved side-scan sonar images (Prior, Coleman, and Garrison, 1979). Excellent examples occur on the Mississippi Delta front, where recently deposited sediment experiences retrogressive liquefaction, subsidence, and flow. The landslides and flows head at collapse depressions which are often floored with large chaotic blocks. These grade into elongate, sinuous landslide gullies. The upper parts of these gullies are scoured and display abrupt scarps on their lateral margins. The landslide gullies and chutes are organized by tributary junctions into an overall dendritic pattern, as described in more detail by Prior and Coleman (1978).

Komar (1979) demonstrated that many similarities exist between Martian outflow channels and terrestrial deep sea channels formed by turbidity currents. The deep sea channels display similar lengths, widths, reliefs, and overall patterns. Moreover, they originate at specific sources, submarine canyons which trap marine sediments that later slump, releasing turbidity currents that carve the channels. Komar (1979) emphasized the analogy for dynamic reasons, to compare the hydraulics of large-scale flows and to investigate the effects of reduced gravity. However, Nummedal (1978) contended that the analogy is genetic. On Mars, large-scale mass flows, such as turbidity currents, mudflows, or debris flows, were presumably initiated by liquefaction in the chaotic terrain regions. From these point sources the liquid moved down gently sinuous, deeply incised single channels with few or no tributaries. The liquid near the channel head is considered to be a viscous, de-

bris-laden slurry of low turbulence and heavy sediment concentration (Thompson, 1979). This liquid produced the single, isolated streamlined hills common near channel heads. However, further downstream this slurry is presumed to have lost much of its debris/sediment concentration. The remaining water was much more turbulent than the slurry, and was responsible for eroding the massive anastomosing patterns and diamond-shaped "islands" that are observed downstream.

Some support for the action of very viscous fluids in shaping the outflow channels has come from fluid stability theory. Thompson (1979) analyzed the requisite conditions for the growth or decay of large longitudinal roller vortex perturbations imposed at the base of various fluids (water, mudflow/debris flow, lava, glacier). The analysis was constrained by the morphology of a headwater reach in Tiu Vallis, where the channel is severely constricted about 30 km downstream from the chaotic terrain source area. Thompson concluded that the pattern of longitudinal grooves through this constriction is best explained by vortex motion in a mudflow or debris-laden slurry. Thompson also noted that his analytical procedure cannot distinguish the influences of various sediment/debris concentrations in the mudflow or slurry, nor can the procedure adjust for the various sediment transport mechanisms that might have prevailed in a Martian flood. Nummedal (1980) concluded that a variety of debris flow processes may be responsible for outflow channels.

While the liquefaction/debris flow hypothesis can explain features in the proximal (headwater) portions of some outflow channels, it does not adequately explain many of the distal relationships. Downstream reaches of the channels show bed forms that seem to require a highly turbulent fluid (Baker, 1978b). These include transverse headcuts (cataracts) at the heads of inner channels, scabland, and scour marks around flow obstacles. Moreover, some of the source and sink arguments that can be overcome with the catastrophic flood hypothesis are not so easily reconciled when proposing mudflows. The im-

mense magnitude of those flows, persisting over 1,000 km along some channels, would be expected to produce immense debris fans. None can be recognized on Mars. Reducing the percentage of water in the fluid proposed for carving the channels greatly increases the difficulty of explaining the general lack of obvious deposition in the Martian outflow channels.

Other questions center on the material that comprises the Martian crust. First, it must behave in a similar manner to materials that occur under unusual terrestrial circumstances: glaciomarine clays and submarine canyon head deposits. Second, it must be extremely common on Mars in the extensive areas with chaotic terrain. Finally, once it is liquefied, it must be able to persist in that transient state for a sufficient time to traverse channels as long as 2,000 km. Even at the peak velocity of 100 km/hr cited for short-term occurrence in some Norwegian liquefaction (Nummedal, 1978), a mudflow would have to retain its excess pore pressure for nearly a day to traverse the required distance. If some mudflows did not traverse the entire channel length, one would expect that their massive deposits would be obvious on the channel floors.

Glaciers. Lucchitta and Anderson (1980) suggested that glacial ice may have sculptured the outflow channels of Mars. Glaciers on Earth produce a number of morphological features that are similar to those in the outflow channels, including anastomosis, scour marks, longitudinal ridges and grooves, and streamlined uplands. The glacial analog for Martian outflow channels has not been fully developed and warrants further study. Some problems with the hypothesis include (1) the mechanics of glacial flow in the Martian environment (see Chapter 5), (2) the lack of cirques at the heads of large outflow channels (chaotic terrain occurs instead), (3) the lack of massive glacial deposits at the mouths of the outflow channels, (4) the difficulties of achieving an atmosphere capable of precipitating considerable snow for glacial ice accumulation preferentially at equatorial latitudes (where most of the outflow channels occur), and (5) the inability of ice to generate turbulent flow (which appears to be required by some landforms in the outflow channels).

Lucchitta, Anderson, and Shoji (1981) have attempted to resolve some of the above difficulties. Their theoretical analysis suggests that ice could have moved on Mars even under present conditions, although a warmer climate in the past would have been more favorable. They envision ice accumulation in source areas largely from spring flow. The seepage would create domes or plugs, which, after building to a critical thickness, would then move as glaciers through the outflow channels and locally generate the distinctive morphology that suggests glacial action. Combinations of water flows, icings, and ice drives are thus an important corollary to the glacial hypothesis. These possibilities will be discussed more fully in Chapter 8.

Lava Channels

An example of the problems of similarity between fluvial and volcanic landforms occurred because of the pattern of high-resolution frames obtained by Mariner 9. A single high-resolution frame from the flanks of Alba Patera showed what appeared to be a dendritic network of gullies (Milton, 1973, Fig. 9). Milton (1973) stated that another possible origin for the features was lava channels, but the lack of other high-resolution pictures precluded a definitive statement. Viking imagery of the region (Carr and others, 1977b) shows that instead of a contributive fluvial drainage network, the Mariner image had shown part of a complex lava tube–channel flow system (Figure 3.13). At

the distal ends of lava tubes in this region, distributary patterns developed which spread the lava over a broad region. Thus, the lava was flowing in the opposite direction to the direction inferred for water in the fluvial hypothesis.

Lava channels have been clearly recognized on the flanks of Martian shield volcanoes (Greeley, 1973). The possibility that the large Martian outflow channels originated through lava flow processes has been pursued by several investigators (Schonfeld, 1976, 1977; Cutts, Roberts, and Blasius, 1978). Their hypothesis is consistent with several observations: (1) lava has no problems of stability in the present Martian atmosphere; (2) lava would also be expected from lithospheric source areas; (3) some large channel systems, such as the Elysium channels (Chapter 4), occur on the flanks

of volcanic constructs; (4) other channels, such as Mangala, Maja, and Kasei Valles are associated with volcanic plains units; (5) channeling in some areas, notably the Chryse Planitia, was nearly contemporaneous with the emplacement of volcanic plains (Carr and others, 1976). Theoretical models of Martian volcanism (Smyth, Huguenin, and McGetchin, 1978) indicate that Martian primary lavas should have unusually low viscosity. Schonfeld (1977) speculated that Martian lavas were of such low viscosity (because of their probable iron oxide and water contents) that flow was turbulent and capable of producing fluvial-like features.

Low-viscosity lava eruptions may result in a variety of channel-like landforms. Such eruptions have been extensively studied on the basaltic shield volcanoes of Hawaii (Figure

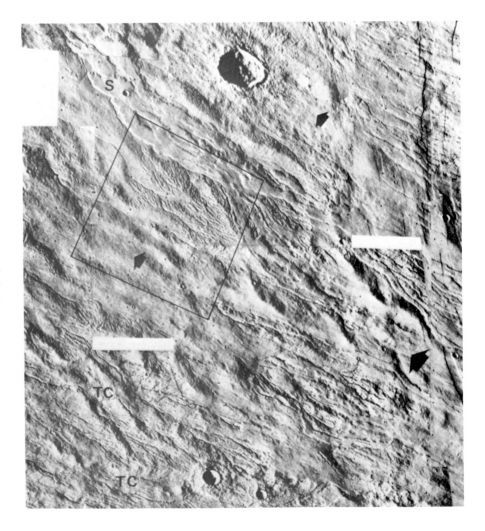

Figure 3.13. Oblique view of the northwest flank of Alba Patera centered at approximately 44°N, 116°W. Lava sheet flows (S) and tube-channel flows (TC) were emplaced by lavas moving in the direction of arrows. The outlined region was previously believed to be a fluvial dendritic drainage pattern, but is now thought to be a pattern of distributary lava tubes. (From Carr and others, 1977b, Jour. Geophys. Res., v. 82, p. 3985–4015, copyrighted by American Geophysical Union.)

Figure 3.14. Braided lava channels that developed in pahoehoe lavas erupted along the southwest rift zone of Mauna Loa, Hawaii. (Photograph by V. R. Baker, August 1979.)

Figure 3.15. Cinder-spatter cones that served as source vents for channel-fed lavas on the northeast rift of Mauna Loa. (Photograph by V. R. Baker, August 1979.)

3.14). Lava channels develop as large quantities of lava are distributed from source vents to outlying areas (Figure 3.15). In aa flows the lava rivers tend to occupy the center lines of flows feeding the actively advancing flow fronts (Figure 3.16). In pahoehoe flows the channels either crust over or develop arches by spatter accretion and periodic flooding (Greeley, 1971; Swanson, 1973). The result is a lava tube that continues to serve as a major artery for erupted lava (Figure 3.17). Many smaller lava tubes branch off the arteries to distribute lava to the distal margins of the flow. The largest tubes and channels serve as semipermanent conduits to continually changing and advancing distributary systems. When the eruption ceases, lava tubes may be drained of lava. The resulting loss of support can produce local collapse along portions of the tube, revealing its presence on aerial photography.

The mare regions of the Moon display numerous examples of winding valleys called "sinuous rilles." These features commonly head at a crater or

Figure 3.16. Aa lava flows on the southwest rift of Mauna Loa, showing complex overlapping relationships and prominent lava channels with flow levees. (Photograph by V. R. Baker, August 1979.)

Figure 3.17. Partly collapsed lava tube developed in the 1919–1920 pahoehoe lavas of Mauna Iki volcano in the Kau Desert, southwest rift zone of Kilauea Volcano, Hawaii. A natural bridge separates two collapsed skylights into the tube. The cinder cones on the horizon are Kamakaiauka (left) and Kamakaiawaena (center). (Photograph by V. R. Baker, August 1979.)

other depression, and they are largely restricted to probable basaltic lava flows. They are commonly attributed to lava tube and channel formation (Kuiper, Strom, and LePoole, 1966), although alternative suggestions range from erosion by water (Lingenfelter, Peale, and Schubert, 1968) to outgassing and fluidization of the regolith (Schumm, 1970). The major problems with the volcanic origin suggested for the lunar rilles are (1) their large size, and (2) the tendency of some rilles to have developed partly in nonvolcanic terrains. Problem (1) is not severe, since the entire scale of mare volcanism is magnified from the terrestrial cases. Lava flows were probably sustained for very long periods, allowing initially small tubes to grow to the dimensions that characterize the larger sinuous rilles. Problem (2) seems to require erosive capability by the sustained lava flows.

Carr (1974b) noted that there have been several terrestrial observations of lava erosion. He used a theoretical analysis to demonstrate lava erosion on surrounding rock. First the lava heats the wall rock of its channel or tube; the wall rock then melts and flows; finally it is incorporated into the lava stream. Carr's theoretical model predicts that sustained lava flow over several years could erode features hundreds of meters deep, assuming an eruptive temperature of 1,200°C and the maintenance of the gravitational slope necessary to sustain flow. Because heat loss from the lava is minimized by the formation of crusts and tube roofs, the erosion can occur over several tens of kilometers.

Erosive lava channels could result in a number of features that would be morphologically similar to fluvial landforms. Repeated eruptions can produce terraces. Channels can branch and rejoin in an anastomosing

TABLE 3.5. Compatibility of Various Fluid-Flow Processes with Certain Features of Martian Outflow Channels

Morphological Features	Wind	Mudflow	Glacier	Lava	Catastrophic Flood
Anastomosis	?	X	X	X	X
Streamlined uplands	X	X	X	?	X
Longitudinal grooves	X	X	X	?	X
Scour marks	X	?	?	?	X
Scabland	?	—	?	—	X
Inner channels	?	X	?	X	X
Lack of solidified fluid at channel mouth	X	—	X	—	X
Localized source region	?	X	X	X	X
Flow for thousands of kilometers	X	—	X	X	X
Bar-like bed forms	?	?	?	?	X
Pronounced upper limit to fluid erosion	—	X	X	X	X
Consistent downhill fluid flow	?	X	X	X	X
Sinuous channels	?	X	X	X	X
High width-depth ratio	X	X	X	—	X
Headcuts	—	?	X	—	X

pattern. "Islands" isolated by this branching might be modified by subsequent erosion to resemble alluvial islands. These and other considerations have led to suggestions that some Martian outflow channels formed by lava erosion (Cutts, Blasius, and Roberts, 1978). Schonfeld (1977) suggested that Martian lavas would probably have had low viscosity when erupted. Their turbulent flow might effectively mimic fluvial morphological features.

Carr (1974b) recognized that, although his model provided a reasonable explanation for lunar rilles, it could not explain many of the Martian channels. Most lava channels have a discrete source of volcanic origin, and they develop a distributary system in their distal reaches. Water channel networks have a tributary system in headwater areas. Water channels also show evidence of lateral movement and deposition, while lava channels remain relatively fixed. Finally, extensive lava accumulations must occur near the mouths of lava channels. Carr (1974b) explained that many of these morphological attributes of lava channels are missing in Martian outflow and "runoff" channels, while the attributes consistent with water flow are present.

Wind, Water, Earth, Ice, or Fire?

Suggested origins for the Martian channels have encompassed the entire range of fluidal processes. Even liquid alkanes have been proposed as the responsible fluids (Yung and Pinto, 1978). Table 3.5 summarizes the compatibility of various fluid-flow processes with certain morphological features of the Martian outflow channels. Every flow system can explain some of the important Martian channel macroforms and mesoforms (marked by X on the table). However, some channel forms are either questionable (?) or extremely unlikely (—) in all the flow systems save one. The analysis identifies the unlikely consequences of all the postulated flow systems. Only glacial ice and catastrophic flooding emerge without severe constraints. Only catastrophic flooding appears capable of producing the complete assemblage (Baker and Milton, 1974; Baker, 1978b).

As with many so-called "results" in science, the above conclusion is quite misleading. Indeed the question "Which fluid?" is overly simple. We are only at a point of departure for much more fascinating questions. If the outflow channels resulted from flood processes, what of the small valley networks and other channel types on Mars? Did they all form by aqueous erosion? Is there other evidence for water- or ice-related processes? What of the details in the outflow channels themselves? Did the hundreds of millions of years that have passed since flooding result in significant channel modification by other processes? Are there landscapes on Earth that formed by catastrophic flooding? How do such landscapes compare and contrast to Martian outflow channels? What of the mechanics of catastrophic flooding in the Martian environments? Were the Martian floods outbursts of clear water, or was this water highly charged with sediment? Was it choked or confined by ice? What are the implications of the Martian channels for the origin of Mars and the evolution of its atmosphere? These and other questions will be considered in the remaining chapters.

Chapter 4
Patterns and Networks
of Martian Valleys

This chapter will consider diverse valley features on Mars in which exogenic processes are the primary agents of formation. Excluded from consideration are troughs which are primarily formed by endogenic processes, e.g., tectonic grabens or volcanic calderas. The distinction is arbitrary, since endogenic troughs are commonly modified by exogenic processes (Sharp and Malin, 1975). This chapter will also refrain from describing the true "channels" of Mars, those troughs where fluid flow has left its obvious imprint on the valley floor. Many of the valleys described here may have also carried fluids, but the evidence is less obvious than in the outflow channels discussed in Chapter 6.

The term "dry valley" is used in terrestrial geomorphology to denote valleys which may have once carried streamflow, but which are now streamless. Such features are common in limestone regions, where the progressive development of subsurface drainage routes has occurred by solution. In limestone areas a system of subsurface solution conduits may eventually replace the system of surface valleys. Another cause of terrestrial dry valleys is climatic change. The climatic fluctuations of the Pleistocene locally altered the conditions that favor surface runoff, generally conditions of precipitation and evaporation. A decrease in the water balance that yields runoff could transform a terrestrial fluvial valley to a dry valley. Mars may have experienced an analogous adjustment in its valley networks.

River valleys and the streams that flow in them display an organized pattern of connection known as a drainage network. A drainage network is analogous to the veins of a leaf (Davis, 1899). Moreover, networks exist in spatially limited systems, known as drainage basins (Figure 4.1). A network of streams in a drainage basin constitutes a system that is adjusted to the hydrologic, geomorphic, and erodibility characteristics of the basin (Schumm, 1977). The network can be described in either of two ways: (1) topologically, considering the interconnections of the system through a scheme of stream ordering (Figure 4.2), and (2) geometrically, considering the lengths, orientations, and patterns of networks and their elements. Many types of drainage patterns occur on Earth, the most common being the dendritic pattern in which the branching of tributaries is tree-like in plan.

It is tempting to assume that the quantitative regularity known to exist in terrestrial drainage networks (Horton, 1945) may be used as a means of identifying the fluvial genesis of a landscape. Unfortunately many of the so-called "laws" of drainage network development derive from probabilistic theory rather than from clearly understood causes (Shreve, 1975). The study of valley networks on Mars requires an initial description of the phenomenon and a search for the broad similarities and differences from terrestrial drainage networks. This chapter will emphasize this initial phase of analysis. It is meant to be an introduction to a rich variety of phenomena on Mars.

Two properties of networks that may be useful for interplanetary comparisons are their junction angles and drainage density. Junction angles are the angles between the main stem (high order) segments and entering tributary (lower order) segments of networks. Pieri and Sagan (1978) compared junction angle statistics in terrestrial and Martian networks. They found that the small valley networks on Mars display very low junction angles in comparison to terrestrial drainage networks. Drainage density is the length of stream channel per unit of drainage area. It is one of the most important properties of terrestrial drainage basins, and it can be envisioned as a measure of basin efficiency at removing excess precipitation inputs as surface runoff (Patton and Baker, 1976).

The classification problems for the valley systems of Mars have been discussed in Chapter 3. The working approach of this review is to distinguish

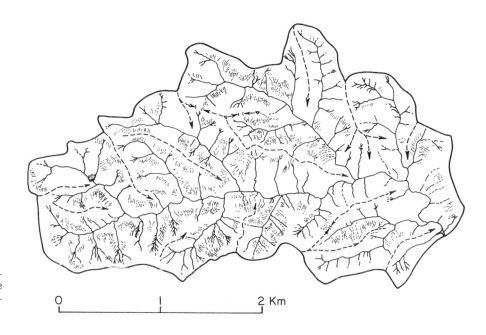

Figure 4.1 Detailed geomorphic map of a terrestrial drainage network developed on limestone bedrock. The heavy dashed lines are ridge crests, and the light hatchure-like lines are discontinuous hill-slope gullies. The heavy line outlines the drainage basin.

0   1   2 Km

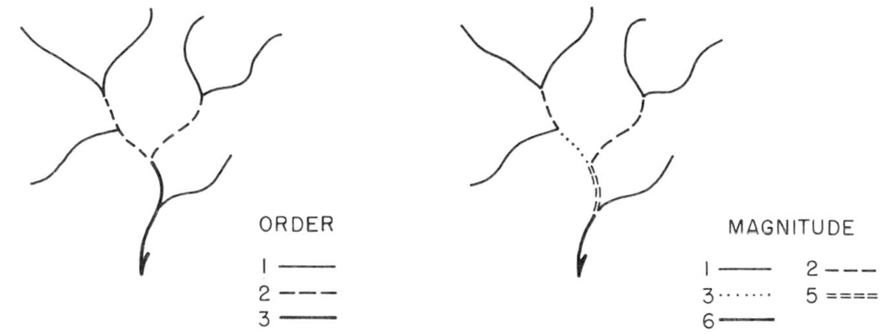

Figure 4.2. Methods of ordering the channels in a drainage network using the systems proposed by Strahler (1957) and by Shreve (1966a), left and right respectively.

ORDER

1 ——
2 – – –
3 ——

MAGNITUDE

1 ——   2 – – –
3 ·······   5 ====
6 ——

the obvious fluid-sculptured troughs (outflow channels) from the valleys which possess relatively little direct morphological evidence of fluid erosion. Pieri and Sagan (1979) adopted the same procedure and further divided the valley network systems into three categories: longitudinal, integrated, and slope valley systems. The longitudinal valleys are those quite large valleys with tributary systems but relatively poor integration of the tributaries into an overall drainage network. Examples are Ma'adim, Nirgal, Ladon, Nanedi, and Al Qahira Valles. The integrated and slope valley systems are clearly a separate class of phenomena. They correspond to the relatively small valley systems of the heavily cratered terrain of Mars. These small valley systems occur in almost all old cratered terrains, but they are absent in plains units equivalent to the Lunae Planum in age and younger (Carr, 1979b; Pieri, 1980b). Wallace and Sagan (1979) estimated that the number of individual valleys on Mars is several hundred thousand.

SMALL VALLEY NETWORKS

The relatively small valley networks of Mars have been given numerous names: "fine channels" (McCauley and others, 1972; Scott and Carr, 1978), "gullies" (Milton, 1973), "arroyos" (Hartmann, 1974a), "filamental channels" (Soderblom and others, 1974), "runoff channels" (Sharp and Malin, 1975), "furrowed terrain" (Mutch and others, 1976), and "dendritic channels" (Masursky and others, 1977). A consensus on the origin of the Martian valley networks has been as elusive as one on a name for the phenomenon (Baker, 1980b).

Despite the fact that the small valley systems have received the least study among the Martian "channel" types, they have been much cited as the best evidence for a denser atmosphere and warmer temperatures in the early history of Mars (Milton,

1973; Sagan, Toon, and Gierasch, 1973; Masursky and others, 1977; Carr, 1979b; Pollack, 1979). The reasons for the lack of study of the small valleys include the problems of image resolution and the complexity of valley development in association with impact cratering and other processes of valley modification. Pieri (1976) attempted the only comprehensive study of small dry valleys depicted on Mariner 9 imagery. He noted the tendency of the valleys to coalesce into low topographic areas, indicating fluid flow. He also noted the wide distribution of dry valleys and their association in geography and age with the heavily cratered upland terrains of Mars. However, problems of resolution prevented more definitive statements until the acquisition of the improved Viking pictures.

From his recent studies of Viking imagery, Pieri (1980a, 1980b) concluded that the small valley networks are very ancient. His crater counts on these channels are equivalent to crater counts on the ancient heavily cratered terrain in which the channels occur. Moreover, the valleys are consistently embayed by younger plains lavas. Carr (1979b) found more variability in crater counts from the small valley systems. He estimated about $6-9 \times 10^3$ craters larger than 1 km diameter per 10^6 km² for relatively fresh appearing valleys. However, other valley systems, such as Ladon Vallis, have a morphology that suggests formation prior to the decline of early, very high impact rates (Carr, 1979a). Thus, some small valley systems could be as old as 4 billion years. The age relationships for valley networks in the heavily cratered terrains are illustrated in Figure 4.3.

The formation of valley networks in the heavily cratered terrain appears to have extended over a considerable period of time. Some networks have a relatively fresh appearance in the Viking pictures (Figure 4.4), while others have a more degraded appearance (Figure 4.3). The cause of the degradation of these valley networks is not known. Possibilities include impact cratering and slope processes on valley sides, possibly influenced by ground ice in the Martian regolith.

The major morphological attributes of the small valley systems have been summarized by Pieri and Sagan (1979). The valleys occur in networks that tend to be parallel or quasi-parallel. Structural control of the pattern is sometimes evident, but the high-density dendritic networks common on Earth are notably absent on Mars. The parallel tributaries have their proximal terminations at theater-like valley heads. These tributaries tend to join the main valleys at mean junction angles that are much lower than in terrestrial drainage networks. Often the tributaries entering a trunk valley are as deep and as wide as that trunk valley. Moreover, the complete valley networks appear quite inefficient at filling space. Large intervalley areas remain undissected with no evidence of sharp drainage divides or competition for drainage (Pieri and Sagan, 1979). The individual valleys appear to have steep, crenulated walls, and they often show a relatively constant width downvalley. All these features are sufficiently different from terrestrial drainage networks that Pieri (1980a, 1980b) concluded that rainfall had little or no direct role in creating the small Martian valley networks.

Figure 4.3. Viking mosaic of heavily cratered terrain centered at about 26°S, 20°W. The scale is only approximate. Note the contrast between the large valleys, such as Ladon (L) and Samara (S), and the small networks in squares A and B. The latter two areas are shown in more detail in Figures 4.7 and 4.8. Ladon Vallis is intersected by several large craters, which presumably were emplaced during the primordial heavy bombardment of Mars by meteors. The small networks drain the lowland plains around the large craters. (Viking mosaic 211-5821.)

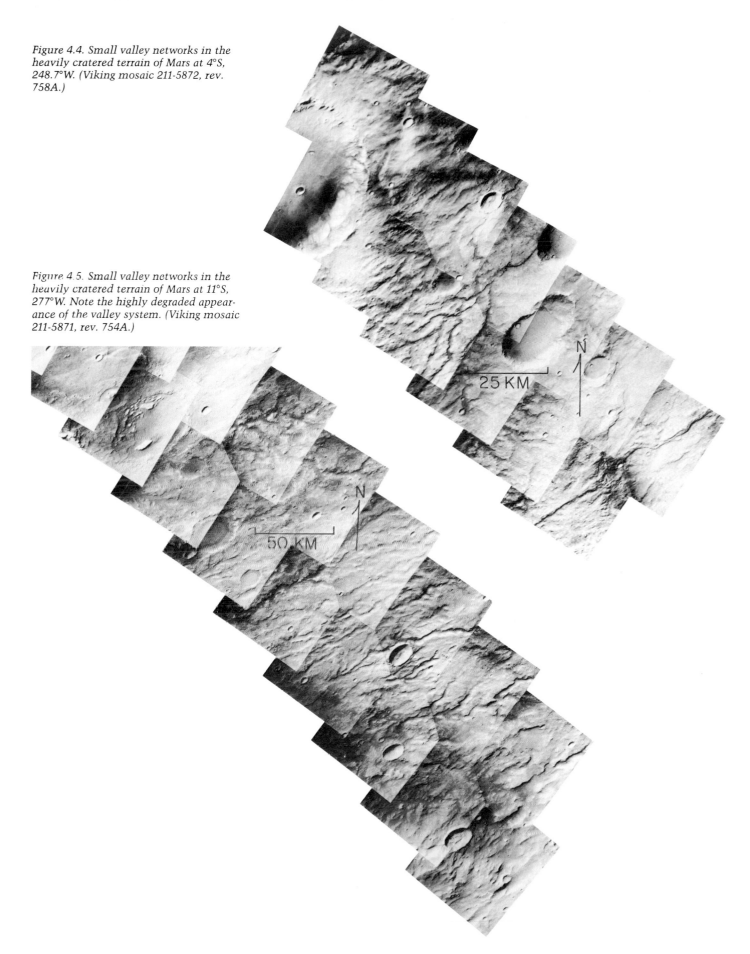

Figure 4.4. Small valley networks in the heavily cratered terrain of Mars at 4°S, 248.7°W. (Viking mosaic 211-5872, rev. 758A.)

Figure 4.5. Small valley networks in the heavily cratered terrain of Mars at 11°S, 277°W. Note the highly degraded appearance of the valley system. (Viking mosaic 211-5871, rev. 754A.)

Pieri and Sagan (1979) considered valley origins by a truth table analysis similar to that presented earlier for outflow channels (Table 3.5). As possible genetic processes they considered the following: fluvial, glacial, lava, debris flow, wind, structural-tectonic, dry sapping, wet sapping, alkane, and liquid or solid carbon dioxide. They found that either fluvial or wet sapping processes could explain the valley distributions, planimetric properties, and morphology. Moreover, many features of the valleys were more compatible with a lithospheric source of water than with a pluvial source. Their conclusion was that the valleys formed by a headward sapping mechanism coupled with downvalley planimetric development by fluvial erosion. The fluids were released from the subsurface at restricted source areas (the valley heads).

Margaritifer Sinus Region

The Margaritifer Sinus region of Mars contains one of the highest densities of small valley systems on the entire planet (Carr, 1979a). The Viking imagery of this region shows a variety of valley morphologies (Figure 4.6) that illustrate the properties of small valley networks. The valleys are developed on two major terrain types: (1) the hilly and cratered terrain, which displays rugged intercratered areas and is believed to be the oldest surface on Mars (Scott and Carr, 1978), and (2) the cratered plateau areas, which show relatively smooth intercrater plains that have buried or partially buried some craters.

Figure 4.7 shows valleys that cross the boundary between the two terrain types. A deeply entrenched system is evident in the hilly and cratered material. The rectangular pattern suggests a fracture control of valley development, and the whole network shows low junction angles and very low drainage density. Several impact craters are superimposed on the network. The valleys terminate in the cratered plains region, but the exact relationships are indistinct. The gradual shallowing of the valleys may result from percolation and/or evaporation of the fluids responsible for their genesis (Masursky and others, 1977). Alternatively, the cratered plateau

materials may be partially burying the valley systems. The occurrence of wrinkle ridges (W in Figure 4.6) suggests that the cratered plateaus may be underlain by extensive lava flows.

Another variety of Martian valley drains into prominent linear troughs in the cratered plateaus (localities T in Figure 4.6). The troughs are enclosed depressions that probably result from a subsidence process concentrated along fractures. They would be classified as endogenic depressions by Sharp and Malin (1975). Long, slightly sinuous valleys drain into these depressions from the surrounding cratered plateau.

One of the highest density valley networks on Mars is illustrated in Figure 4.8. Even here the pattern is not a true dendritic one, but rather subparallel with tributaries entering at low junction angles. Pieri, Malin, and Laity (1980) classify this pattern as "digitate." The superimposed craters indicate considerable age for this system. The network indicates fluid flows toward a complex lowland, showing a pattern of isolated flat-topped hills similar to the fretted terrain of the highlands-lowlands planetary boundary (Figure 4.9). Fretted terrain development is generally attributed to permafrost-related processes (Sharp, 1973b). The fretted terrain zone has a lower crater density than the valley terrain tributary to it, suggesting that the fretting process occurred after the valley network development.

The regional setting of this valley system is sketched in Figure 4.10. The mapping distinguishes large valleys with relatively broad floors ("major channels") from small parallel valleys ("filamentous channels"). The latter types are concentrated along crater rims and appear to represent patterns of outward fluid drainage. Figure 4.6 shows that similar crater-rim drainage systems are common around the larger craters. Thus, these small valley systems clearly postdate the period of massive planetary bombardment that is considered responsible for the largest craters. Other crater types associated with the small valley systems show various degrees of crater rim degradation and interaction with valley development. These are classi-

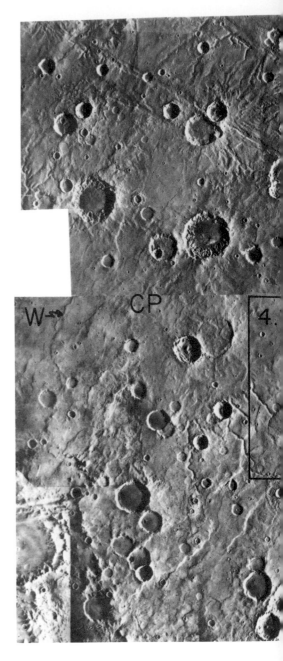

Figure 4.6. Viking mosaic of heavily cratered terrain in the Margaritifer Sinus region centered at about 30°S, 15°W. Compare to Figure 4.3. The scale is approximate. The boxes outline the approximate regions shown in Figures 4.7, 4.8, and 4.9. Note the contrast in crater density between the hilly and cratered areas (HC) and the cratered plateau (CP) areas as separated by the dashed line. (Viking mosaic 211-5207.)

100 KM

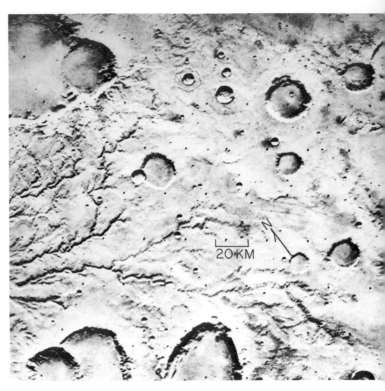

Figure 4.7. Valley systems along the boundary between hilly and cratered terrain (HC) and cratered plateau area (CP). Note regional setting from Figure 4.6. (Viking frame 084A06.)

Figure 4.8. Subparallel or "digitate" valley network development in the Margaritifer Sinus region of Mars. This valley system is also known as Parana Valles. (Viking frame 084A47.)

Figure 4.9. Cratered plateau lowland serving as baselevel for valley system shown in Figure 4.8. A portion of the valley network is visible at V, and fretted terrain (F) occurs at the valley mouths. Area is located at approximately 21°S, 10°W. (Viking frame 084A46.)

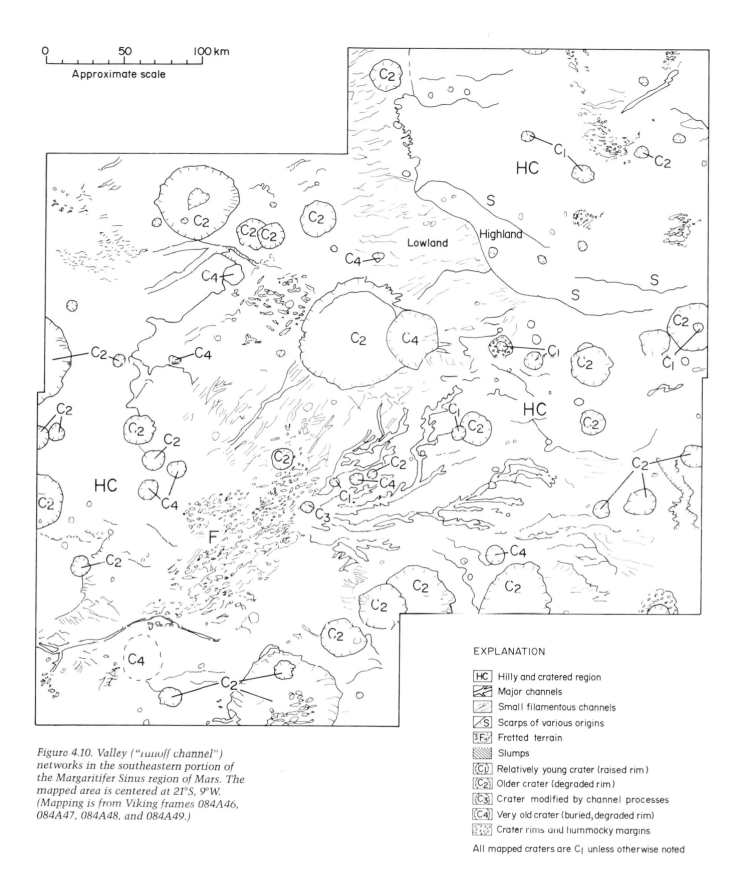

Figure 4.10. Valley ("runoff channel") networks in the southeastern portion of the Margaritifer Sinus region of Mars. The mapped area is centered at 21°S, 9°W. (Mapping is from Viking frames 084A46, 084A47, 084A48, and 084A49.)

EXPLANATION

HC	Hilly and cratered region
	Major channels
	Small filamentous channels
S	Scarps of various origins
F	Fretted terrain
	Slumps
C₁	Relatively young crater (raised rim)
C₂	Older crater (degraded rim)
C₃	Crater modified by channel processes
C₄	Very old crater (buried, degraded rim)
	Crater rims and hummocky margins

All mapped craters are C_1 unless otherwise noted

fied in Figure 4.10 as follows: (C₁) young craters, with raised rims; (C₂) older craters, with degraded rims; (C₃) craters that have been modified by valley formation; and (C₄) very ancient, highly degraded craters. Crater types C_1 and C_2 occur superimposed on the valley networks.

LONGITUDINAL VALLEY SYSTEMS

Nirgal Vallis

Nirgal Vallis is a striking valley system centered at approximately 29°S, 40°W. Its western half displays a tributary pattern, while the eastern half is a tightly sinuous, deeply entrenched valley (Figure 4.11). The valley debouches into a very wide, slightly sinuous trough named Uzboi Vallis. Another sinuous channel-like feature lies 80 km northwest of Nirgal Vallis. It flows about 100 km along a similar trend to Nirgal before disappearing abruptly along a zone of structural lineaments. Nirgal Vallis has a total length of about 700 km.

Nirgal Vallis is developed in a cratered plateau upland with Uzboi Vallis acting as a baselevel for the valley system (Figure 4.12). Wrinkle ridges on the cratered plateau upland suggest that mare-like basalts occur near the surface. Probably the basalt flows are underlain by impact breccia of the megaregolith. This is implied by the partially buried craters of the cratered plateau and by the exposure of probable megaregolith in the lower portions of canyon walls in the Valles Marineris, about 500 km to the north.

Milton (1973) recognized an unusual dual nature for the morphology of Nirgal Vallis. The downstream half of the channel system displays interlocking tight bends with sharp-pointed inner cusps and has an apparently low width-to-depth ratio. The downstream reach also widens irregularly. Milton suggested that all these qualities are more similar to the lunar sinuous rilles than to terrestrial meandering rivers. However, the upstream portion of the system has a hierarchical pattern of tributaries, not found in lunar rilles. Milton noted that this pattern somewhat resembles that of entrenched arroyo systems in the

southwestern United States. Because of the internal coherence of the Nirgal system, Milton rejected a twofold hypothesis of lava tube development followed by aqueous erosion. He speculated that the system may have developed by a process of partial melting in the Martian permafrost zone.

Schumm (1974) noted that the core of one meander bend in the middle reach of Kasei Vallis is a sharp projection rather than the usual rounded shape. Unlike most terrestrial rivers, this sharp core has an apex that coincides with the apex of the outer meander bend. Schumm suggested that Nirgal Vallis is a large tension fracture that has been or is propogating westward toward the unnamed sinuous depression located to the northwest. Sharp and Malin (1975) were not convinced that a purely endogenic hypothesis could explain the total pattern and configuration of Nirgal Vallis. They noted a strong resemblance between Nirgal's upper reach and terrestrial sapping and runoff features and concluded that fluidal activity was the principal agent of genesis.

Weihaupt (1974) examined the meandering pattern of Nirgal Vallis in an effort to determine whether it was consistent with what was expected of meandering rivers on earth. The three properties he measured were (1) the channel width, (2) the meander wavelength, i.e., the spacing of two alternating bends, and (3) the radius of curvature, i.e., the radius of a circle whose arc will fit the meander bends. For terrestrial rivers the ratio of wavelength to channel width ranges from 7 to 11 and the ratio of wavelength to radius of curvature is about 4.7 (Leopold, Wolman, and Miller, 1964). Weihaupt found that the respective ratios in the meandering reach of Nirgal Vallis ranged from 1.4 to 2.6 (wavelength/width) and 3.3 to 4.4 (wavelength/radius of curvature). Despite these discrepancies with the properties of terrestrial rivers, Weihaupt employed empirical relationships from terrestrial rivers to calculate expected mean annual discharges of a terrestrial river with comparable dimensions to Nirgal Vallis.

As recognized by Sharp and Malin (1975), Weihaupt's conclusions are suspect because of his use of valley

width, rather than true fluid-channel width, in the various empirical equations. Weihaupt also stated that discharges and flow velocities can be compared on Earth and Mars independently of the differing gravity terms. Komar (1979) gave a complete discussion of the significant influence of gravity on the calculation of discharge and velocity. The origin of valley systems on Mars cannot be established by the simple application of empirical equations.

Figure 4.13 is a detailed morphological map of Nirgal Vallis compiled from high-resolution Viking images. These pictures show that incision is relatively shallow upstream (Figure 4.14). Tributaries are very short and terminate at steep-walled, cuspate valley heads. Such valleys are sometimes called "amphitheater-headed valleys," indicating a similarity to classical theaters because of their flat floors and steep surrounding slopes. Cirques, which form at the heads of glaciated valleys, are a terrestrial example, but amphitheater-headed valleys are also well known in tropical landscapes, such as Hawaii (Hinds, 1925; Wentworth, 1928). The terminology is not precise, because a true amphitheater would be a closed oval, not a feature open at one end. A more appropriate term would be "theater-headed valley" (A. Bloom, written communication, 1980).

The "meandering reach" of Nirgal Vallis (Figure 4.15) is considerably wider than the 2 km average upvalley width. Width in this reach averages 5 km. The valley is deeply incised, and short, stubby tributaries occur along the steep valley walls. The downstream portion of Nirgal Vallis has a valley width of up to 15 km, and it is clear that much of the valley widening has occurred by landsliding (Figure 4.16). A hanging valley in this section shows that entrenchment was favored on one local branch of the valley system. The hanging valley is highly degraded in appearance, while the main valley shows more pristine wall morphology.

Nirgal Vallis is clearly a very different kind of valley from the common dendritic valleys of fluvial landscapes on Earth. On Mars it probably represents a relatively well-preserved

Figure 4.11. Regional setting of Nirgal
Vallis (N). Note the wide valley, Uzboi
Vallis, connecting the craters Holden and
Bond. An unnamed valley (W) is tributary to a system of linear troughs (T). Ladon Vallis (L) occurs in the heavily cratered terrain to the east.

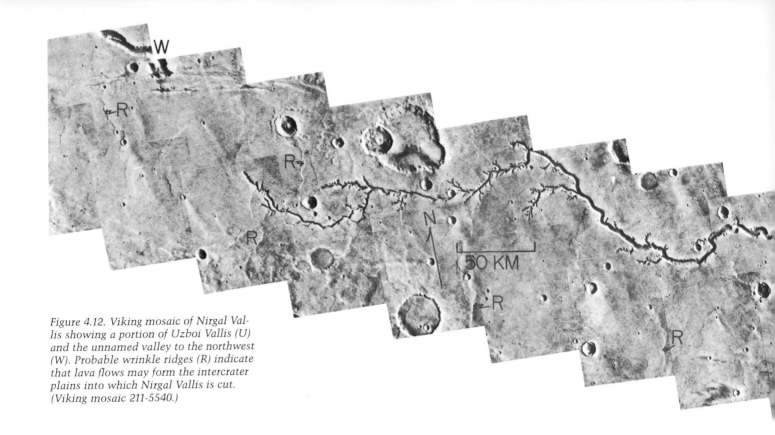

Figure 4.12. Viking mosaic of Nirgal Vallis showing a portion of Uzboi Vallis (U) and the unnamed valley to the northwest (W). Probable wrinkle ridges (R) indicate that lava flows may form the intercrater plains into which Nirgal Vallis is cut. (Viking mosaic 211-5540.)

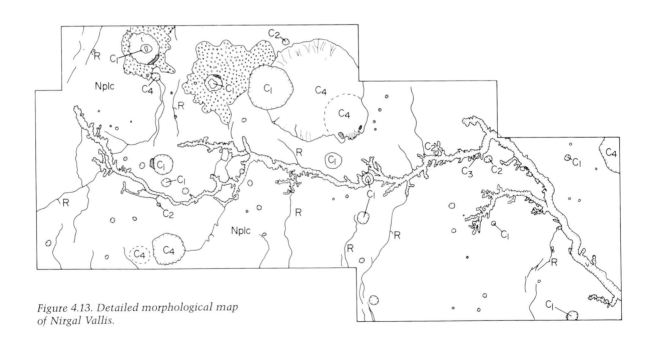

Figure 4.13. Detailed morphological map of Nirgal Vallis.

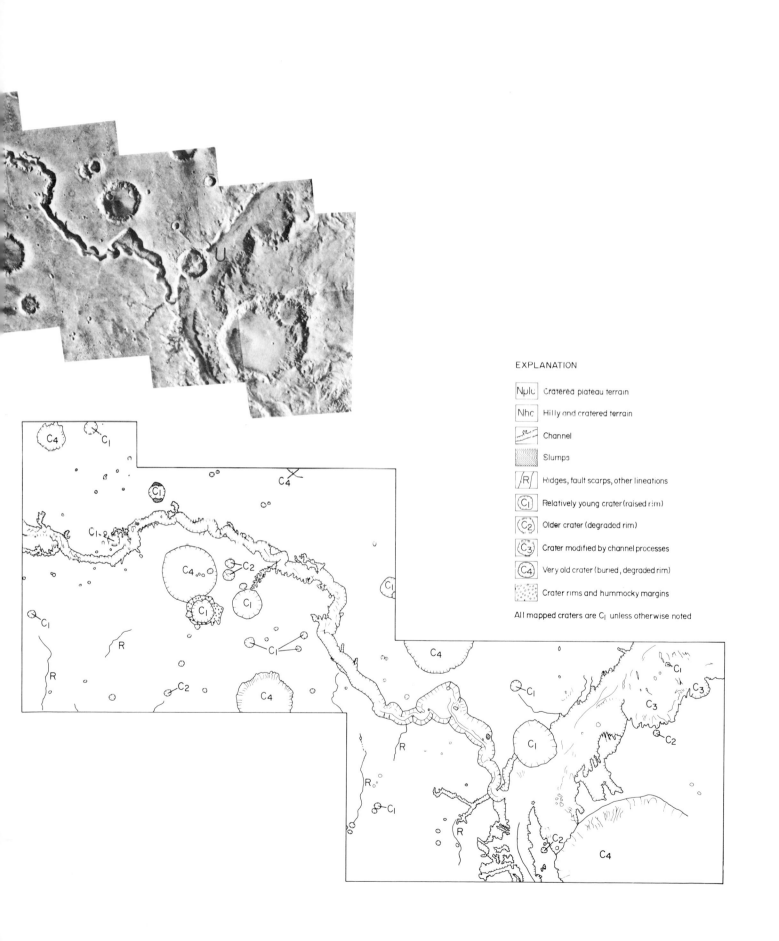

EXPLANATION

Nplc	Cratered plateau terrain
Nhc	Hilly and cratered terrain
	Channel
	Slumps
/R/	Ridges, fault scarps, other lineations
(C₁)	Relatively young crater (raised rim)
(C₂)	Older crater (degraded rim)
(C₃)	Crater modified by channel processes
(C₄)	Very old crater (buried, degraded rim)
	Crater rims and hummocky margins

All mapped craters are C_1 unless otherwise noted

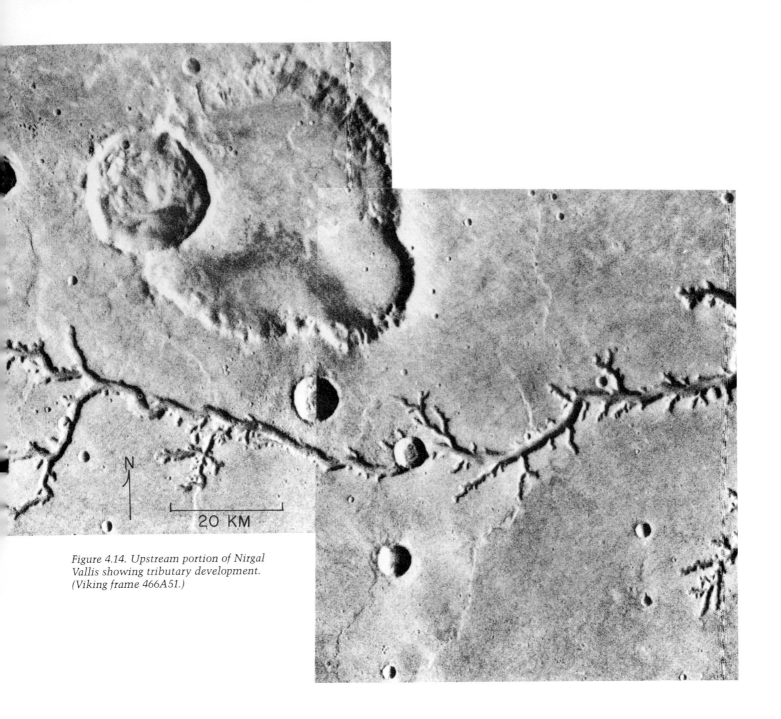

Figure 4.14. Upstream portion of Nirgal
Vallis showing tributary development.
(Viking frame 466A51.)

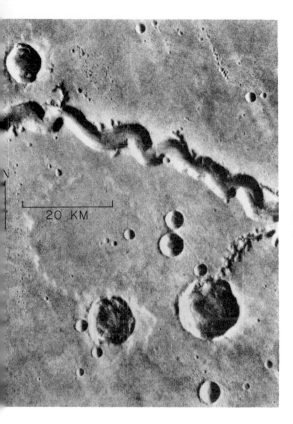

Figure 4.15. "Meandering reach" of Nirgal Vallis. (Viking frame 466A59.)

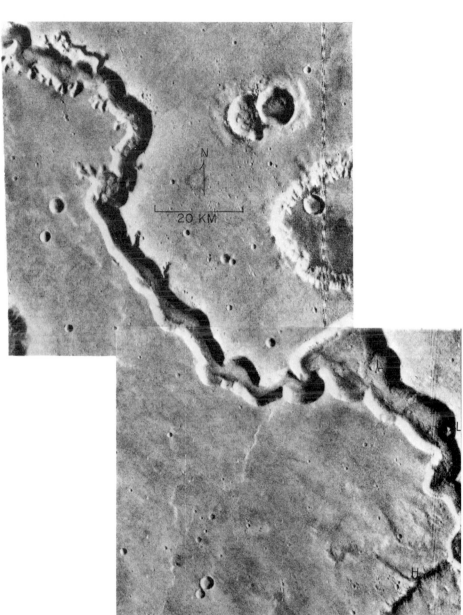

Figure 4.16. Downstream portion of Nirgal Vallis showing landslide deposits (L) and hanging valley (H). (Viking frames 466A61 and 466A64.)

end member of a spectrum of elongate valley types that are deeply incised and have theater-headed tributaries (Pieri, 1980a). The lack of dissection of adjacent uplands by these valley systems, the relatively short accordant tributaries with width and depth equal to the main trunk channels, and various morphometric properties provide strong arguments against the formation of such valleys by rainfall and surface runoff (Pieri, 1980b; Baker, 1980c).

Nanedi Vallis

Nanedi Vallis is a system of long, narrow valleys located on the cratered upland between Shalbatana Vallis and upper Maja Vallis. The system has two prominent, long branches (Figure 4.17) and is centered at about 7°N, 48°W. The length is about 500 km, and maximum width is about 4 km at the northern end (Sharp and Malin, 1975). The valley is highly sinuous, deeply entrenched, and has only a few short tributaries. Structural control is evident in several reaches, and the whole system more strongly resembles Nirgal Vallis than it does any other Martian channel types. Sharp and Malin (1975) concluded that Na-

nedi Vallis is a runoff channel possibly fed by seepage along its course.

Masursky, Dial, and Strobell (1980) found that Nanedi Vallis was a relatively ancient network. Its floor showed 3600 ± 250 as the cumulative number of craters greater than or equal to 1 km diameter per 10^6 km². This implies an age of 3.4 billion years on the lunar curve. The ridged plains into which the valley is cut have a crater number of 10,000 ± 2,000, which implies an age of 3.8 billion years on the lunar curve.

Bahram Vallis

Bahram Vallis is a long, sinuous valley located at about 21°N, 57°W, nearly midway between Vedra Vallis and lower Kasei Vallis (see Chapter 6). Unlike its neighboring valleys, Bahram does not show features that classify it as an outflow channel. It consists of a single trunk valley that is deeply entrenched into a heavily cratered upland surface (Figure 4.18). The upstream parts of the valley are cut in Lunae Planum materials (Theilig and Greeley, 1979). The channel walls display pronounced scalloping in this section. Theilig and Greeley (1979) attributed the scalloping and associated

blocky valley-side slopes to mass wasting of the valley walls, perhaps facilitated by undercutting. Their geological mapping shows easily erodible crater plateau material probably underlying Lunae Planum materials (lava flows). The relationships are similar to those at Nirgal Vallis, and the same processes may have formed both valley systems.

Bahram Vallis does display some streamlined erosional topography where fluid flows apparently diverged when they debouched from the valley mouth into Chryse Planitia (Theilig and Greeley, 1979). Immediately south of Bahram Vallis are several true outflow channels to be discussed in Chapter 6: Vedra, Maumee, and Maja Valles. These channels, like Bahram, transect Lunae Planum and cratered plateau materials, and eventually debouch into the Chryse Planitia. However, unlike Bahram, they display extensive evidence for fluid flow on their valley floors. Bahram may have developed by a sapping mechanism similar to that invoked for Nirgal, but the flow divergence features at Bahram's mouth indicate a fluid role in its development. The lack of appropriate upstream channel morphology

Figure 4.17. High-resolution Viking frames showing the western member of the two large valleys comprising Nanedi Vallis. Note the general similarity in morphology to Nirgal Vallis. (Viking mosaic 211-5540.)

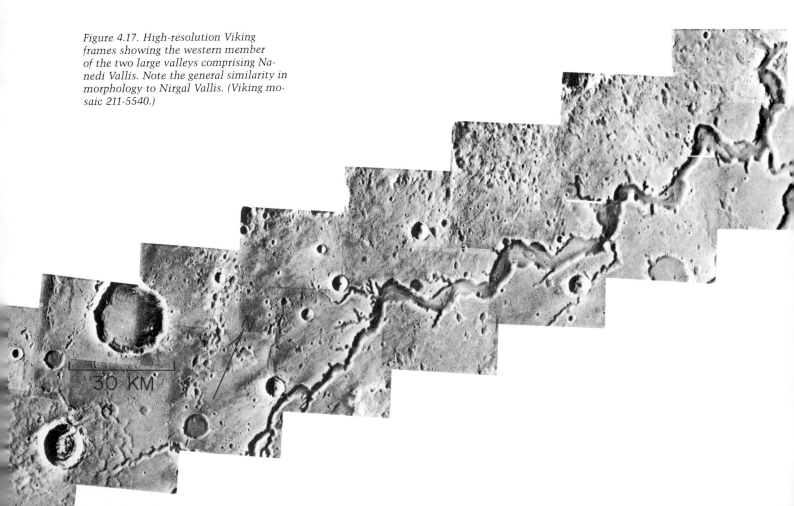

30 KM

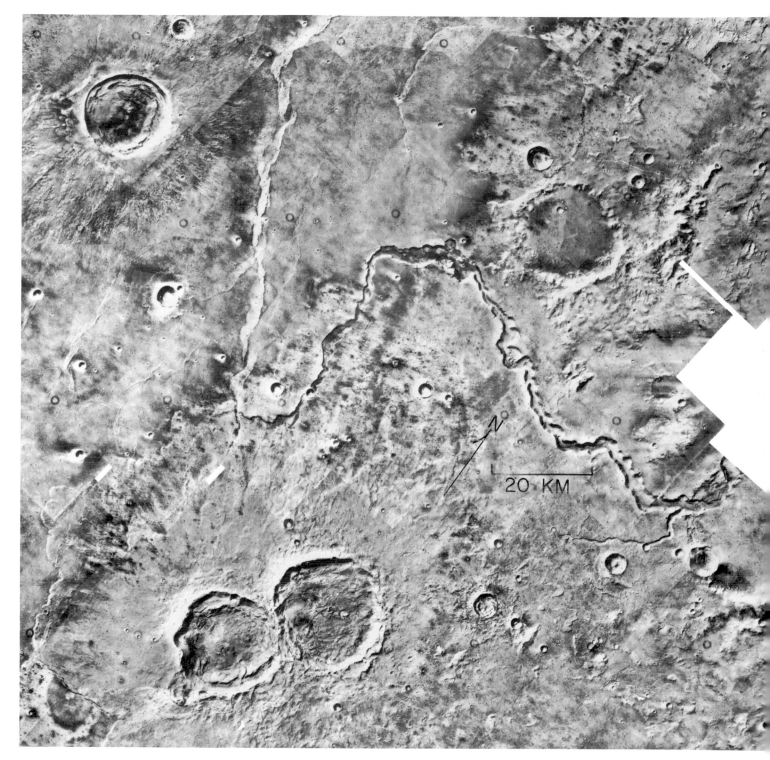

Figure 4.18. Viking mosaic of Bahram Vallis. Mass-movement processes have altered the primary morphology of the valley. The valley is incised into ancient cratered terrain and into the younger volcanic plains of Lunae Planum. (Viking mosaic 211-5189.)

rules against the large-scale flooding argued for outflow channels.

Crater counts on the valley floor of Bahram give a crater number of 2,900, in contrast to a crater number of 12,000 on the adjacent upland surface (Masursky and others, 1977). The system may be slightly older than nearby outflow channels, which have crater numbers of 1,500–2,500 (Carr, 1979b). Bahram Vallis probably formed by a sapping mechanism, similar to that invoked for Nirgal. However, Bahram occurs in a region that was later affected by massive outflows of erosive fluids, as discussed in Chapter 6.

Ma'adim and Al Qahira Valles

The Ma'adim channel system extends for about 700 km between south latitudes 29° and 16° and west longitudes 184° and 181°. The average width is about 15 km, widening to about 25 km near the mouth (Sharp and Malin, 1975). Among large Martian channels the system is unusual in several respects. Its upstream reaches (Figure 4.19) display long, branched tributaries. However, the gently winding downstream reach has only relatively short "box canyons" as tributaries. The system terminates at a smooth-floored 30-km diameter crater. Although the crater floor may be alluviated, a problem exists in accounting for the material eroded from the extensive valley system. Ma'adim also is unusually old for a large channel. Masursky, Dial, and Strobell (1980) found that its floor displayed 3,400 ± 200 cumulative craters per 10^6 km^2 that were greater than or equal to 1 km in diameter. This implies an age of about 3.4 billion years on the lunar curve.

Because of the tributary development near its headwaters, Sharp and Malin (1975) classified Ma'adim as a runoff channel. They suggested that headwater seepage areas probably provided the fluids for channel excava-

tion. However, the immense size of the untributaried downstream reach does not seem consistent with the relatively small headwater zone. Possibly the valley walls have been considerably widened by hillslope retreat that followed deep incision into the heavily cratered surface. The apparent great age for the feature suggests that long spans of time may have been available for channel modification.

Another large valley system, somewhat similar to Ma'adim, is Al Qahira Vallis, located about 800 km to the west at 18°S, 197°W. The system also displays a broad, flat-floored main valley, indistinct termination, and

heavy cratering. Tributaries are not as prominent as for Ma'adim. The great age of Al Qahira and Ma'adim Valles and their termination within the heavily cratered terrain suggests that both systems are related to processes within the cratered terrain and not to the highlands-plains boundary. Perhaps these large valleys began as small valley systems, but were subsequently enlarged by wall retreat in their deeply incised lower courses. Factors consistent with this hypothesis might be a greater concentration of ground ice at their locations and/or a greater length of time for modification processes to operate.

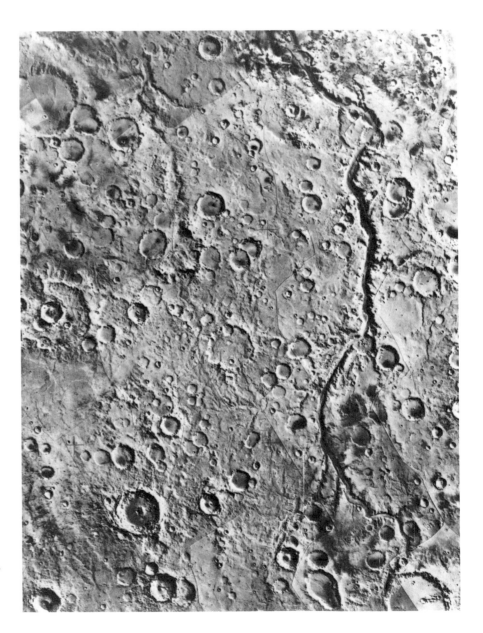

Figure 4.19. Ma'adim Vallis. Note the tributary development in the upstream reach and the heavily cratered nature of the upland surface. This mosaic of Viking orbiter frames shows an area of approximately 500 × 750 km. North is at the top of the image.

SLOPE SYSTEMS

Ius Chasma

Valleys or gullies marginal to the Valles Marineris have been the subject of considerable speculation. The pattern is angular and dendritic, displaying the hierarchical pattern observed in fluvial tributary valley systems. The valley width also tends to decrease headward as one would expect in fluvial valleys. As recognized by Sharp (1973a), however, headward sapping by water or ground ice is a more likely explanation for this pattern than is river action. Carr and Schaber (1977) noted that many of the gullies lack identifiable catchment areas, and each tributary terminates in a cirquelike valley head. The high-resolution Viking frames show V-shaped valley cross sections probably formed by talus slopes extending from both walls. Many valleys have a medial stripe along their center that may indicate downvalley movement of the talus, perhaps facilitated by interstitial ice. Some valley systems, such as those near Ius Chasma (Figure 4.20) assume an orientation along probable fracture sets. All these features seem best explained by headward sapping of the valleys along structural lines of weakness. Additional valley modification probably occurred by mass movement processes.

Seepage Gullies

Some evidence for wet seepage on Mars was presented by Sharp (1973a). He described a tableland of layered deposits on the floor of Gangis Chasma at about 7.5°S, 49°W. A somewhat similar tableland occurs in Ophir Chasma (Figure 4.21). The lower slopes that flank this tableland are dissected by closely spaced parallel, linear, unbranched, and largely unintegrated gullies as much as 7–8 km long and 300–400 m wide (Sharp and Malin, 1975). The gullies all head at a uniform level on the slope along the outcrop of a prominent dark stratum.

Many gullies abruptly terminate headward in "box canyons," and some gullies die out partly down the 12° slope (Sharp and Malin, 1975). The gully pattern seems best explained by fluid seepage along a distinct zone. Some gullies in the Valles Marineris appear to have been modified by eolian processes.

OTHER VALLEY SYSTEMS

Hellas and Argyre Valleys

Several long, unnamed valley systems are tributaries to the Hellas Planitia (Figure 2.2). Because of the relatively low resolution Mariner 9 imagery of this region, these channels have not been given much study. One channel (Harmakhis Vallis) heads in a quasicircular depression on the flanks of Hadriaca Patera at 34°S, 267°W and follows a sinuous course for 500 km to the Hellas Planitia. The valley appears to narrow downstream, suggesting a possible volcanic origin (Mutch and others, 1976).

Malin (1976), attempting to interpret low-contrast Mariner 9 pictures resulting from poor illumination conditions, concluded that the Hellas channels were mantled with debris. One high-resolution Mariner 9 B-

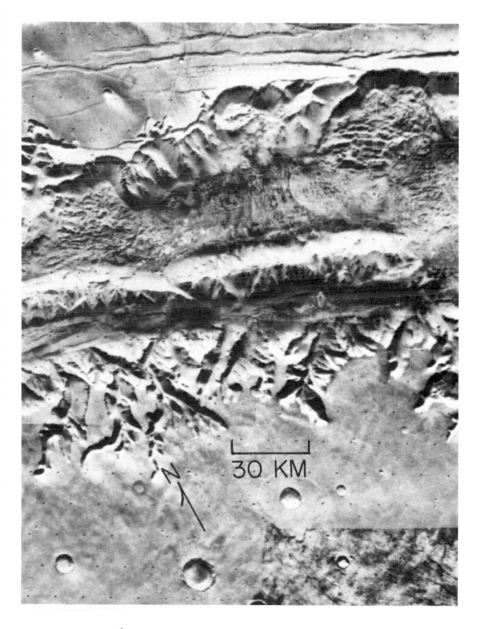

Figure 4.20. Portion of Ius Chasma showing tributary development on the southern canyon walls. (Viking mosaic 211-5208.)

Figure 4.21. Gully development on the flanks of a tableland in Ophir Chasma (3°S, 74°W). (Viking mosaic 211-5891.)

Figure 4.22. Upper reach of Reull Vallis, a long valley tributary to the Hellas Planitia. The picture is centered at about 41°S, 257°W. Prominent debris mantles surround isolated uplands, and the valley floor also appears to be filled with debris, probably introduced by periglacial processes (Carr and Schaber, 1977). The imaged scene is 233 × 279 km. (Viking frame 97A62.)

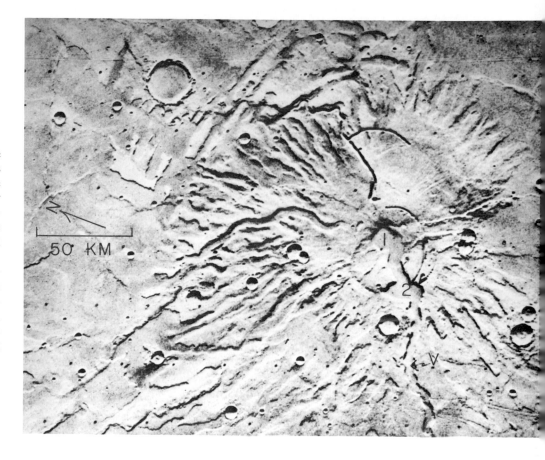

Figure 4.23. Tyrrhena Patera with its radial system of troughs (22°S, 253°W). Note the two calderas (numbered 1 and 2) and the long valley (V) that connects them to the volcano flanks. This radial volcano is representative of several other ancient patera on Mars. (Viking frame 087A14.)

frame image showed lineations on the channel floor and slump-like features along the walls. Viking imagery of the upper portion of one Hellas valley (Reull Valles) shows that debris mantling and mass movement processes have indeed extensively modified these channels (Figure 4.22).

Argyre Planitia, another huge impact basin on Mars (Figure 2.2), also has valleys which breach its surrounding ring of mountainous terrain (Figure 2.1). Pieri (1980a) described the breaching of Argyre's surrounding massifs at at least five major points. The inward-directed valley systems, centered on this basin, are an example of a radial centripetal pattern of valley development (Pieri, Malin, and Laity, 1980).

Tyrrhena Patera

Tyrrhena Patera is located about 2,000 km northeast of Hellas Planitia at 23°S, 253°W. Like many ancient patera features on Mars it displays a central caldera, ring-like fracture systems, and an outward-radiating system of troughs (Figure 4.23). The largest trough is about 5 km wide and can be followed from the central caldera through a second caldera and another 200 km down the volcano flanks. This trough is broad and flat-floored, and it narrows slightly with distance from the central caldera zones. Based on a study of Mariner 9 imagery, Carr (1974b) concluded that the radial pattern of troughs required erosion, most likely by lava flows. The downchannel decrease in width is consistent with this hypothesis. However, the valley is phenomenally large compared to terrestrial examples of lava erosion.

High-resolution Viking imagery of Tyrrhena Patera shows that the various troughs formed over a prolonged time period (Figure 4.24). The distal troughs widen downgradient and are transitional into isolated remnants of

the volcano (F), reminiscent of fretted terrain. Several of these troughs have distinct theater-like valley heads, i.e., they terminate headward in steep slopes, like "box canyons" (A). Several large rampart craters overlie the troughs, and others are apparently eroded by the trough-forming process. The heads of the longest troughs appear to be overlain by a mantling deposit (M); note especially point 1, where a prominent trough seems to be partially buried. This mantling unit is, in turn, cut by a deep, narrow canyon (C) and by the broad, shallow valley (V) that connects the two calderas. Additional troughs are developed along an inner ring fracture system (I) and an outer ring system (O). These temporal relationships are confirmed by the outer ring system cutting the older radial trough system at point R.

These relationships are not easily resolved by a simple hypothesis of lava erosion combined with tectonic collapse. The radial troughs have no apparent volcanic source, and their

apparent dissection of the old volcanic form also is not consistent with lava erosion. The development of fretted terrain and theater-like valley heads is similar enough to the fretted channels that a similar process may be inferred. The hypothesis of sapping erosion, probably associated with permafrost, is more consistent with the morphology. The sapping may have largely occurred before an explosive eruption phase that produced the mantling unit. The two troughs cutting this mantling deposit are headed at the ring fracture complex or in the central calderas.

The Elysium Channels

The northwestern flanks of Elysium Mons and nearby Hecates Tholus (Figure 2.2) show a series of very unusual sinuous valleys (approximately 37°N, 220°W). Like the outflow channels (Chapter 6), these valleys head at discrete source areas (Figure 4.25), contain streamlined uplands (Figure 4.26), and erode diverse terrains. However, their source areas are structur-

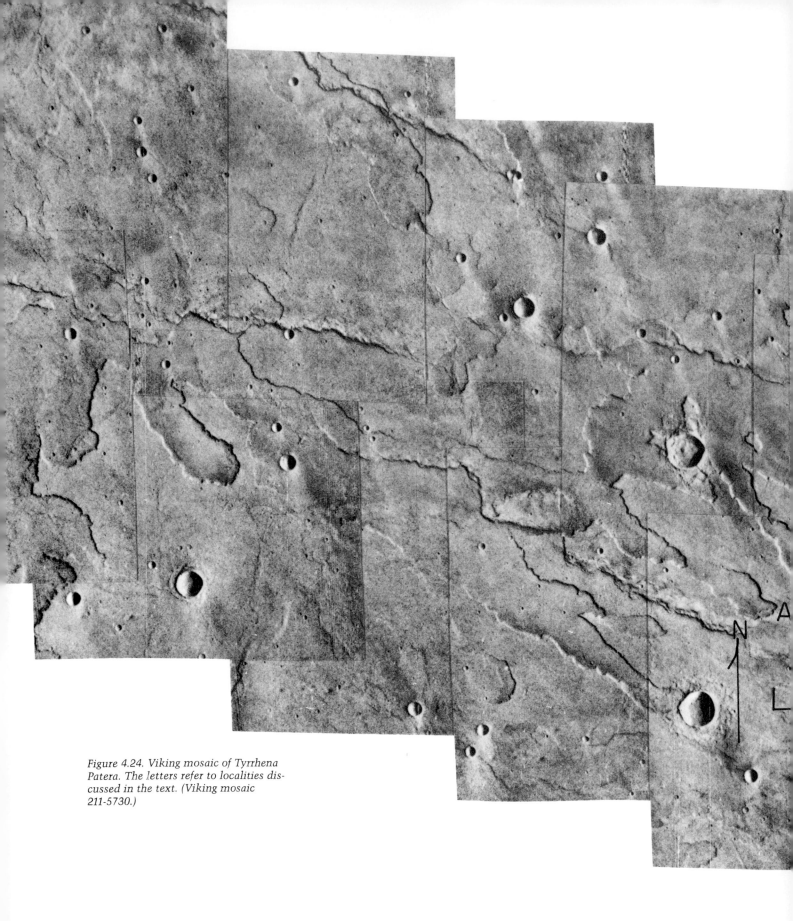

Figure 4.24. Viking mosaic of Tyrrhena Patera. The letters refer to localities discussed in the text. (Viking mosaic 211-5730.)

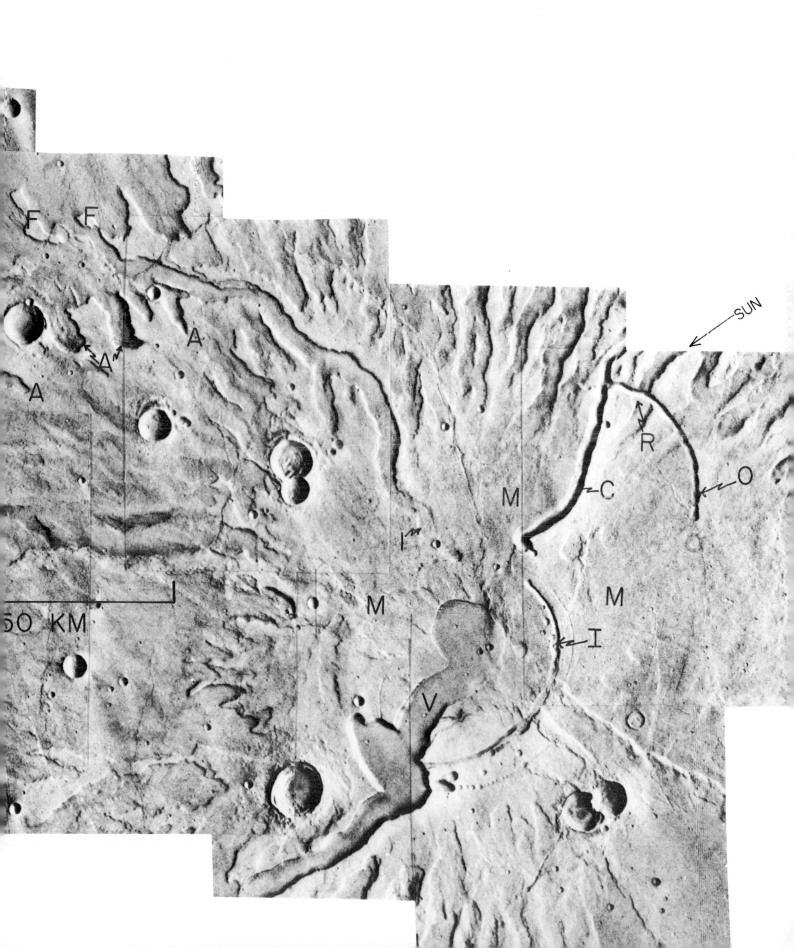

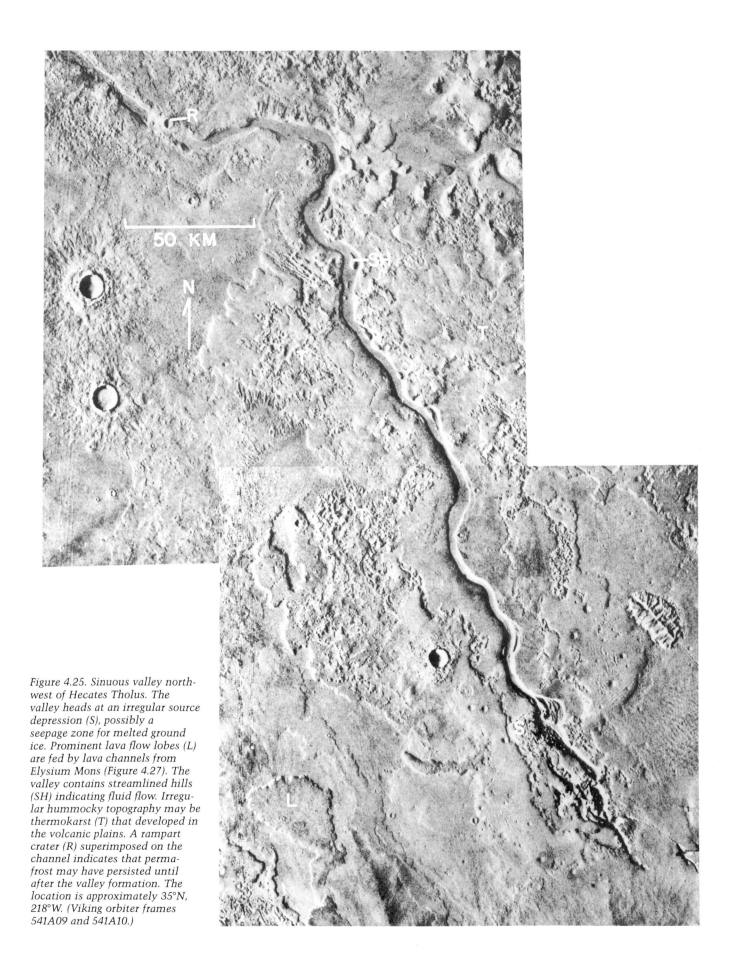

Figure 4.25. Sinuous valley north-west of Hecates Tholus. The valley heads at an irregular source depression (S), possibly a seepage zone for melted ground ice. Prominent lava flow lobes (L) are fed by lava channels from Elysium Mons (Figure 4.27). The valley contains streamlined hills (SH) indicating fluid flow. Irregular hummocky topography may be thermokarst (T) that developed in the volcanic plains. A rampart crater (R) superimposed on the channel indicates that permafrost may have persisted until after the valley formation. The location is approximately 35°N, 218°W. (Viking orbiter frames 541A09 and 541A10.)

Figure 4.26. Sketch map of streamlined hills in an unnamed channel west of Elysium Mons (locality B in Figure 4.27). Arrows show the postulated fluid flow directions along large erosional residuals (with secondary channels) and smaller streamlined forms.

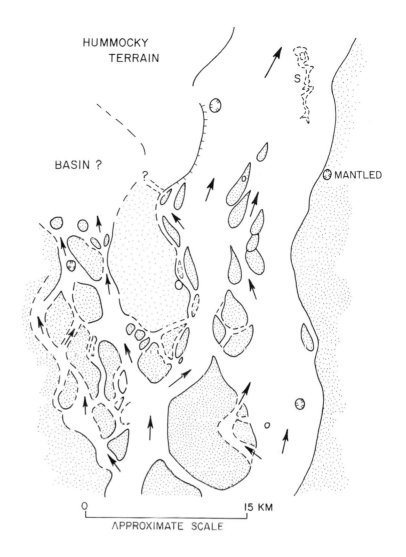

HUMMOCKY TERRAIN

BASIN ?

MANTLED

0   15 KM

APPROXIMATE SCALE

ally controlled zones of collapse on volcano flanks, not chaotic terrain zones as in the common outflow channels. The Elysium channels locally comprise anastomosing complexes (Figure 4.26).

Carr (1974b) considered that the Elysium channels were of probable volcanic origin, but he did not elaborate. Malin (1976) presented the most detailed description of the channels based on Mariner 9 data. He noted that these channels are unique among those observed in Mariner 9 imagery because (1) they occur wholly in volcanic plains units, outside the heavily cratered terrains; (2) they head in broad, deep, elongate depressions that Sharp and Malin (1975) interpreted as volcano/tectonic features; and (3) their cross-sectional relief is markedly subdued. Malin (1976) suggested that the subdued appearance of the channels might result from burial by a blanket of weakly consolidated material, perhaps ash from the nearby Elysium volcanic field.

Viking imagery of the Elysium channel area (Figure 4.27) shows that the channels are associated with a transition from constructional lava plains to fretted, mantled, and channeled plains. Fretted terrain, rampart craters, and chaotic terrain all indicate that ice-rich permafrost may have been important in the development of these distal plains.

The northern flanks of Hecates Tholus also show evidence that the material beneath its lava flows may have unusual physical properties (Figure 4.28). The margins of the shield volcano terminate in a highly irregular scarp. In several huge alcoves the upland terrain is broken into large angular blocks. Smaller blocks or mesas occur on the lowland plains to the north. The terrain appears somewhat similar to the chaotic terrain south of the Chryse Planitia, but there are important differences. Most apparent

is the small size of the rotational slump blocks and the apparent small amount of vertical subsidence involved. Rather, the blocks appear to have dispersed by lateral displacement toward the plains to the north.

The surface of Hecates Tholus south of the "chaotic terrain" shows numerous channels about 1 km wide. The larger of these channels originate near the volcano summit. They lack levees and do not coalesce as would be expected for lava channels.

The enigmatic association of fluvial-like and volcanic features is not limited to the Elysium volcanic province. Sharp and Malin (1975) described channels on Ceraunius Tholus, near the northeast end of the Tharsis Ridge at about 24°N and 97.6°W. Ceraunius Tholus is an elliptical dome 140 × 100 km. Numerous small channels radi-

ate from just below the rim and erode the hummocky surface of the volcano. As with the channels on Hecates Tholus (Figure 4.28), these channels lack levees, are noncoalescent, and have discontinuous sections. One much larger channel heads at the summit caldera, which measures about 12 km across. This channel is about 40 km long and 2 km wide. Unlike a lava channel, this feature shows no evidence of accretion, and it debouches into another volcanic caldera, forming a delta-like deposit. However, an extensive zone of lava accumulation is not present. This provides a strong argument against lava tube collapse or lava erosion. Sharp and Malin (1975) speculated that heated volcanic water, possibly rich in freezing-point depressants, might have created the Ceraunius Tholus channels. The caldera de-

Figure 4.27. Regional setting of the Elysium channels. Tectonic troughs (T) of the Elysium Fossae are local source areas. Elysium Mons volcano (E) is surrounded by ring fractures (RF), and lava channels (L) locally feed lava flow lobes. Beyond the lava-covered volcano flanks is the complex topography cut by large channels (localities A, B, and C shown in Figures 4.25, 4.26, and 4.28 respectively). Fretted terrain (F), rampart craters (R), and chaotic terrain (C) indicate probable permafrost-related processes. The mantling of some channels (M) may result from explosive volcanic eruptions. (Viking mosaic 211-5643.)

pression at the channel mouth could easily hold the necessary volume of eroded fluvial sediments, but it could not mask the immense volume of lava that would be required in the alternative hypothesis.

Reimers and Komar (1979) agreed that the slope channels on Ceraunius Tholus and Hecates Tholus have morphologies inconsistent with lava erosion, lava tube collapse, or tectonic fracturing. They believed that the radial, noncoalescent, parallel pattern of these channels resulted from explosive volcanic density currents. These currents were presumably generated when rising magma intersected a thick layer of ice-rich permafrost. The melting of ground ice provided water for base surge activity. Collapsed terrain on the northwest flank of Hecates Tholus (Figure 4.28) might be a result of this melting process. The length of the explosive phase of volcanism for Martian volcanoes would therefore be limited by the depletion time of the local permafrost reservoir. Volcanoes that continued to erupt beyond their explosive phase might bury older ash flow tuff deposits beneath younger lava flows.

A hypothesis for the Elysium channels emerges from the above considerations. The region probably experienced a history of equilibrium/disequilibrium in ice-rich permafrost induced by volcanism. Permafrost dissipation may at times have been relatively slow, producing the chaotic terrain margins of the original plains and the remnant upland of fretted terrain (Figure 4.28). Alternatively, local zones of permafrost may have occa-

sionally melted catastrophically or perhaps released water from confined aquifers (Carr, 1979b). The model is certainly speculative when based on the evidence in the local region. However, additional evidence for permafrost-related features on Mars (Chapter 5) is so pervasive that judgment should be reserved until the planet-wide story has been considered.

TERRESTRIAL VALLEY DEVELOPMENT BY SAPPING PROCESSES

Several varieties of Martian valley networks have been described in this chapter. Some Martian networks have been mapped in detail for comparison to terrestrial drainage networks that result from the runoff of precipitation. Despite the superficial similarity of some patterns, several important differences are apparent. Unlike most terrestrial drainage networks, the Martian valley networks display the following: (1) a relatively low drainage density, (2) an abrupt headward termination of valleys at scarps, forming theater-like or cirque-like valley heads, (3) abnormally long high-order "trunk" valleys contrasting with very short low-order tributaries, (4) the local occurrence of hanging valleys, (5) a morphology of relatively smooth valley floors contrasting with steep valley walls which often show prominent mass movement features, and (6) tributary junction angles lacking correlation to position in the network, as occurs in terrestrial dendritic networks (Pieri, 1980a, 1980b). The conclusion is that these features are more consistent with various subsurface sapping processes than with surface runoff (Baker, 1980b). The same conclusion was reached for several of the valley networks by Sharp and Malin (1975) and applied more generally to all the non-outflow type "channels" by Pieri and Sagan (1979) and by Pieri (1980a, 1980b).

Terrestrial sapping processes have received relatively little attention in recent geomorphic research. The following review is merely meant to illustrate the existence of possible terrestrial analogs to the Martian dry valley networks. An exhaustive analysis will have to await further research.

Terrestrial scarp recession by sapping occurs most readily where a massive, resistant lithology overlies a relatively weak, incompetent one. Undermining of the resistant layer occurs by weathering and ground-water flow along the contact between the two lithologies. Where scarp steepness is maintained, scarp recession will leave a smooth, nearly flat surface, often termed a "pediment." Because sapping is favored along certain zones of structural weakness, scarp recession is often differential, forming great embayments. Residual areas of upland may be isolated to form outliers, or "inselbergs," in the pediplain that is produced by scarp retreat.

Dunne (1980) outlined a complete model for terrestrial channel network evolution by spring sapping. The initial conditions require a water table with a regional slope toward a hydraulic sink provided by a depressed region. Water emerging along a spring line would then foster chemical weathering, thereby increasing the porosity of the seepage zone, reducing the local rock tensile strength, and rendering the weathered zone more susceptible to piping. Piping is the intraformational erosion of rock or soil by the mechanical action of ground-water flow (Parker, 1963; Kälin, 1977). Local zones of heterogeneity in the rock will result in some zones achieving the critical conditions necessary for piping before other zones achieve them. Joints, faults, and folds serve this function. These critical zones then experience piping, which quickly excavates a spring head and locally lowers the ground surface. Once initiated, this process becomes self-enhancing because the ground-water flow lines converge on the spring head. The increased flow accelerates chemical weathering which leads to further piping at the same site.

The further a spring head retreats, the greater the flow convergence that it generates, thereby increasing the rate of headward erosion. Headward sapping proceeds faster than valley widening because the valley head is the site of greatest flow convergence (Dunne, 1980). However, headward growth may intersect other zones that are highly susceptible to piping. A particularly favorable zone will re-

Figure 4.28. Chaotic terrain on the northern margin of the volcanic plains (V) of Hecates Tholus at 33°N, 213°W. The fretted plains to the north (F) probably developed by the dissipation of ground ice. The prominent rampart crater (R) also indicates permafrost processes and is 13 km in diameter. Small channels (C) are part of a radial pattern about Hecates Tholus that may have originated by explosive volcano-permafrost interactions (Reimers and Komar, 1979). (Viking mosaic 211-5274.)

sult in a tributary which also experiences headward growth, and which may generate tributaries of its own. Thus, sapping which develops in a zone of jointing or faulting will develop a pattern aligned with those structures. It will, however, be organized by the hydraulic controls on piping.

This process of piping, headward retreat, and branching eventually will form a network of valleys. The developing network works to counteract the self-enhancing effect of flow concentration mentioned above. As spring heads migrate to the neighborhood of one another, their demands for the available ground water compete with each other. Eventually an equilibrium will be achieved at some optimum drainage density.

On Earth in humid climates we can assume that the ground-water system is continually replenished by rainfall. The drainage systems can therefore proceed to the equilibrium state, provided sufficient time is available. On Mars, however, drainage maturity was apparently not achieved, probably because the subsurface reservoir was not replenished as efficiently as on Earth. Hence, Martian networks like Nirgal Vallis display extremely low drainage density.

Valley development by sapping is especially prominent in the sandstone plateau country of southeastern Utah (Figure 4.29) (Laity, 1980). This region has many examples of headward recession by steep walled valleys because of spring sapping at the base of massive sandstone caprocks (Figure 4.30). Angular patterns develop where jointing is prominent, and gently sinuous valleys occur in more homogeneous rock terrains. The valleys have long high-order segments and terminate headward in steep, cirque-like scarps. They also display hanging tributaries which enter the main canyon at waterfalls. Mass movement processes characterize the steep valley walls. Broad valleys develop by the recession of valley-wall cliffs. Laity and Pieri (1980) note that grain release by chemical weathering is an important process in the Colorado Plateau sandstone units. The granular disintegration of the sand-

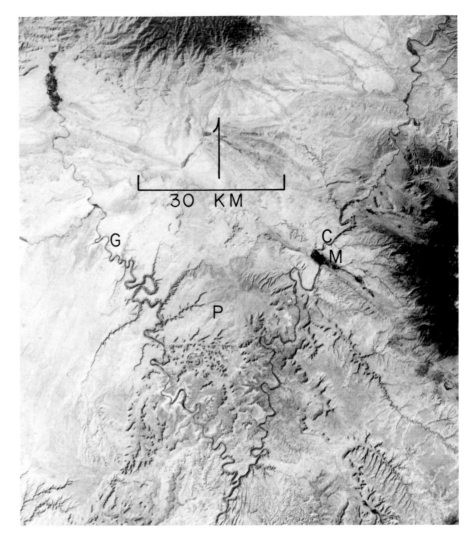

Figure 4.29. Landsat image of sandstone plateau terrains in eastern Utah. The two major rivers are the Colorado (C) and Green (G). Moab (M) is the largest town. Sandstone plateau areas (P) are being dissected by headwardly eroding valleys. (Landsat Image E-1409-17303-5.)

stone results in a lack of talus in valleys that are extending themselves by headward sapping.

VOLCANIC VALLEYS

The problem of limited water in Martian valley development might be effectively modeled by terrestrial sapping landscapes that were limited by time in their development. Young volcanic landscapes meet this requirement and also offer similar rock materials to those that probably occur in many Martian situations.

Valley development on terrestrial volcanoes has received surprisingly little geomorphic scrutiny. In the Hawaiian Islands, valley development by fluvial action is inhibited by the extremely high permeability of the lava flows comprising individual shield volcanoes. Volcanic ash and deep weathering may allow the initiation of some fluvial dissection, as along the Hamakua Coast of Hawaii. Penetration of this low-permeability layer, however, will again result in rapid infiltration by surface water into the highly permeable lavas.

The youngest Hawaiian volcanoes, Kilauea and Mauna Loa, generally lack surface streams. The relationship of fluvial valley development to former lava tubes and lava flow-unit boundaries is very pronounced in this early phase of development.

An important aspect of the "fluvial" valley systems in Hawaii is the role of ground water in their development. Major valleys grow at the expense of others because they tap larger supplies of ground water. Chemical disintegration is favored at the water table, and this leads to steep valley walls (Figure 4.31) (Wentworth, 1928). The characteristic valley form for the older Hawaiian volcanoes is U-shaped in cross section and ends in an abrupt theater-like head (Figure 4.32). Valley sides are steep-walled and sub-

jected to soil avalanches, slumps, and other mass movement processes (Wentworth, 1943). Because of the very high permeability of the lava flows, only the deeper valleys maintain perennial streams, even in high-rainfall regions. Headward valley growth occurs by spring sapping. Hanging valleys may be created when deep, headwardly receding valleys intersect and capture perched fluvial valleys (MacDonald and Abbott, 1970).

On the island of Maui, chemical weathering and volcanic ash falls have allowed considerable fluvial dissection of valleys. West Maui shows the huge theater-headed valleys that characterize the more mature erosional stage of Hawaiian volcanoes (Figure 4.33). The most interesting of these, from the viewpoint of com-

parative planetology, are those with large springs at their heads. The springs form in especially deep valleys that have eroded back into the dike complex of the volcano to tap ground water held in interdike compartments. The sapping of these springs results in rapid headward valley growth. This rapid erosional rate combines with a dominance of mass wasting and soil avalanches to yield very steep valley walls.

On Haleakala volcano (eastern Maui) a large caldera has been breached by two theater-headed valleys (Figure 4.34). The relationship of the valleys to the caldera is somewhat similar to that described for Tyrrhena Patera on Mars. On Haleakala the valleys were partly filled with lava flows by subsequent eruptions of the volcano (Figure 4.35). This relationship

Figure 4.30. Headwardly eroding valleys in the southeastern Utah sandstone plateaus. The U-shaped, steep valley heads are maintained by sandstone caprocks that are being undercut by spring sapping and slope retreat. (Photograph by V. R. Baker, August 1979.)

Figure 4.31. U-shaped valleys cut in the Pololu volcanic series on the northeastern side of Kohala volcano, island of Hawaii. The steep valley walls are subject to soil slips, soil avalanches, and waterfalls. (Photograph by V. R. Baker, August 1979.)

Figure 4.32. Amphitheater-headed valley on the southern flank of Haleakala volcano, island of Maui. Headward recession of such valleys occurs by spring sapping and/or valley head waterfall activity. (Photograph by V. R. Baker, August 1979.)

Figure 4.33. Dissected slopes of the western Maui shield volcano. (Photograph by V. R. Baker, August 1979.)

Figure 4.34. Summit caldera of Haleakala volcano, showing the breaching of the crater by Kaupo Gap, a former amphitheater-headed valley now partially filled with lava flows. (Photograph by V. R. Baker, August 1979.)

Figure 4.35. Kaupo Gap, Haleakala volcano, showing lava flows of the Hana volcanic series that poured seaward through the erosional valley that had breached the central caldera. (Photograph by V. R. Baker, August 1979.)

is also instructive for Martian situations. The Haleakala lavas partly fill a valley created by nonvolcanic erosional processes, and similar interactions between constructional volcanic processes (lava flows and ash falls) and erosional processes (sapping, permafrost wastage, and mass movements) probably account for the unusual relationships observed for the Elysium channels (Baker, 1980d).

SUMMARY

The discovery of branching, coalescing valley networks on Mars, first depicted on Mariner 9 images, led to immediate speculation that these were analogous features to terrestrial drainage networks. The paleoclimatic implications of surface runoff from rainfall are profound for Mars, but in retrospect, this conclusion was premature.

Pieri (1980a, 1980b) demonstrated that valley networks on Mars have theater-headed tributaries, short accordant tributaries with anomalous width and junction angle statistics, little or no intervalley dissection of upland terrains, pronounced structural control, and interior filling by mass-wasting debris. These attributes led him to conclude that the valleys formed by basal sapping in which ground water acted to undermine support in layered rocks. The networks developed by headward extension and bifurcation of the seepage zones, such as occur in valleys of the Colorado Plateau. There is some local evidence that surface flows of fluid occurred on valley floors and especially at valley mouths. Crater statistics and stratigraphic relationships indicate that this was an ancient process. Very little valley formation can be found that postdates the formation of the Lunae Planum.

Very large valleys occur on the margins of older volcanoes and patera landforms on Mars. These also appear to have formed by sapping because of the many similarities to theater-headed valleys that dissect the older Hawaiian volcanoes. The sapping agent was probably water released from local concentrations of ground ice. The role of ground ice is critical to landform evolution on Mars; Chapter 5 will review the geomorphic evidence and discuss its implications.

Chapter 5
Ice and the
Martian Surface

The Viking imagery of Mars has revealed a great variety of landforms which strongly resemble terrestrial features common in areas of cold climate. Terrestrial geomorphologists often apply the term "periglacial" to these landforms because their optimum development occurred adjacent to the borders of the great Pleistocene glaciers (French, 1976; Washburn, 1980). The term can be extended, however, to refer to phenomena produced in a distinctive climatic environment. Such an environment is characterized by very low annual temperatures, by many fluctuations above and below freezing, and often by very strong wind action. In this extended definition the term is applicable on Mars, even though that planet may never have had true glaciers on its surface.

Another term commonly applied to Mars is "permafrost." By strict definition this term is applied to terrestrial rock or soil material that has remained below 0°C for more than two years (Ferrians, Kachadoorian, and Greene, 1969). The term is independent of water content. In this strict sense water within a permafrost should be termed "ground ice." Nevertheless, most of the papers concerned with ground ice on Mars have used the general term "permafrost" as a synonym. Judging from the abundant morphological evidence on Mars for landforms that form in terrestrial regions of ground ice, it is likely that much of the Martian permafrost contained considerable water in the past, if not also at present. This review will describe evidence for an ice-rich permafrost for Mars. Although a true proof of this ice-rich zone requires actual samples of the deep regolith, the photographic evidence is so overwhelming that the conclusion is inescapable.

The presence of permafrost and periglacial features on Mars was suspected even before the recognition of diagnostic landforms on Mariner imagery (Anderson, Gaffney , and Law, 1967; Lederberg and Sagan, 1962; Wade and deWys, 1968). Following the success of Mariner 9, many analogies were suggested between terrestrial and Martian cold-climate terrains (Anderson, Gatto, and Ugolini, 1972, 1973; Gatto and Anderson, 1975; Lucchitta, 1978a; Sharp, 1973b). Studies of Viking imagery have supported these hypotheses and extended the list of processes that may have been operative (Carr and Schaber, 1977; Lucchitta, 1978b). A list of the possible periglacial and permafrost indicators on Mars is presented in Table 5.1. Although not all these landforms will be fully discussed in this chapter, the full list is presented with references to more extended discussions. This chapter will show that some problems remain in explaining some of the individual features, especially their immense size. However, the total list of features is exactly what would be expected in a well-developed terrestrial cold-climate landscape with abundant ground ice.

GLACIATION?

No features of unequivocal glacial origin have yet been recognized on Mars. This is perhaps surprising because, as will be discussed further, there is overwhelming geological evidence for a variety of processes involving ice. However, nearly all those processes involve ground ice, i.e., ice within the Martian regolith.

Northwest of Arsia Mons volcano (one of the Tharsis Montes) an immense lobe-shaped feature extends approximately 350 km from the base of the volcano (Figure 5.1). The lobe mainly consists of a hummocky blanket, but its outer margin is marked by continuous, subparallel curvilinear ridges. Individual ridges can be traced hundreds of kilometers around the lobe. Most interesting is the apparent superposition of these ridges right across craters and lava flow topography without apparent deflection (Figure 5.2). Carr and others (1977b) explained this lobe as an immense landslide shed from the lower flanks of the volcano. They did note that the outer margin of curvilinear ridges is difficult to explain by the landslide hypothesis. The ridges might result from compression in front of the slide, but this mechanism fails to explain the peculiar superposition of ridges over undeformed craters. Lucchitta (1978b) presented the alternative idea that both the ridges and the hummocky terrain could be explained by a former ice cap on the volcano. The ridges may have been superposed on the terrain as moraines that developed from a mixture of glacial ice and debris.

Postulating ancient glaciers on Mars requires overcoming some severe limitations posed by the present planetary environment (Sharp, 1974). Glacial flow, erosion, and deposition would be difficult to achieve because of the intense cold. Ice experiences a drastic drop in plasticity with lowering temperature (Glen, 1955). Glacial flow would require a warmer climate during a past epoch of Martian history. Alternatively, Lucchitta and Anderson (1980) calculated that geothermal heating might facilitate glacial flow on Mars.

Lucchitta, Anderson, and Shoji (1981) suggested glacial flow might even be achieved under present Martian conditions if the ice could attain certain critical thicknesses and then generate frictional heating at local zones of steep gradient. These authors described some striking similarities between erosional features in Martian outflow channels and features sculpted by terrestrial ice streams. Ice streams related to the present Antarctic and former Laurentide ice sheets produced grooved terrain and streamlined uplands of similar scale to those observed in the outflow channels.

Figure 5.1. Detail of lobate ridges on the western flank of Arsia Mons. The ridge pattern at top center (north) is reminiscent of that displayed by terrestrial glacial moraines. The mosaic frames are 210 km across. (Viking mosaic 211-5324.)

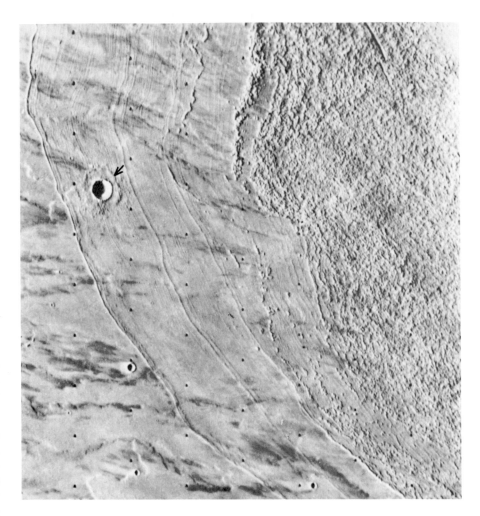

Figure 5.2. Viking mosaic (approximately 200 × 200 km) showing the western flank of Arsia Mons. The granular texture at the right of the mosaic is part of a larger lobe-shaped feature that is marked by curvilinear ridges on its outer margin (left). The ridges pass undeflected across a crater and its ejecta (arrow). This area occurs immediately west (to the left) of that shown in Figure 5.1. (Viking mosaic 211-5317.)

PATTERNED GROUND

Extensive portions of the northern plains of Mars display a polygonal pattern of fractures (Figure 5.3). Carr and Schaber (1977) noted that the distinctive crack pattern implied a contraction mechanism of some kind. The two most probable mechanisms were thought to be ice wedging, such as occurs in permafrost terrains, and cooling-contraction cracking, such as occurs in lava flows. Because of similar mechanics (Lachenbruch, 1962), the two processes probably cannot be distinguished by the crack pattern alone. The interpretation of the northern plains as volcanic is consistent with the cooling-contraction hypothesis (Morris and Underwood, 1978). However, the question can be raised as to why these lava plains among many Martian lava plains are the only ones to display the crack patterns.

The pervasive development of polygonal crack patterns on the northern plains implies a latitudinal control of the responsible process. Lucchitta (1978b) notes that polygons close to the ice cap have upturned edges around the cracks in a manner similar to actively developing terrestrial polygons of the ice-wedge type (Péwé, 1975). These observations appear most consistent with an origin that relates to permafrost processes. The major difficulty with this explanation is the immense size of the polygons. The "cracks" are really troughs hundreds of meters wide, with average spacings of 5–10 km (Pechmann, 1980). By contrast, terrestrial ice-wedge polygons generally vary from 1 to 100 m in diameter (Washburn, 1973). The huge scale is also inconsistent with cooling-contraction in lava flows, in which the

possible terrestrial analogs are even smaller than the permafrost polygons.

Pechmann (1980) proposed that the polygonal patterns of troughs on the northern plains of Mars resulted from extensional fracturing or graben formation. He invoked uplift or planetary expansion to generate regional, near-surface horizontal tensional stresses. Pechmann noted that the depths of cracks determine the scale of polygonal crack patterns in terrestrial lavas, desiccating sediments, and permafrost. Observations on earth suggest a lower limit on crack depths of one-tenth the average crack spacing. Thermal cracking in permafrost must occur near the surface, since ice can creep at depth to relieve stresses generated by long-term cooling (Pechmann, 1980).

If the Martian polygonal terrain formed by an ice-wedging process, the immense scale requires that the pro-

cess must have been very different than what is observed on Earth. The smaller terrestrial examples develop with diurnal or annual freeze-thaw cycles (Black, 1976). It is intriguing to speculate that the Martian polygonal cracks may have formed in a very thick permafrost layer, perhaps several kilometers thick. The upper "active" zone of this layer was then subjected to freeze-thaw cycles of extremely long period, perhaps on the order of a hundred thousand or even a million years. Mars has a rather effective mechanism for generating such cycles through insolation changes induced by its orbital mechanics, particularly its obliquity (W. R. Ward, 1973). Using a secular temperature variation of 1.2×10^5 years, Coradini and Flamini (1979) showed that the active permafrost layer on Mars could extend to about 100 m. Presumably the very thick active zone could promote

Figure 5.3. Fractured plains in the
Cydonia region, southeastern Acidalia
Planitia, Mars, at about 44°N, 18°W. The
polygonal fracture pattern resembles
crack patterns that develop from ice
wedging, desiccation, and lava cooling-
contraction. However, the Martian crack
pattern has an immensely magnified
scale. These polygons are up to 20 km
across. (Viking frame 32A18.)

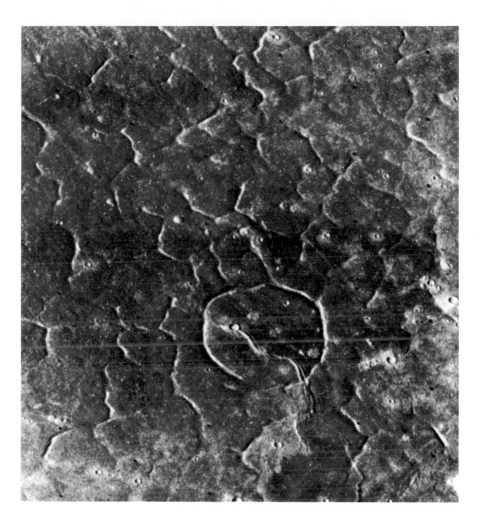

the development of immense tensional cracks during the secular freeze-
thaw cycles. Another proposal suggested that large-scale fracturing in a
thick (200 m) frozen silt layer might
be achieved by upward doming (Helfenstein, 1980). The doming would result from the progressive downward
freezing of an underlying aquifer.

Small-scale patterned ground may
manifest itself as a network of small
troughs at the Viking 2 landing site
(Figure 1.10). High-resolution images
obtained late in the Viking mission
show small-scale patterned ground
features outlined by ridges rather
than troughs (Evans and Rossbacher,
1980). The features occur near Lunae
Planum and have diameters of 400 m
to 1,000 m.

The northern plains of Mars also
display unusual curvilinear patterns
of sinuous ridges and troughs. The
possible terrestrial periglacial analog
is striped ground (Carr and Schaber,
1977). However, this suggestion is
limited by the fact that terrestrial examples tend to occur on slopes greater
than 3° and are no more than a few
meters wide (Washburn, 1973). The
Martian examples occur on what appear to be level plains, and the stripes
are up to several hundred meters wide
(Carr and Schaber, 1977).

THERMOKARST

Thermokarst is the process of melting
ground ice to produce local collapse of
the ground surface (Muller, 1944). The
extent of thermokarst development
depends on the ice content of the
ground material and on the degree
and rate of disruption of the thermal
equilibrium in the permafrost. The
process is most effective in materials
with high ice contents, such as the
eolian silt deposits of Siberia, which
contain up to 90 percent ice by volume (Czudek and Demek, 1970). This
ice may occur as segregated ice in layers, films, lenses, seams, and pods
(Péwé, 1975). It also occurs as wedge
ice in thermal contraction cracks
(Lachenbruch, 1962).

In addition to vertical collapse, thermokarstic development also proceeds
by a backwasting process (termed
"back-wearing" by Czudek and Demek, 1970). This process is commonly triggered by fluvial, lacustrine, or
marine erosion. For example, a spring
flood may undercut a riverbank and
allow relatively warm water to create
erosional niches at river level. Once a
vertical exposure is made in ice-rich
ground, the process becomes self-enhancing. Water from thawed ground
ice can be heated by the sun. If retained at the base of the exposure, it
will undercut the face, thereby inducing collapse and exposing fresh
ground ice to melting. The result is
extensive cliff retreat by the headward
recession of scarps exposing the
ground ice. Often the scarp retreat is
localized, perhaps by higher water
content, to produce a scalloped cliff
line known as "thermocirque" topography (Czudek and Demek, 1970).

The backwasting process may produce broad depressions with steep
slopes and flat floors. Some vertical
deepening of these features, called
"alases," may occur because of heat
retention by ponded meltwater on
their floors. Extensive alas development occurs in Siberia where the features range up to 15 km in diameter
(Washburn, 1973).

Sharp (1973b) cited backwasting
processes and ground ice exposed on
scarp faces as probable origins of the
fretted terrain on Mars. Either melt-

ing or sublimation of the ice could induce undermining and retreat of escarpment lips. Sharp also discussed the role of ground ice in forming the chaotic terrain of Mars. The principal problem is one of scale: several chaos areas are well over 100 km in diameter, and their bounding escarpments are 2–3 km high. An immense quantity of material had to be removed from these regions. If the material were magma, one would expect surficial manifestations of the volcanism. The association between many chaotic terrain regions and outflow channels is so pronounced that the two landscapes are almost certainly linked in origin.

Gatto and Anderson (1975) compared thermokarst terrain in the northern Alaska coastal plain to fretted terrain on Mars. Specially enhanced Landsat images of the Alaskan terrain show a pattern of irregularly shaped, flat-floored pits, basins, and valleys that is remarkably similar to fretted terrain shown on Mariner 9 B-frame imagery. One important difference is the larger scale of the Martian terrain features.

Carr and Schaber (1977) described probable thermokarstic features in the vicinities of both Viking lander sites. In the southern Chryse Planitia they interpreted irregular depressions and scalloped scarps of a tableland as alases and thermocirques respectively (Figure 5.4). However, the Martian example results in an extensive planation surface, whereas terrestrial alas development does not produce such large lowland surfaces. Near the Viking 2 landing site in the Utopia region, the terrain displays numerous extremely irregular depressions at a variety of scales. A channel-like feature, Hrad Vallis, appears to have developed by a coalescence of these depressions, perhaps as an alas valley that subsequently enlarged by receiving a more sustained meltwater flow.

Theilig and Greeley (1978) described a high albedo mantling unit overlying the heavily cratered terrain of the Lunae Planum region. This mantling unit, which is interpreted as eolian sediment, has locally been eroded to form steep-sided, flat-floored depressions averaging 8 km in diameter. Knobs of material along the inner de-

pression walls appear to be slump blocks from a backwasting process (Figure 5.5). The correspondence to terrestrial alas valleys appears correct even to the scale.

The heavily cratered terrain provinces of Mars also have features that appear to be of thermokarstic origin (Figure 5.6). Unfortunately, orbital images alone do not suffice to establish an unequivocal origin for various flat-floored depressions that are remarkably common erosional features on Mars.

HILLSLOPE MORPHOLOGY

The term "mass movement" is applied to various kinds of debris movements on hillslopes under the influence of gravity. It includes the phenomena commonly referred to as "landslides," but true sliding mass movements involve shear failure along two or more surfaces. Free fall, toppling motions, and flow are additional varieties of mass movement. Because these movements are part of a continuous series and because they often occur together

TABLE 5.1. Possible Periglacial and Permafrost Features on Mars

Process	Feature	Size on Mars	Size of Terrestrial Analogs	Reference
Patterned ground processes	Polygons	20 km diameter	1–100 m diameter	Carr and Schaber (1977)
	Stripes	1–2 km wide; 1–2 km spacing	0.1–1.5 m wide; 3–4 m spacing	Carr and Schaber (1977)
Thermokarst	Alases	10 km diameter	Several km diameter	Theilig and Greeley (1978)
	Coalescing alases (alas valleys)	10–100 km long	Several 10's km long	Carr and Schaber (1977)
	Fretted terrain	Scarps 1–3 km high	Relief of 10–100 m	Gatto and Anderson (1975)
	Chaotic terrain	100 km diameter	Several km diameter	Sharp (1973b)
	Sapping and ground ice collapse features	Several km high; several 10's km long	Variable, but smaller than Martian counterpart	Sharp (1973b)
Hillslope processes	Spur-and-gully topography	Several km high	Up to 1 km high	Lucchitta (1978a)
	Hillslope chutes	Several km high	Up to 1 km high	Sharp (1973a)
Mass movement	Debris lobes	10–50 km long	Variable, up to several km long	Squyres (1978, 1979)
	Landslides	1–100 km wide; 1–180 km long	1–10 km wide; 1–20 km long	Lucchitta (1978c, 1979)
	Solifluction	10's km long	Variable, up to several km long	Lucchitta (1978a)

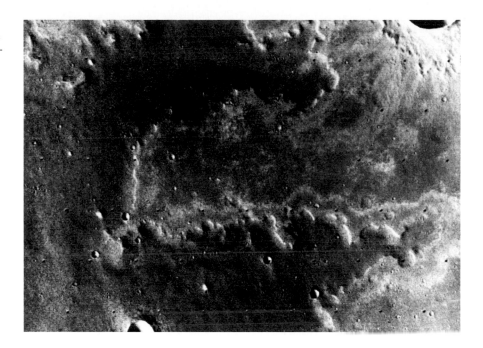

Figure 5.4. Possible thermocirque and alas development in the southern Chryse Planitia, as interpreted by Carr and Schaber (1977). The depicted area is about 45 km wide. (Viking frame 8A70.)

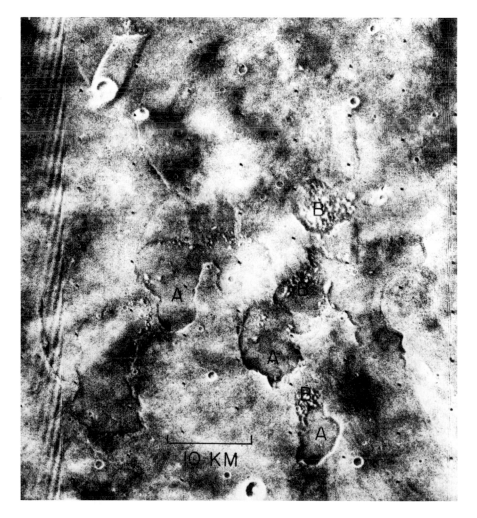

Figure 5.5. Steep-sided, flat-floored depressions that appear to have developed in an eolian mantle that locally overlies the Lunae Planum (Theilig and Greeley, 1978). The depressions are interpreted as alas valleys (A), and they locally contain knobs and blocks of preserved wall material (B). This area is centered at 8.8°N, 59.2°W. (Viking frame 074A09.)

Figure 5.6. Possible thermokarstic development in the heavily cratered terrain of Mars. The large flat-floored trough (T) appears to have developed by backwearing, consuming the surface craters (C) by scarp retreat. The drainage pattern of small valley networks suggests that the depression may occupy the summit of a broad upwarp aligned along the indicated fold axis. Another irregular depression (D) appears to act as the base level for a centripetal (inward-directed) drainage pattern. (Viking frame 097A28.)

in the same complex "landslide," a rigorous classification of mass movement phenomena is not possible.

Mars has numerous steep escarpments and hillslopes. The walls of the Valles Marineris system rise as much as 7,000 m from the adjacent canyon floor (Wu, 1979). It is not surprising then that these hillslopes display an immense variety of mass movement features, some of which are very difficult to interpret and classify even given the immense quantity of experience with terrestrial mass movements.

Many forms of mass movement involve debris that has been separated from intact rock by erosion. Sharp (1968, 1974) presented a strong case for the importance of freeze-thaw processes in shaping the details of Martian topography. He noted that if liquid water were currently available on Mars, frost shattering would be intense, creep would be widespread, and various periglacial landforms might be prolific. Data from the Viking landers seem to indicate that the present-day Martian surface is somewhat colder than what was assumed theoretically when Sharp made this prediction. However, the key point of the argument was that various periglacial processes had probably operated during past epochs when liquid water was available at the Martian surface. Presumably these would also have to have been relatively warmer epochs as well, involving prolonged periods of diurnal atmospheric temperature fluctuation across the freezing point of water.

Frost shattering may even be possible in the present cold environment of Mars if salt-rich frosts are invoked. Malin (1974) used thermodynamic

considerations to show that some salt solutions could exist in liquid phases during part of the modern Martian diurnal cycle, at least in the temperate latitudes.

Slope morphology in the Valles Marineris region often displays a very distinctive form, first recognized by Sharp (1973a). A sharp brink separates the trough walls from the upland surface, and the steep upper walls are characterized by U-shaped chutes. The chutes may lead directly to talus slopes (Figure 5.7) or to a complex of small ridges separated by open swales. The latter topography, which is quite striking on Mars (Figure 5.8), is termed "spur-and-gully topography" by Lucchitta (1978a). The spur-and-gully topography often grades directly into talus slopes. The entire talus slope may be several thousand meters in height and usually has a slope angle near 30° (Lucchitta, 1978b).

Sharp (1973d) noted that the size and abundance of Martian hillslope chutes exceeds anything now recog-

nized on Earth. He speculated, based on the available terrestrial analogs, that the dry avalanching of loose granular or fragmental materials was required for their formation. Martian avalanche chute development is consistent with the hypothesis that much of the near-surface material exposed in the canyon walls is highly fragmental debris, perhaps a megaregolith of impact-shattered rock.

Where larger chutes have developed along valley walls, talus cones and small debris fans develop locally at their base (Figure 5.9). These larger accumulations of debris are made possible because of the larger source area contributing frost-shattered rock.

DEBRIS APRONS AND ROCK GLACIERS

Viking pictures of the fretted terrain regions along the highlands-plains boundary of Mars revealed spectacular piedmont aprons around residual uplands (Carr and Schaber, 1977).

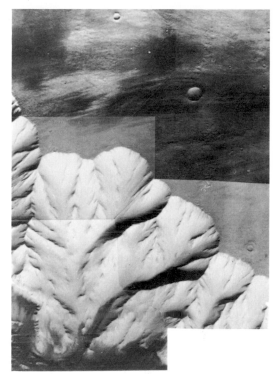

Figure 5.8. Spur-and-gully topography in the Valles Marineris. This example is on the Melas Chasma escarpment (approximately 13°S, 70°W). (Viking mosaic 211-5819.)

Figure 5.7. Talus slopes (center) and a prominent debris flow (right center) in northern Kasei Vallis (27°N, 67°W). The frame depicts a scene approximately 30 km wide. The debris flow is similar to terrestrial examples in which water contributes to mobilizing the sediment. (Viking frame 665A24.)

Figure 5.9. Talus cone and debris fan development in northern Kasei Vallis (27°N, 66°W). The frame is approximately 30 km wide. (Viking frame 665A28.)

Figure 5.10. Geomorphic maps of debris aprons in the fretted terrain of Mars. Note the extensive debris aprons surrounding residual upland plateaus and along upland margins. Dark streaks indicate depressions associated with ridge and furrow systems developed on the debris apron surfaces. Dotted regions are the debris accumulations. (a) Deuteronilus Mensae region, centered at 46°N, 321°W (from Viking frames 058B55–57); (b) Protonilus Mensae region, centered at 44°N, 314°W (from Viking frame 058B61).

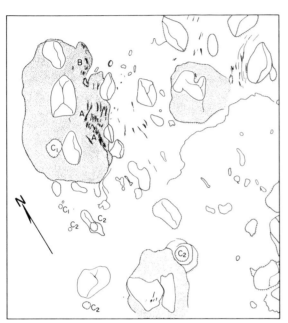

These are well illustrated by examples in the Deuteronilus and Protonilus regions (Figure 5.10). The images suggest an outward flow of debris radially from the upland massifs. Flow lines show that the debris was locally deflected around obstacles (Figure 5.11). The lobate form and convex profiles of the debris aprons are additional evidence of flow by the debris. A different pattern occurs where the residual uplands are closely spaced, as in the Nilosyrtis region (Figure 5.12). Here abrupt escarpments rise to about 1 km in height and slope 10–40° (Squyres, 1978). The valleys between the upland surfaces contain debris that is locally marked by prominent longitudinal ridges. The overall pattern is reminiscent of terrestrial glaciation. Some valleys have cirque-like heads, and there is

one instance of a distinct lobate debris bulge where a valley closed at one end opens onto the lowland plain (Figure 5.13). However, this may be the only clearly demonstrable case of through-valley flow (Squyres, 1978). It is probable that the longitudinal valley floor lineations resulted from compressive forces generated as debris flowed toward the valley centers from the valley walls.

Carr and Schaber (1977) interpreted these features as the result of frost creep and gelifluction, processes that involve ice action in terrestrial regions of cold climate (Washburn, 1973). Frost creep involves surface layers of the regolith that alternately heave by frost action and settle by thawing. Since the heaving is normal to the slope and the settling is nearly vertical, the process produces a net down-

slope movement. Gelifluction is the slow downslope movement of water-saturated debris in a region of considerable ground ice. Because the permafrost table inhibits drainage into the regolith, the near-surface layers become saturated during seasons of thaw. While these mechanisms are very effective in terrestrial cold-climate regions, their application on Mars requires certain assumptions. Foremost is the need to achieve repetitive thaw cycles, which appears impossible in the present climatic regimen. The implication is that the processes occurred under different climatic conditions in the Martian past. Preliminary crater counts on the Martian debris aprons do suggest considerable antiquity, perhaps a billion years or more (Carr and Schaber, 1977).

Squyres (1978) suggested an alterna-

Figure 5.11. Debris aprons surrounding fretted terrain uplands in the Protonilus region of Mars, 44°N, 315°W. Note the radial flow lines and local deflection around obstacles (arrow). No craters occur on the debris aprons, indicating a very young surface by Martian standards. (Viking mosaic 211-5266.)

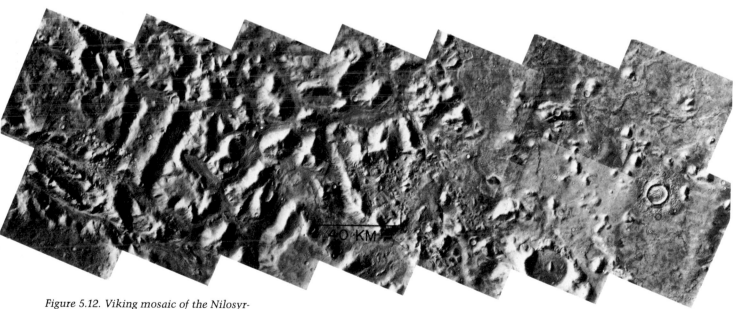

Figure 5.12. Viking mosaic of the Nilosyrtis region, 34°N, 282°W. Fretted terrain uplands are surrounded by valleys that appear to be filled with debris from adjacent slopes. Subparallel ridges and grooves on this debris imply flowage. Craters are generally absent on the debris surfaces. (Viking mosaic 211-5207.)

Figure 5.13. A portion of the mosaic in Figure 5.12 showing the details of valley morphology, including cirque-like valley heads (C), longitudinal ridges (L), tributary junctions (T), and debris fans (F) from adjacent slopes. At point R a debris filling of a valley terminates in a distinct lobate bulge. (Viking frames 84A73 and 84A74.)

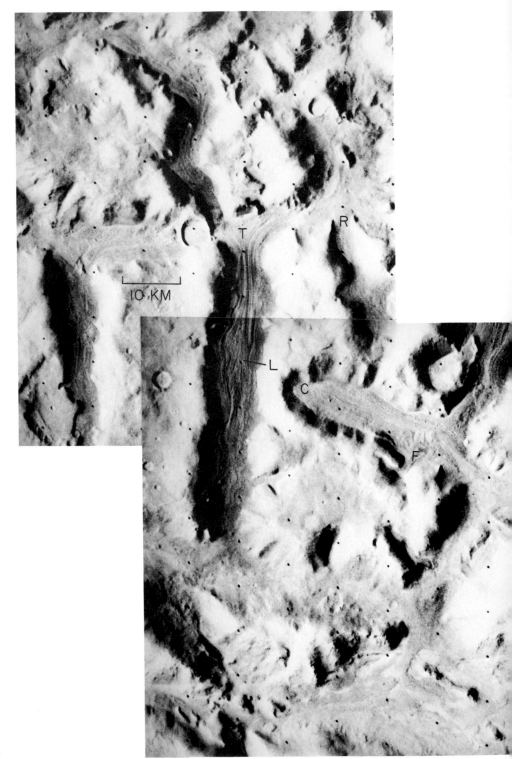

tive model for the debris aprons on Mars. He drew the analogy to the terrestrial environment of rock glacier development. Debris is presumed to be loosened from escarpments and transported downslope by mass wasting processes. At the escarpment base this debris incorporates seasonal accumulations of ice from the atmosphere. The resulting mixture of debris and ice behaves similarly to a rock glacier. The mass deforms in response to continual loading of fresh debris, flowing away from the escarpment in order to maintain an equilibrium between basal shear stress and the yield strength of the material. When an obstruction or another flow is encountered, lineations are generated by the compression. The lack of craters on the debris surfaces indicates that this process has acted relatively recently in Martian history. Lineated floor materials occur in large Martian valleys and in craters (Figure 5.14).

LANDSLIDES

Sharp (1973a) described the huge landslides revealed by Mariner 9 images of the Valles Marineris. He noted that landsliding was probably one of the major processes in shaping the great trough areas of Mars. Sapping processes along the margins of the troughs were considered to be the long-term cause of the landsliding.

Viking imagery of the Valles Marineris system provided considerable new data on the Martian landslides (Blasius and others, 1977). Lucchitta (1978c) described twenty-five large slides with deposits ranging from 40 km² to 7,500 km² in area. Some of the largest landslides occur in Ius Chasma (Figure 5.15), where one measures about 100 km wide and is recessed 30 km into the escarpment wall. Eleva-

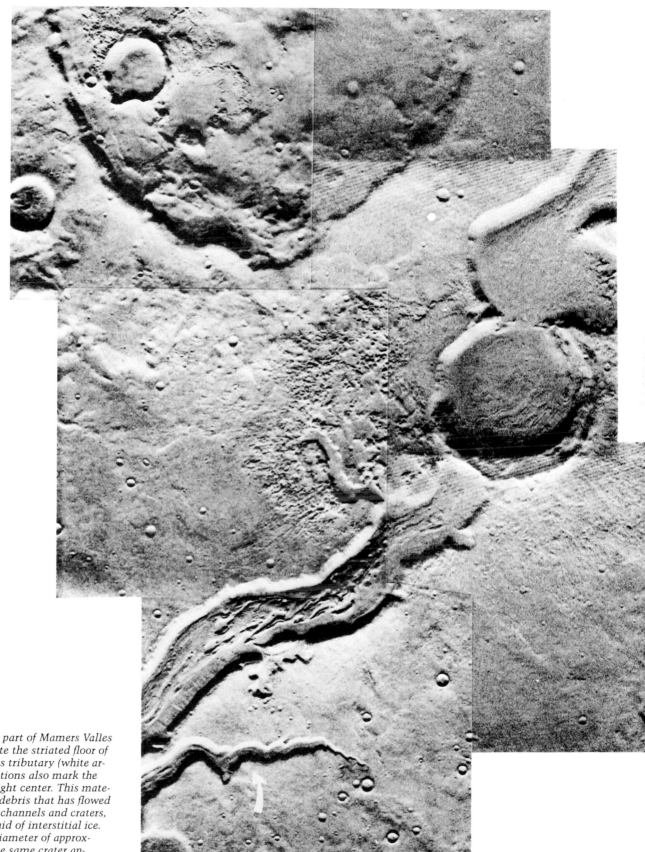

Figure 5.14. Upper part of Mamers Valles
(42°N, 338°W). Note the striated floor of
the channel and its tributary (white arrow). Curved striations also mark the
floor of crater at right center. This material appears to be debris that has flowed
on the floor of the channels and craters,
perhaps with the aid of interstitial ice.
The crater has a diameter of approximately 10 km. The same crater appears at P in figure 3.3. (Viking mosaic
211-5726.)

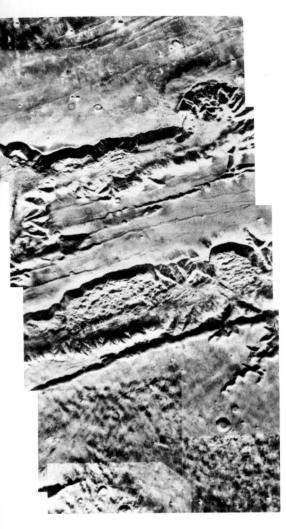

Figure 5.15. Landslides in Ius Chasma (center) and Tithonium Chasma (top) in the Valles Marineris system. The imaged scene is about 250 km wide. (NSSDC picture P-17872.)

tion differences of as much as 7,000 m occur between the escarpment lip and the lowest point on the landslide deposit (Wu, 1979). These immense scarp heights probably resulted from the lack of pluvial erosion on Mars, and their great free, unsupported faces were the ultimate cause of the immense landslides. Lucchitta (1978c) speculated that liquid water from permafrost ice buried beneath the scarp face contributed to scarp instability. Because many of the scarps occur along fault traces (Blasius and others, 1977), earthquakes could easily have acted as a triggering factor. Preliminary crater counts on the landslide deposits (Lucchitta, 1978c) yield a crater number of about 600, relatively young by Martian standards, but certainly hundreds of millions of years old assuming the lunar impact flux rate.

One especially striking landslide (Figure 5.16) has been discussed in detail by Lucchitta (1978d). The landslide has a distinctive morphology consisting of (1) mountainous debris near its head, (2) hummocky deposits further out, and (3) a vast apron of longitudinally ridged material extending far out on the valley floor. Unit 1 appears to consist of immense slump blocks, apparently of the relatively resistant layered rock that comprises the escarpment crest. This rock is most likely lava flows of the type common in the Lunae Planum. Unit 2 is probably composed of the material that crops out below the inferred lava flows on the visible landslide scar. This material is presumably the impact-generated megaregolith, and behaved as a loose, friable material. Unit 3 consists of two lobes separated by a channel that may indicate water drainage from the landslide base. This unit has prominent longitudinal ridges that diverge in a fanlike pattern. The distal part of unit 3 has overtopped several transverse ridges on the canyon floor, indicating a high momentum for the flow of debris. This debris has moved to about 60 km from the headcut of the landslides. Similar large landslides commonly occur in the eastern portion of the Valles Marineris system (Figure 5.17).

The Tharsis volcanoes are another region where large-scale landslide topography occurs. The lobe-shaped

Figure 5.16. Viking mosaic of a large landslide on the south wall of Gangis Chasma (9°S, 44°W). The view is to the south. The landslide scar is about 25 km wide. Note the older landslide to the east (left) that was partly buried by the large slide. (NSSDC picture 16952.)

feature on Arsia Mons was discussed earlier as a manifestation of either landsliding or glaciation. A more varied range of mass movement features occurs along the basal scarp of Olympus Mons (Figure 5.18). Using Mariner 9 imagery, Blasius (1976) interpreted these as slumps, slides, and flows. The Viking imagery of the scarp (Carr and others, 1977b) also shows huge lobate masses varying from 18 to 180 km in length and up to 90 km in breadth (Figure 5.19). These lobate masses have a hummocky interior zone and a marginal zone that consists of a terminal ridge or of a series of ridges and troughs. Carr and others (1977b) interpreted all these lobate masses as huge landslides derived from the scarp, possibly moving over cushions of compressed gas as suggested for some terrestrial landslides by Shreve (1968).

Lucchitta (1979) recognized the Tharsis volcano debris flows as morphologically distinct from the Valles Marineris landslides. She suggested that ice was probably involved in their emplacement because of morphological similarities to rock glaciers and to the Malaspina Glacier in southeastern Alaska. Consistent with this hypothesis is the extent of some debris flows out onto the mountains of aureole material surrounding Olympus Mons. Curvilinear ridges of the debris are superposed without deflection on this irregular terrain. This suggests a lowering of the debris apron onto the aureole mountains by loss of some intervening material. In terrestrial cases ice would be considered to be the most likely material to be subsequently lost by melting or evaporation.

Lucchitta (1978d) emphasized the similarities between the longitudinally ridged Martian slide and a similarly ridged slide generated on Sherman Glacier, Alaska, by the 1964 earthquake. On the Sherman landslide, debris is arranged into longitudinal

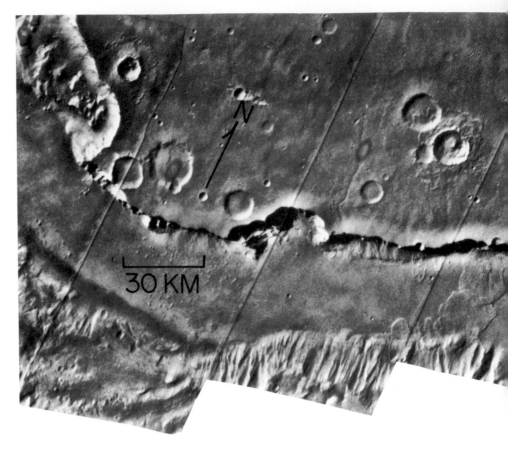

stripes that may develop by shear between debris trains moving at different speeds (Shreve, 1966b). Additional noteworthy features include an overall fan-like shape to the runout deposits on plains areas, the relatively thin nature of the runout deposits in comparison to their lengths, and cones of debris covering large blocks of surface of the slide mass. The Sherman slide may have developed its distinctive morphology because of flow across a relatively smooth bed of glacial ice. The similar morphology of the Martian slide could result from ice underlying the canyon floor over which it moved or from the incorporation of dispersed ground ice in the moving debris (Lucchitta, 1978b). Unfortunately the morphological evidence alone is not sufficient to exclude other possible mechanisms.

The immense scale and detailed morphology of the Martian landslides make them broadly similar to terrestrial phenomena known as "catastrophic rockslides and rock avalanches" (Voight and Pariseau, 1978). The character of these slides is illustrated by the largest of historic examples, the Mayunmarca slide along the Mantaro River, a tributary of the Amazon in the Andes Mountains of central Peru. Like many catastrophic landslides, the Mayunmarca slide occurred in a region of highly oversteepened slopes caused by rapid uplift and deep river incision. A mass of well-bedded arkosic sandstone with intercalated siltstone began to move by sliding along polished bedding surfaces that dipped steeply toward the Mantaro River. Very quickly this material became a debris flow of disintegrated rock that flowed into the river valley. Kojan and Hutchinson (1978) give the awesome dimensions of the Mayunmarca slide: 1.9 km of vertical height, 8.0 km of horizontal movement, and approximately 10^9 m^3 of debris. The duration of the seismic disturbance induced by the slide indicates an average velocity of 130 km/

hr. The mass surged across the Mantaro River and approximately 400 m up the other side of the valley.

The movement of catastrophic landslides can be analyzed with some relatively simple physical calculations. The important parameters for a landslide are its total vertical fall (h), the slide mass (m), its velocity of movement (v), the horizontal distance of movement (x), its maximum runup (r), and the average slope over which the slide moves (angle α). A very simple method of analysis would be to assume that the kinetic energy of movement is expended in creating the potential energy indicated by the runup. The computation can be made from the following equation, in which g is the acceleration of gravity:

$$\frac{mv^2}{2} = mgr.$$

Solving for v shows that the velocity of the idealized landslide is simply a function of the runup:

$$v = \sqrt{2gr}.$$

Of course, this is only the *minimum velocity* required because some of the potential energy is also transformed into frictional loss.

Although frictional loss is very difficult to describe for a complex landslide, it is possible to estimate an overall coefficient of friction (Scheidegger, 1973). Provided that the slide starts from rest, the average coefficient of friction μ is simply the total vertical height of the path of the slide (h) divided by the total horizontal reach (x):

$$\mu = h/x.$$

By these equations the Gangis Chasma slide on Mars achieved velocities of 27–39 m/sec and had a coefficient of friction of 0.03 (Lucchitta, 1978d). Terrestrial catastrophic slides, by comparison, have achieved velocities of up to 78 m/sec but friction coefficients only as low as about 0.09. Thus, the Martian slide appears to have been extremely efficient in moving long distances over relatively flat terrain. This problem of very high

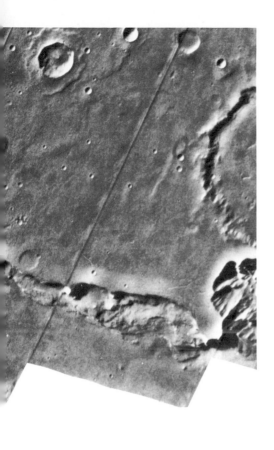

mobility in large landslides has still not been resolved for terrestrial examples, and several investigators have felt that only some unusual mechanism of movement can account for the low ratios of vertical fall to horizontal runout.

Shreve (1966b, 1968) proposed an ingenious explanation for the above problem involving air-layer lubrication. His hypothesis is that a certain class of catastrophic landslides override and trap a cushion of compressed air during their initial fall. This air cushion then allows sliding motion of the debris at high velocity and extremely low friction. Shreve (1968, p. 38) listed several specific attributes thought to be characteristic of air-layer landslides: extremely high velocity, local homogeneity of debris, a distal wedge of transported debris, lateral and distal rims and ridges, indications of air launch of the slide mass, extensive dust clouds, and a three-dimensional "jigsaw-puzzle" effect in the fracture of large rock fragments in the slide mass. In addition there is often a distinctive surface pattern to the slide mass.

Voight and Pariseau (1978, p. 28–32) argued that the criteria cited by Shreve do not demand the presence of an air-layer. Instead they contend that each criterion can be explained by alternative models, most of which involve extensive flow, rather than sliding, in the debris motion. Probably it is this question of sliding versus flowing movement that is most basic in the debate concerning catastrophic landslide mechanics. Shreve (1966b, 1968) emphasized the sliding motion because he observed wedges of distinctive material which apparently had remained during prolonged transport at the base of catastrophic landslides as they overrode sources at their proximal ends. However, Hsü (1975) argued that Shreve's reasoning can only exclude viscous flow from explaining the preservation of original sequential order in catastrophic landslides. He postulated that debris avalanches are characterized by internal deformation that changes a slid-

Figure 5.18. Lobate landslide mass (upper right) along the northern escarpment of Olympus Mons. The tongue-shaped mass is approximately 15 km wide. The striated topography at center and lower left is probably lava flow units that spilled over the escarpment. The imaged scene is approximately 84 × 99 km. (Viking frame 048B08.)

Figure 5.19. Large landslide or debris flow lobe on the west basal scarp of Olympus Mons at 22.5°N, 138°W. This huge mass movement probably originated from the basal scarp, flowing or sliding across the adjacent lowland. This scene measures about 84 × 99 km. (Viking frame 048B04.)

ing block into a sheet of flowing debris. Hsü essentially reconfirmed the concept of Heim (1882, 1932) that attributed the fluidization of large avalanches to the exchange of mechanical energy between colliding blocks in the moving debris stream. This concept lowers internal friction by relatively elastic mechanical collisions rather than by the buoyant effect of compressed interstitial gas.

The Martian landslides are pertinent to the general problem of catastrophic landslides because they probably achieved their low coefficients of friction without the aid of a dense atmosphere. The only possibility for a dense atmosphere on Mars appears to have been in the planet's primordial phase, certainly in the first billion years of its history. The landslides are much too young for that possibility. Rather than a gas phase, the Martian slides must either have been dry or have achieved low friction factors with the aid of water (from melted ice). The bulk of the evidence is consistent with the latter hypothesis.

WEATHERING PROCESSES

The foregoing discussion has shown that steep, high escarpments are common on Mars and that they have experienced a variety of processes that generate large particles of debris. The processes described thus far fall in the realm of physical weathering, that is, the disintegration of intact rock by physical processes. From terrestrial studies of hillslopes, we know that unless the mantle of coarse weathered debris is removed by some transport process, new rock cannot be exposed to further weathering. The hillslope will cease to be active, choked by its own production of rock waste.

Because many of the Martian scarps and hillslopes appear to be (or relative-

ly recently to have been) geologically active, some mechanism must occur on the planet that reduces the large particles to sizes small enough for removal. The removal mechanism is most likely to have been wind. Although the efficiency of eolian processes on Mars is well documented by theory and observation, as reviewed by Mutch and others (1976, p. 235–261), an effective size-reduction process is necessary to produce the necessary fine-grained particles. Orbital images are of little aid in discovering the responsible process or processes.

Some indication of weathering processes on Mars was provided by the Viking landers. The soil beneath the lander pads is very fine-grained and weakly cohesive. It is thought to have originated by the chemical alteration of a mafic igneous rock to produce clay minerals and salts. The clay

mineral with the most appropriate chemical composition to match the inorganic chemistry results is the iron-rich variety of smectite known as nontronite. The nontronite is believed to have formed by the interaction between iron-rich magma and subsurface ice.

The inorganic chemistry results from the Viking landers show unusually high concentrations of sulfur in the soil. Slightly higher concentrations of sulfur occur in a duricrust layer than occur in the loose soil. This duricrust of cemented soil occurs at both landing sites and appears to be about 2–3 cm thick. The sulfur distribution and its abundance imply that sulfate salts are responsible and that these act as a binding agent in the Martian soil (Toulmin and others, 1977). Salt crust formation also implies the upward migration and subsequent evaporation of water containing the dissolved salts. This is the manner of formation of weathering crusts in many terrestrial deserts. The responsible salt minerals probably include the very soluble magnesium sulfate, kieserite.

These data support the concept that salt weathering may be an important surficial process on Mars. This suggestion was made by Malin (1974), who compared the Martian weathering environment to that of the Antarctic dry valleys, where salt weathering is believed to be especially active (Selby and Wilson, 1971). Requirements for Martian salt weathering include salt minerals, sufficient water, and strong winds to remove weathered detritus. Data from the Viking project have confirmed these requirements.

The disintegration of surface rock by salt action in terrestrial deserts has long been recognized by geomorphologists (Walther, 1891). Salts can disrupt rock by (1) crystal growth, (2) hydration expansion, (3) thermal expansion, and (4) osmotic pressure (I. S. Evans, 1969, Winkler, 1975). Crystal growth is generally considered to be the most effective of these processes (Wellman and Wilson, 1965; Bradley, Hutton, and Twidale, 1978). The process is capable of breaking down rock material to micron-sized particles. Rather than producing a distinctive landform, however, salt weathering seems to mainly be a means of reducing the size of debris produced by other processes. On Mars the process could be very effective in reducing the size of blocks generated by fluvial erosion and hillslope processes. The fine-grained weathering products might then be dispersed by the wind, thereby exposing a new surface to erosion and/or weathering.

THE WATER SOURCE

The interpretation of Martian landforms as related to ground ice in a very thick permafrost layer leads to the problem of accounting for the required volumes of water. Terrestrial ground ice forms from meteoric water that has percolated into the ground. A similar mechanism cannot operate in the present Martian atmosphere. The Viking landers demonstrated that

TABLE 5.2. Comparison of Atmospheric Constituents on Earth and Mars

Constituent	Mars	Earth
Carbon dioxide	96%	0.03%
Nitrogen	2.5%	78%
Oxygen	0.1%	21%
Argon 40	1.5%	0.9%
Argon 36	4 ppm	32 ppm
Water vapor	0–85 ppm	up to 5%

Mars has an extremely thin atmosphere with a pressure ranging from about 7 to 10 mb, or only about 1 percent of the terrestrial value. Mars is also extremely cold. The landers showed that peak summer temperatures only reach −30°C by day and plummet to −80°C at night. Moreover, this cold, thin atmosphere is nearly all composed of carbon dioxide, as established by the mass spectrometers on the two landers (Table 5.2). From a computer model of a similar atmosphere, Leighton and Murray (1966) showed that at most only a few tens of meters of ice could have accumulated planet-wide during Martian history. The thick permafrost implied by the geological data requires an unusual mechanism of formation, perhaps either (1) an ancient Martian atmosphere very different from the present one, or (2) accumulation of water as ground ice after migration upward from the planet's interior.

Prior to the receipt of atmospheric data during the Viking mission, the case for a denser atmosphere in the Martian past was hotly debated. Sagan (1971) and Sagan, Toon, and Gierasch (1973) presented arguments for a denser atmosphere with moister conditions during the Martian past. The problems of generating such an atmosphere and then dissipating it, perhaps repeatedly, made it appear to others to be unreasonable and unlikely (Murray, 1973; Murray and Malin, 1973). The alternative to the dense ancient atmosphere was the idea that Mars may have lacked any significant atmosphere until relatively recently. The planet might have evolved its present thin atmosphere during volcanism late in its history. Fortunately the Viking atmospheric

analysis has provided some hard data that set limits on various speculations.

A principal indicator of atmospheric evolution is the abundance of the inert gas argon. Pre-Viking estimates of argon in the Martian atmosphere ranged as high as 30 percent. Such a high value would indicate that Mars had outgassed an immense amount of volatiles during its history, possibly losing much of this inventory by escape into outer space (McElroy, 1972; Sagan and Mullen, 1972). The molecular instrument package on the landers, however, demonstrated that actual atmospheric argon is only 1–2 percent. Even more significant is the ratio of the isotopes argon 36 to argon 40. The Martian ratio is but 10 percent of the terrestrial one and the abundance of argon 36 on Mars per unit mass is only 1 percent of that on Earth (Owen and Biemann, 1976). This argon ratio strongly implies that Mars has outgassed many fewer volatiles than has Earth. The argon data correlate to a theoretical estimate of ancient Martian atmospheric pressure based on probable crustal abundances of various elements (Anders and Owen, 1977). The data suggest ancient atmospheric pressures of no more than 140 mb, about 14 percent of the terrestrial value. The total amount of degassed water might be equivalent at most to a planet-wide layer a few tens of meters deep. This value is approximately 5 percent of what would be expected if Mars had the same volatile inventory and degree of outgassing as the Earth (Fanale, 1976).

The Viking nitrogen analysis provided a surprise. Mars does have considerable nitrogen, about 2.5 percent of its atmosphere. This value does not carry strong indications of total Martian outgassing, but it does imply that considerable water must also have been outgassed from Mars. The water cannot now be present in the very thin atmosphere, and only a minor percentage can be accounted for at the polar caps. The missing water seems likely to have been largely stored in the subsurface permafrost. A corollary to the nitrogen story is the very high ratio of nitrogen 15 to nitrogen 14 that was observed during the Viking lander entry experiments. Enrichment in nitrogen 15 strongly implies the loss of the lighter isotope nitrogen 14 to outer space. The degree of enrichment suggests that the Martian atmosphere was formerly much more dense than at present (McElroy, Yung, and Nier, 1976).

While the various Viking data have not produced a definitive picture of the ancient Martian atmosphere, they do point to a moderate planetary history of degassing and a probable, very ancient period of greater atmospheric pressure. The conservative view, based on calculations from crustal abundances, implies that this ancient atmosphere had perhaps 14 percent of the terrestrial density. Considerations based on exospheric escape of gases (McElroy, 1972) seem to imply a very dense ancient atmosphere with considerable water. More speculative arguments can be used to envision very high primordial atmospheric pressures, perhaps 500 mb or more, if greenhouse effects can be generated by reducing compounds (see Chapter 9). The water from this warmer, denser ancient atmosphere could have been the meteoric source for ground ice.

A possible scenario for the Martian permafrost has been outlined by Mutch and others (1976). They argued that Mars experienced a phase of catastrophic outgassing early in its history. This phase would correspond to planetary accretion over 4 billion years ago. During this "warm" phase of planetary history, Mars may have outgassed the equivalent of 1 km of water spread over its surface, assuming sufficiently volatile-rich accretionary materials. As this early dense, water-rich atmosphere cooled, it eventually allowed equilibrium between liquid and vapor phases. The liquid phase during this ancient period may have been responsible for developing the small valley networks of the ancient cratered terrain (Carr, 1979b). Water would be expected to percolate through the extremely porous surface materials of these terrains to develop extensive ground-water systems. Because the atmosphere continued to cool, this liquid ground water was never allowed to re-enter the atmosphere. Instead it was trapped by subsurface freezing.

Is the above model compatible with the extremely low water content and low atmospheric pressure of the modern atmosphere? The present Martian atmosphere has probably slowly evolved for billions of years since its early dense phase. More volatiles were added by episodic volcanic activity, but Mutch and others (1976) considered these subordinate in volume. It is likely that much of the nitrogen and water of the later atmosphere was lost by exospheric escape (McElroy, 1972). The great reservoir of ground ice generally was prevented from adding to the atmosphere because of the extremely low temperature. However, various processes may have temporarily released melted ground ice or confined water beneath that ice to the Martian surface over the past three billion years or so. The results of this release would be many of the various channel-forming and periglacial processes implied by Martian landforms.

The view that considerable Martian permafrost might form from lithospheric water (Sharp, 1974) definitely avoids the problem of precipitating water from the ancient atmosphere. The extreme cold of the Martian environment may have allowed considerable ground ice to form within the crust by the freezing of juvenile water ascending from the degassing planetary interior. Sharp (1974) noted that this water might lie deep enough within the crust to be sealed off from free contact with the atmosphere.

The nature of the Martian permafrost depends largely on the porosity of the subsurface formations and on the internal heat flow of the planet. Porosity beneath the ancient cratered uplands of Mars is probably very high. The first 500 million years of Martian history is generally thought to have been marked by an extremely high rate of impacting. Perhaps this was the final stage in the sweeping up of debris from the primordial nebula that produced the planets. Whatever the cause, this high early cratering rate also affected Mercury and the Moon. By saturating the planetary surface with large craters, this high impact rate produced a megaregolith of pulverized debris several kilometers deep (Hartmann, 1977).

The nature of the lunar megaregolith, which is probably similar to that on Mars, can be inferred from seismic evidence. Latham and others (1971) proposed that the near-surface materials are loosely consolidated and that seismic velocities increase with depth because of self compaction that continues to depths of about 25 km. Because of the higher gravity on Mars, Carr (1979b) proposed that the depth of self compaction, and therefore sealing of the subsurface pore space, is probably in the 10–20 km range for Mars. Porosities in the rock above this zone certainly can be as high as 0.1 because of the fragmental nature of the megaregolith and the volcanic nature of near-surface rock (Sharp, 1974; Carr, 1979b).

The ultimate thickness of the Martian permafrost is probably limited more by thermal considerations than by porosity. Öpik (1966) suggested that the thermal regimen of the Martian crust would allow ground ice development to a depth of 1–3 km. This estimate correlates well with data on recurring escarpment heights that suggest a permafrost layer to a depth of 1–2 km throughout much of Mars' geologic history (Soderblom and Wenner, 1978). Moreover, current mean annual surface temperatures on Mars imply that, for probable values of internal heat flow, the 0°C isotherm occurs at a depth of 1 km at the equator and several kilometers at the pole.

The upper surface of the permafrost depends on the stability of ice relative to the atmosphere. At latitudes lower than approximately 40°, water ice is unstable at the surface; it would sublimate and dissipate if exposed to it (Fanale, 1976). Ground ice at the equatorial latitudes would only persist if it were blocked from contact with the atmosphere by the overlying materials. This condition favors the dry ground-ice sapping process proposed by Sharp (1973b) for fretted terrain development. If ground ice were exposed in the face of an open fracture, steep slope, or cliff, it would sublimate, thereby disaggregating ice-cemented materials and undermining resistant caprocks.

Ice in the Martian permafrost may locally melt to produce near-surface liquid water, probably with the aid of local salt concentrations to act as freezing point depressants. Huguenin and Clifford (1980) described several regions on Mars where this might occur, including Solis Lacus (25°S, 85°W) and Noachis-Hellespontus (20°S, 310°W). The presence of such "oases" is indicated by observations of anomalous water condensates forming in and around these regions throughout much of the Martian year. Zisk and Mouginis-Mark (1980) found that these same regions are characterized by coincident very high radar reflectivities and unusual smoothness. The best explanation for the anomalies appears to be liquid water within 50–100 cm of the surface. The interpretation is strengthened by the fact that radar reflectivity in these areas tends to increase from Martian spring to summer. This exciting result appears to confirm the geomorphic evidence for water in the Martian regolith.

SUMMARY

Mars has an abundant variety of landforms which on Earth are known to be associated with cold climates (periglacial regions) and/or ice-rich permafrost. The most prominent examples are thermokarst features, hillslope morphologies, debris aprons, and landslide features. Scale differences between terrestrial and Martian periglacial landforms pose difficulties for their comparison, especially in the case of the large-scale polygonal troughs of the northern plains. A possible explanation of the immense scale of some Martian periglacial landforms is that the relevant time scale for freeze-thaw cycles on Mars is secular, rather than diurnal or seasonal as on Earth.

The geomorphic data for Mars support the concept of an ice-rich permafrost zone extending to a depth of perhaps 1–3 km. Although the amount of water required by this hypothesis is not consistent with some models of Martian atmospheric volatile abundances, other considerations can resolve this apparent discrepancy (Chapter 9).

The foregoing analysis suggests that indeed Mars is a water-rich planet, although most of its water lies hidden in a great subsurface reservoir of ground ice. What wasn't hidden from the imaging systems of Mariner 9 and Viking was the erosion and deposition by the fluids released from that reservoir. Chapter 6 will discuss the Martian landscapes created by massive fluid flows

Chapter 6
The Outflow Channels

The outflow channels are certainly the most spectacular of the fluvial-like features on Mars. They are much larger than the valley systems described in Chapter 4. Channel widths range from 10 to 100 km, and some channel complexes (Maja, Ares, and Kasei Valles) can be traced 2,000 km or more. These channels appear full-born at localized sources, usually collapse zones marked by chaotic terrain. They transect terrains of varying age, generally emanating from sources within the heavily cratered terrain, and traversing the great "planetary dichotomy" between the northern plains and the heavily cratered uplands.

The largest outflow channels flow toward the Chryse Planitia from the surrounding plateaus and cratered uplands (Figure 6.1). These channels include Kasei, Maja, Shalbatana, Simud, Tiu, and Ares Valles. All of these constitute true channels, rather than valleys, because the erosive fluid produced a variety of bed forms that often mark the entire valley width. This chapter will discuss the detailed mapping of portions of these channel systems. The purpose is to lay the observational groundwork for considering the similarities of outflow channel morphology to the Channeled Scabland of the northwestern United States (Chapter 7) and for considering the responsible fluid flow processes (Chapter 8).

THE SOUTHERN CHRYSE PLANITIA

The Chryse Planitia is one of the lowest regions on the Martian surface, lying 2–3 km below the mean surface elevation of the planet. The surrounding topography and channel systems (Figure 6.1) are related to the first-order elevation features of Mars (Mutch and others, 1976). The Tharsis Ridge, highest region of the planet, lies to the west. A general eastward topographic gradient leads from Tharsis to Chryse and to a related trough (Chryse Trough) extending southward from Chryse to Margaritifer Sinus and Ladon Vallis (Figure 2.2). Kasei Vallis, Maja Vallis, Nanedi Vallis, Shalbatana Vallis, Nirgal Vallis, and the eastern Valles Marineris are all developed along this approximate eastward gradient off the Tharsis "bulge." The topography also rises to the east of the Chryse Trough. Ares Vallis and part of Tiu Vallis are oriented with respect to this eastward rise. Much of Tiu Vallis and Simud Vallis trend northward along the axis of the Chryse Trough toward the great depression of Chryse Planitia. This striking topographic setting is not duplicated elsewhere on Mars, and is probably an important factor in the concentration of outflow channels around the Chryse Planitia.

The outflow channels that enter the southern Chryse Planitia show an excellent progression of morphology from their proximal (headward) reaches to their distal reaches. The proximal relationships are well displayed by Ravi Vallis (Figure 3.12), a tributary of Hydraotes Chaos and Simud Vallis. The line map of Ravi Vallis (Figure 6.2) shows that the collapse process producing chaotic terrain was spread in time. Several collapse zones downstream from the channel head occurred after the main channeling event responsible for the grooved channel floor.

In the headward reaches of Tiu Vallis (Figure 2.16) there is a progressive downstream change in morphology from chaotic terrain to remnants of chaotic blocks that were apparently eroded by the release of fluids from upstream. These relationships indicate that some outflow channels developed by a process of headward channel growth (Baker, 1977, 1978b). The working hypothesis is that the headward growth occurred by the progressive transformation of cratered terrain underlain by fluid-saturated, fractured rock into chaotic terrain. With continued headward extension, the downstream zones of chaotic terrain would have no relation to channel volume, which is partly produced by collapse over earlier fluid-release zones and subsequently modified by flows produced at new collapse zones upstream. Downstream zones could have been preflood lows on the topography of collapse-generated troughs that subsequently were modified by flooding to produce the characteristic assemblage of flood bed forms on their floors.

Ares Vallis

Ares Vallis is the easternmost of the southern Chryse channels. It emanates from Iani Chaos at about 0° latitude, 17°W (Figure 6.3). Upstream from the chaos zone the fluid-modified zone is about 100 km wide (Figure 6.4). A deep, gorge-like channel connects Aram Chaos to Ares Vallis at about 3°N, 18°W. A broad channel enters from the east at 4°N. A broad, shallow channel appears to cross the cratered divide between Hydaspis Chaos and Ares Vallis, joining the latter at about 8°N, 23°W.

The eastern channel of the Ares Vallis system transects two medium-sized craters (Jb and Jd in Figure 6.4). The surface between the eastern branch and the main valley appears to be scoured, and much of the upland adjacent to the eastern channel has a scabland-like topography (Sharp and Malin, 1975). These relationships seem best explained by catastrophic flooding (Baker and Milton, 1974). Sharp and Malin (1975) suggested that the great depth of the main channel of Ares Vallis (about 1.5 km) may have initially formed by chaotic terrain collapse. This trough was then modified by catastrophic flooding.

The following is a somewhat speculative outline of events. Converging and diverging flow patterns are typical of a river's attempt to create an alternating pattern of energy dissipa-

Figure 6.1. Map of the Chryse Basin, outflow channel systems, and related features of equatorial Mars between longitudes 0° and 90°W. The contour interval is 1 km as mapped by the U.S. Geological Survey (1976).

THE OUTFLOW CHANNELS **109**

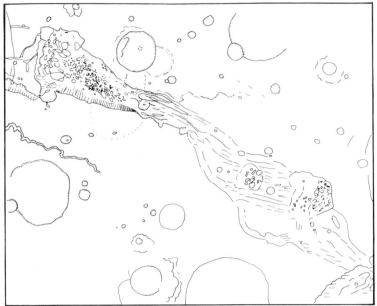

Figure 6.2. Sketch map of Ravi Vallis, located at approximately 0° latitude, 42°W. Hydraotes Chaos is at the lower right of the map. The mapped region measures 250 × 200 km. Zones of chaotic terrain (Aromatum Chaos) occur at the channel head (upper left) and served as a source for catastrophic fluid release. Other chaotic terrain (right, center) truncates the grooved and streamlined channel topography created by upstream fluid release. Compare to Figure 3.12.

Figure 6.3. High-resolution Viking images of Iani Chaos near its transition to Ares Vallis (upper right). Each frame depicts an area about 30 km across. (Viking mosaic 211-5608.)

Figure 6.4. Sketch map of Ares Vallis showing the effects of fluid flows on ancient cratered terrains and the relation of channel heads to zones of chaotic terrain. Craters Jb and Jd were breached by fluid flows (arrows) as discussed by Malin (1976). Streamlined uplands (S) are indicated by shading. A possible expansion bar complex that was later streamlined by waning flows occurs at B. Other possible flood bars are indicated by the dotted pattern at locations where deposition of flood load would be expected. A possible inner channel cataract complex (C) occurs where relatively shallow flows spilled into Ares Vallis after crossing the interchannel divide from Hydaspis Chaos (which mainly drained to the west). The complete terrain relationship is geomorphically unique to catastrophic flood erosion. However, this argument derives from considering terrestrial analogs.

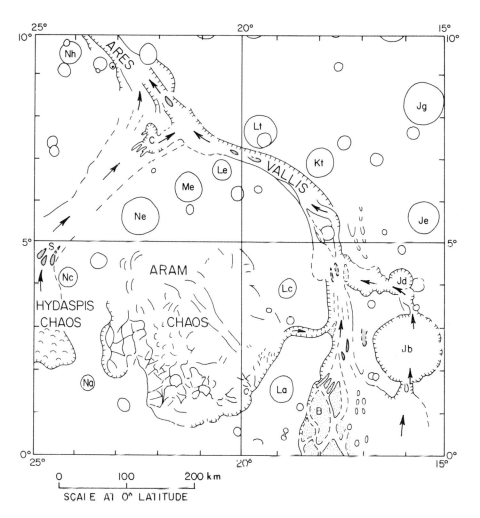

tion. Given sufficient time, terrestrial rivers will eventually scour at the convergences to create pools. With continued discharge the pool-and-riffle pattern gradually will evolve to the equilibrium morphology of a meandering or braided river (Keller and Melhorn, 1973). Catastrophic flood channel-ways only display the incipient stages of the process (Baker 1978c). Channel equilibrium is rarely achieved. As Martian flows emanated from headwardly growing chaotic terrain source areas, they traversed an irregular surface of raised crater rims, intervening plains, and occasional structurally controlled troughs. To move as a sheetflood across an irregular preflood surface would have been highly inefficient relative to energy dissipation. Therefore, the fluid eroded into the existing troughs of the topography, between the larger craters and along pre-existing structural lows. This produced an anastomosing system of secondary channels separated by residual hills of the preflood topography. The individual channels were then relatively efficient compared to the original land surface, but the complete development was inhibited by the short duration of the flow event (or events). In some areas the fluids were ponded above obstacles, such as crater rims. Ponding continued until sufficient depth was achieved to breach the rims and form patterns such as shown in Figure 6.4.

Similar relationships are shown in portions of Maja and Kasei Valles, to be discussed later.

High-resolution Viking images of upper Ares Vallis reveal details of features on the channel floor (Figure 6.5). These show that modifications have occurred since the major fluid releases of the "diluvian" epoch. The postdiluvian modifications include mass-movement processes along the channel walls and probable eolian erosion of the channel floor materials. Further downstream a highly constricted canyon section of Ares Vallis shows evidence of high-level fluid erosion on the upland surface (Figure 6.6). The erosion occurs only to a well-defined upper elevational limit and no higher. The high-level erosion zone was probably abandoned as the fluid cut the deep canyon, or when a downstream zone of headward channel growth receded past this point.

The distinct level of fluid erosion and the spilling relationship of fluid to craters (Figure 6.4) indicates that the primary erosive fluid had a free surface. This would allow erosion by water, ice, or lava but not by eolian processes.

An interesting clue to the nature of channel floor materials in Ares Vallis has come from the infrared thermal mappers (IRTM) on the Viking orbiters. These instruments can determine the thermal inertia of the land surface. Thermal inertia is dependent primarily on conductivity, which, in turn, is closely related to particle size. Although some problems remain in correlating thermal inertia values directly to grain size, the data do appear consistent with inferences from Viking images of the terrain. Christensen and Kieffer (1979) showed that the floor materials of both Ares Vallis and Kasei Vallis have much higher ther-

Figure 6.5. Mosaic of high-resolution Viking pictures showing the upper reaches of Ares Vallis. Features shown include a tributary from Aram Chaos (T), eroded remnants of chaotic terrain (C), streamlined bar-like topography (B), and postdiluvian landslide (L). The highly elongate ridges on the channel floor (Y) may be yardangs derived by postdiluvian eolian erosion of relatively soft materials on the channel floor. (Viking mosaic 211-5608.)

mal inertias than surrounding upland regions. The interpretation is that the material is sand-sized (0.2–1.0 mm), possibly windblown accumulations. The eolian deposition probably occurred long after the outflow channel erosion.

Shalbatana Vallis

Shalbatana Vallis is the westernmost of the southern Chryse outflow channels. It emanates from a zone of chaotic terrain at 0° latitude, 46°W. The channel maintains a surprisingly uniform width of about 50 km along a low-sinuosity northward course for 650 km through the heavily cratered terrain. It then splits into two narrower, shallower distributaries. These valleys are about 12–15 km wide and they eventually empty on the Chryse Planitia.

High-resolution images of Shalbatana Vallis show that the valley is quite deep. The channel walls have the irregular scalloped appearance so common in all deep Martian valleys. Mass movement processes have probably contributed to considerable postdiluvian modification of this outflow channel. The relatively small extent of chaotic terrain in the proximal reaches of Shalbatana Vallis suggests that, like Ares Vallis, its volume probably resulted from a headward growth process rather than from simple fluid erosion.

Residual Uplands

The distal portions of the Chryse outflow channels occur where the channel fluids debouch from constricted canyons in the heavily cratered terrain out on to the open lowland of Chryse Planitia. This region was originally proposed as the prime Viking 1 landing site (Figure 6.7), but it had to be abandoned when the Viking orbiter cameras revealed rugged "scabland" topography within the landing ellipse (Figure 3.10).

The southern Chryse Planitia displays an excellent variety of residual uplands within and separating broad channels. The largest uplands have steep flanking scarps and preserve an upland surface that has not been modified by fluid flows. An example is the large irregular upland at the center of the mapped region in Figure 6.8. Be-

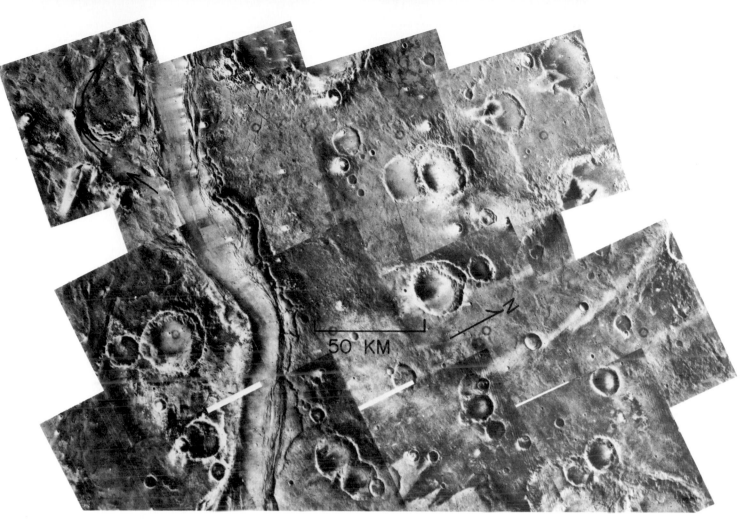

Figure 6.6. Viking mosaic of a constricted reach of Ares Vallis at approximately 10°N, 22°W. The arrows show a zone of high-level fluid erosion. Note the relatively smooth appearance of the channel floor in this reach in contrast to that shown in Figure 6.4. (Viking mosaic 211-5238.)

Figure 6.7. Sketch map of the outflow portion of Ares and Tiu Valles in the southern Chryse Planitia. The "Viking Landing Ellipse" is for a proposed site that was abandoned during the site verification process. The geological interpretation was prepared from Mariner 9 data prior to the Viking mission. Viking data show that the streamlined forms are erosional and that the cratered plain is mantled with duricrusted fine sediment and rocks.

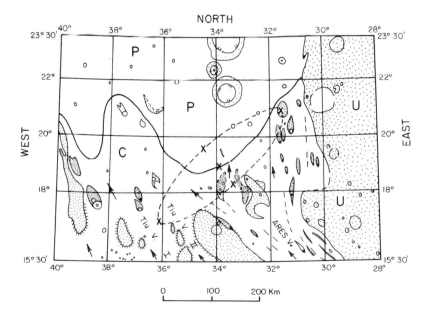

Figure 6.8. Geomorphic map (right) prepared from Viking images 3A11, 3A12, 3A13, 3A14, 3A15, and 3A16 (left). The region occurs along the eastern margin in the Chryse Planitia, where fluid flows emanated from Ares Vallis. The large crater at the bottom (named "Shawnee") acted as an obstacle in the flow field, favoring preservation of the terraced upland (T) on its downstream margin. The resulting streamlined "island" is approximately 100 m high. A low scarp separates unmodified preflow terrain in the eastern one-third of the mapped region from the channelized zones (CHg) to the west. The erosive fluid apparently crossed low troughs in the preflow topography to create the recurving, narrow channels along the eastern margin of the channelized zones.

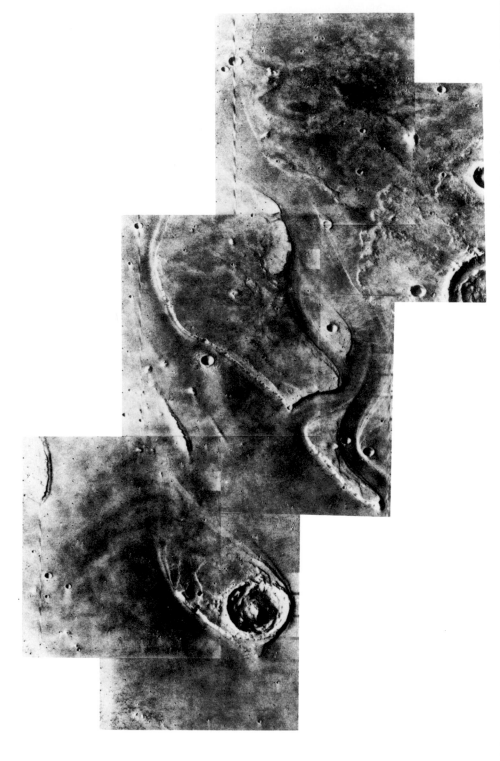

cause of the extensive zones of unmodified antediluvian terrain at their crests, these uplands probably constituted "islands" in the fluid flows that shaped the channels. They acted as elements of the overall channel pattern, separating the anastomosing channels. The scarped margins of the uplands are often scalloped or smoothed into recurving shapes. Longitudinal grooves on the adjacent channel floor often conform to these patterns, demonstrating their origin by fluid erosion. Zones of "scabland" scour also occur on the adjacent channel floor and conform to probable fluid-flow directions (Figure 6.9).

A second variety of residual upland is characterized by secondary channels that cut through the antediluvian terrain at the upland crest. These secondary channels often constitute hanging valleys because they do not erode so deeply as the primary channel floor. They probably formed as fluid crossed low divides along the crests of residual uplands. An example is the upland at the left of the mapped region in Figure 6.10. This form is transitional from the unmodified residual upland and is morphologically similar to it except for the secondary channels. Both types of residual upland sometimes have a quadrilateral shape, similar to the rhombic or diamond shape of eroded braid bars in terrestrial braided rivers.

The third variety of residual upland is that showing a pronounced stream-

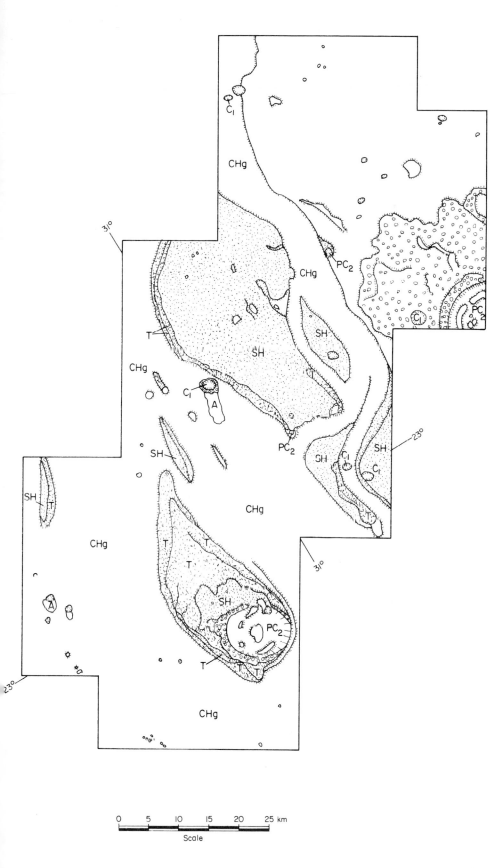

0 5 10 15 20 25 km
Scale

lined shape. Streamlining is a dynamic quality, generally defined as a shaping that prevents the separation of flow along a fluid boundary. Martian uplands tend to achieve maximum streamlining by a planimetric configuration that approximates the mathematical form of a single lemniscate loop (Baker and Kochel, 1978b). An example is the prominent teardrop-shaped upland at the bottom center of the mapped region in Figure 6.11. Antediluvian craters appear to have relatively resistant rims that form the upstream ends of these uplands and may account for the preservation of antediluvian terrain on their downstream ends. In map view the streamlined hills thus show rounded prows that point upchannel and tapering tails that point downchannel. The tails sometimes display terrace-like benches, apparently formed as layered rocks were differentially stripped by the fluid flows.

The action of the fluid flows on crater ejecta and rim materials is quite instructive. The ejecta materials appear to have been easily removed (Figure 6.12), consistent with the hypothesis that they consist of loose, impact-shattered debris. However, the crater rims are often preferentially preserved; the rims apparently diverted the fluid flows. The implication is that the rims are resistant to erosion, perhaps in a manner similar to the cores of deeply eroded terrestrial impact craters, such as Gosses Bluff in central Australia (Milton and others, 1972).

All the residual uplands are interpreted as remnants of more continuous antediluvian terrains. The three classes of upland appear to represent a continuous series of progressive modification by fluid flows. With continued fluid erosion, the larger residual uplands may be cut by channels and eventually reduced to smaller streamlined hills of minimum flow resistance. The lemniscate-shaped hills tend to be preserved because they have reduced fluid drag to a minimum (Baker and Kochel, 1978b).

I have suggested (Baker, 1978b) that the residual uplands in Martian outflow channels may have formed in a manner similar to residual uplands in the Cheney-Palouse Scabland tract of

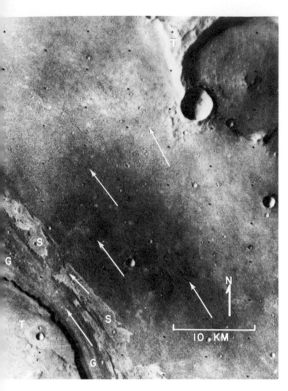

Figure 6.9. Scabland (S) on the channel floor at the mouth of Tiu Vallis, 18°N, 36°30'W. Portions of two terraced residual uplands (T) are also shown. Arrows indicate the probable fluid flow directions. Longitudinal grooves occur at G. (NSSDC picture 16828.)

Figure 6.10. Geomorphic map of the mouth of Tiu Vallis prepared from Viking images 3A18, 3A20, 4A57, and 4A76. The map shows several features that are also characteristic of catastrophic flood erosion in the Channeled Scabland. The streamlined uplands (SH) are probably remnants of preflood terrain that have been shaped by flood erosion. The uplands were preserved downstream from craters (PC₁ and PC₂). Other craters postdate the flood flows (C₁). Low points on the remnant uplands were sometimes scoured to form oblique channels (left center) in a manner similar to the divide crossings of the Channeled Scabland. Terraced margins (T) on the streamlined uplands and irregular scabland zones (S) on the channel floors are analogous to forms produced by the differential plucking erosion of the interlayered Columbia Plateau basalt and sediments by deep, high-velocity Missoula flood flows. Erosional grooves on the channel floor indicate the presumed paleoflow streamlined for the channel. The larger crater at the lower left of the map area is also shown in Figure 6.12.

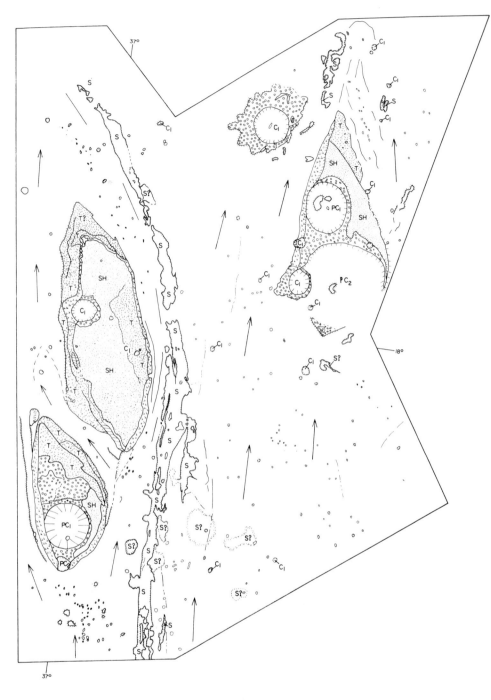

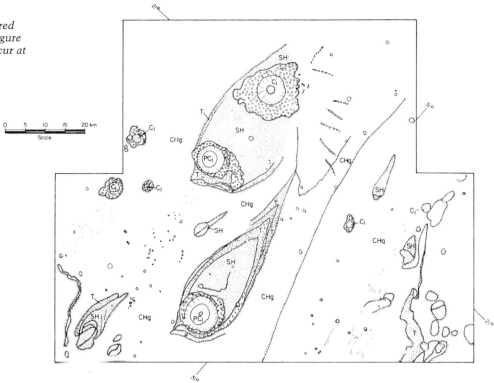

Figure 6.11. Geomorphic map prepared from Viking frames 4A50–4A54 (Figure 3.8). These streamlined uplands occur at the mouth of Ares Vallis.

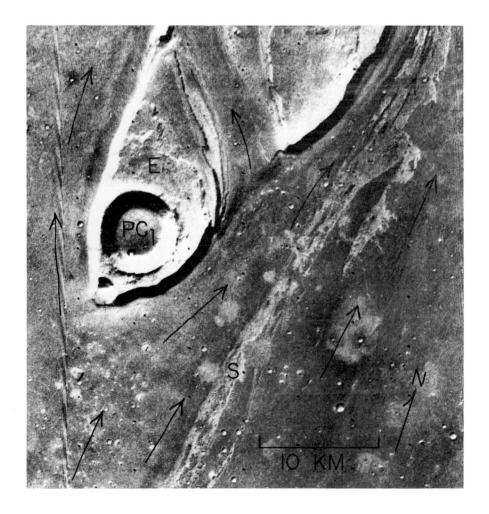

Figure 6.12. Eroded crater (PC₁) at the mouth of Tiu Vallis (Figure 6.10). Note the preservation of ejecta (E) on the downstream side of the crater. Scabland (S) occurs on the channel floor. Arrows indicate the probable fluid flow directions.

eastern Washington. The Cheney-Palouse Scabland originated when catastrophic flooding from Pleistocene Lake Missoula eroded the cover of loess over basalt in eastern Washington (Patton and Baker, 1978a). The residual hills of loess and basalt probably behaved in analogous fashion to the longitudinal bars of braided streams. Initially, the flooding spread widely over the rough planetary surface in relatively wide, shallow channels. As these widened by bank erosion, they encountered increasing boundary resistance. In order to maintain a sufficient velocity for sediment transport, the channel was forced to divide into relatively narrow and deep secondary channels. The residual uplands were preserved between these secondary channels.

In the southern Chryse Planitia, isolated uplands may have existed prior to the imposition of catastrophic fluid flows. The isolated, little-modified uplands are remnants of the continuous heavily cratered upland that abruptly gives way to the more sparsely cratered northern plains. The boundary between the two terrains is often marked by the isolated mesas and buttes that Sharp (1973b) named "fretted terrain." It is likely that the outflows of erosive fluid were imposed on a zone of fretted terrain at the southern margin of the Chryse Planitia. For some of the antediluvian fretted uplands the fluids merely reshaped the bounding escarpments. For other, smaller uplands the fluid was able to cut across low divides. Extreme reshaping produced the beautifully streamlined lemniscate shapes. The mechanics of this streamlining process is discussed in Chapter 8.

MAJA VALLIS

The Maja Vallis system (Figure 6.13) has been well imaged by the Viking orbiters. The system originates at Juventae Chasma, a 250 × 100 km depression. The southern half of Juventae Chasma is mostly enclosed by a high escarpment, and its relatively flat floor is similar to portions of the Valles Marineris (Chapter 2). The

Figure 6.13. Sketch map of the Maja Vallis system. The flow features extend northward from Juventae Chasma, a depression bound by a steep inward-facing escarpment and partially filled with chaotic terrain. The relatively old hilly and cratered terrain (HC) separates the younger volcanic plains of Chryse Planitia and Lunae Planum.

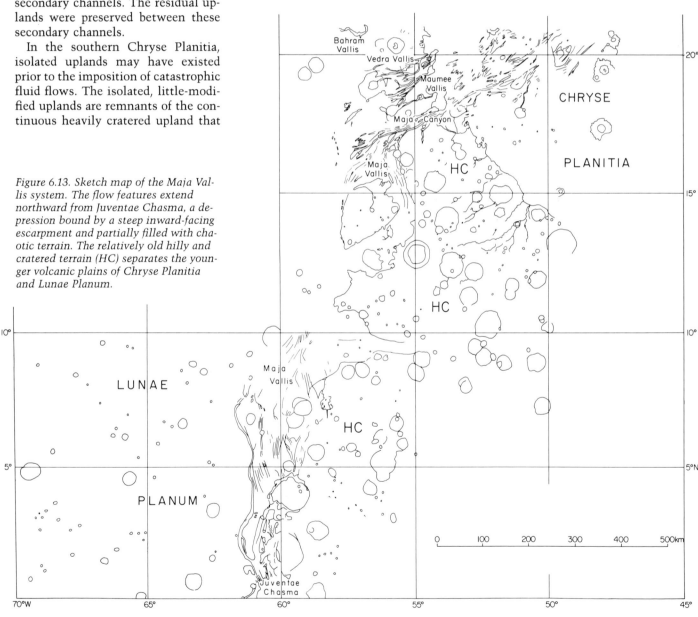

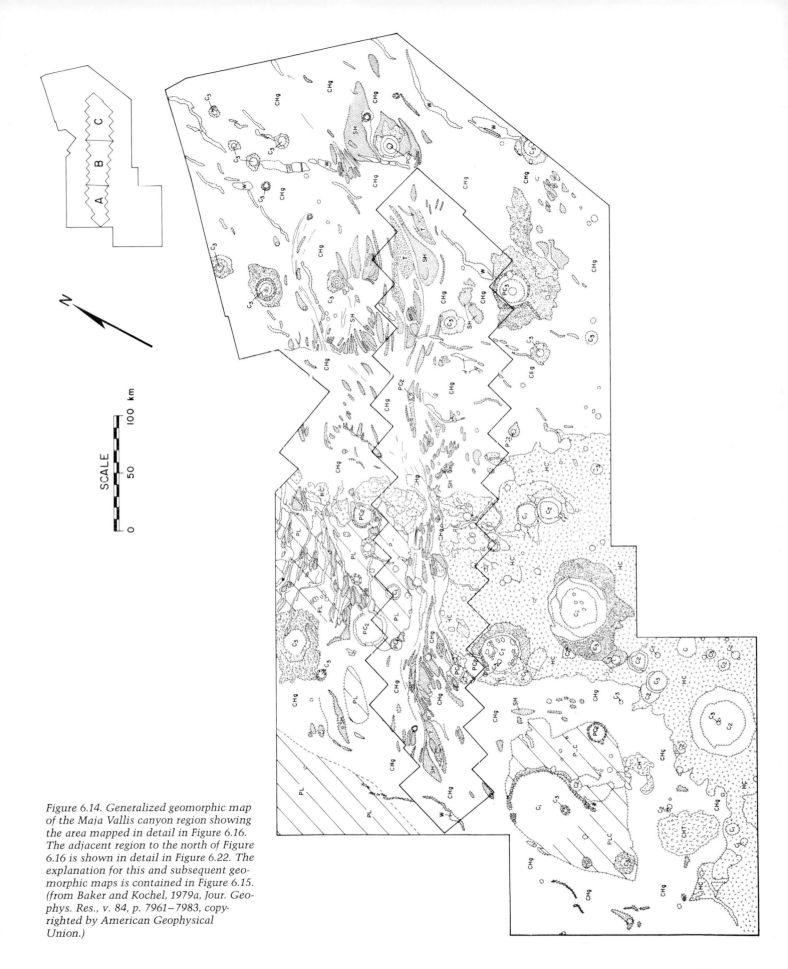

Figure 6.14. Generalized geomorphic map of the Maja Vallis canyon region showing the area mapped in detail in Figure 6.16. The adjacent region to the north of Figure 6.16 is shown in detail in Figure 6.22. The explanation for this and subsequent geomorphic maps is contained in Figure 6.15. (from Baker and Kochel, 1979a, Jour. Geophys. Res., v. 84, p. 7961–7983, copyrighted by American Geophysical Union.)

Figure 6.15. Legend for Figures 6.14, 6.16, and 6.22. The mapping units are discussed in the text. (From Baker and Kochel, 1979a, Jour. Geophys. Res., v. 84, p. 7961–7983, copyrighted by American Geophysical Union.)

northern half of Juventae Chasma consists of chaotic terrain. The channeled terrain begins to the north of the chaos zone and closely follows the boundary between the old cratered terrain (HC in Figure 6.13) to the east and the younger Lunae Planum to the west. This channeled terrain continues for about 1,000 km northward from Juventae Chasma. Between latitudes 15°N and 20°N the channels converge at several deeply incised canyons through the old cratered terrain (Maja Canyon, Maumee Vallis, and Vedra Vallis in Figure 6.13). The channeling on Chryse Planitia shows divergence of the fluid flow features and eventual disappearance.

Despite its immense size, the Maja Vallis system was only poorly known from Mariner 9 A-frame images. On Mariner images the channel system appeared discontinuous, its northward path interrupted by the volcanic plains of Lunae Planum. Milton (1975) mapped these boundaries as imprecise. Malin (1976) suggested that plains units of the eastern Lunae Planum were locally burying the channel. The resurgence of channel landforms was attributed to incomplete burial by thin volcanic units. The Viking images show the continuity of the channel system, although erosion is more pronounced in some reaches than in others.

The canyon section of Maja Vallis, with its approach reach and its outflow reach on the Chryse Planitia (Figure 6.14), provides an excellent illustration of outflow channel morphology. Morphological maps have been prepared for this region by Baker and Kochel (1979a). This mapping (Figures 6.15 and 6.16) will now be discussed in some detail.

Channel Host Terrains

Maja Vallis is incised into older terrain surfaces that are extensively preserved in the uplands west of Chryse Planitia. Remnants of the antedilu-

	DESCRIPTION	INTERPRETATION
	Plateau material typical of the Lunae Planum.	Basaltic lava flows underlain by less consolidated material, perhaps sediments, impact debris, palagonite, or volcanic ash.
	Hilly and cratered region, heavily cratered.	Probably predates large-scale outflow events.
	Channels and inferred flow direction.	Approximate direction of catastrophic flood flow across this surface.
	Dark-appearing lineations that parallel inferred flow direction.	Longitudinal grooves eroded parallel to flood flows by large roller vortices (macroturbulence).
	Inner channel headcuts.	Headwardly-eroding inner channel cataracts developed during flooding.
	Etched zones in which the upper 25–75 m of channel floor has been stripped away.	Erosive plucking by macroturbulent flood flows produced butte-and-basin scabland.
	Streamlined hills and uplands. Prows point upstream, and downstream ends have tapering tails.	Erosional remnants, often downstream from flow obstructions, whose streamlined shape results from high-velocity fluvial activity.
	Terrace-like benches along edges of streamlined hills and plateaus.	Lithologic units of varying resistance to plucking erosion that have been differentially stripped by catastrophic flood flows.
	Bars?	Possible accumulation of flood debris downstream from flow obstructions.
	Talus accumulations.	Blocky talus piles resulting from back wasting.
	Fans, often possessing radiating flow lines and a lobate front.	Portions of the talus slopes that have flowed across adjacent channel floor.
	Slumps.	Rotational landslides.
	Chaotic terrain.	Zones of collapse resulting from melted ground ice and/or fluid release.
	Light-appearing zones developed behind craters.	Eolian deposits.
	Lineations oblique to inferred flow direction.	Joints, faults, and other bedrock structures accentuated by erosion.
	Mare-like ridges.	Wrinkle ridges, many of which have interacted with channeling processes.
	Sinuous depressions oblique to inferred flow direction.	Rilles or downfaulted areas.
	Crater with poorly preserved rim morphology.	Crater rim is highly degraded and partly buried but not modified by catastrophic flood processes.
	Crater with poorly preserved rim morphology that has interacted with channeling processes.	Highly modified pre-flood crater that was modified by catastrophic flood flows.
	Crater with raised rim that has not interacted with channeling processes.	The rim is commonly incomplete and the crater highly degraded. Antediluvian crater.
	Crater with raised rim that has interacted with channeling processes.	Impact crater that created a flow obstacle or was modified by catastrophic flood flows.
	Crater with distinct, raised rim and fresh-appearing morphology.	Relatively young crater, probably postdiluvian.
	Crater rims and hummocky zones with lobate raised margins.	Crater ejecta, partially deposited by fluid debris flows released when meteors impacted the Martian cryolithosphere.

All the mapped craters are C_3 unless otherwise noted.

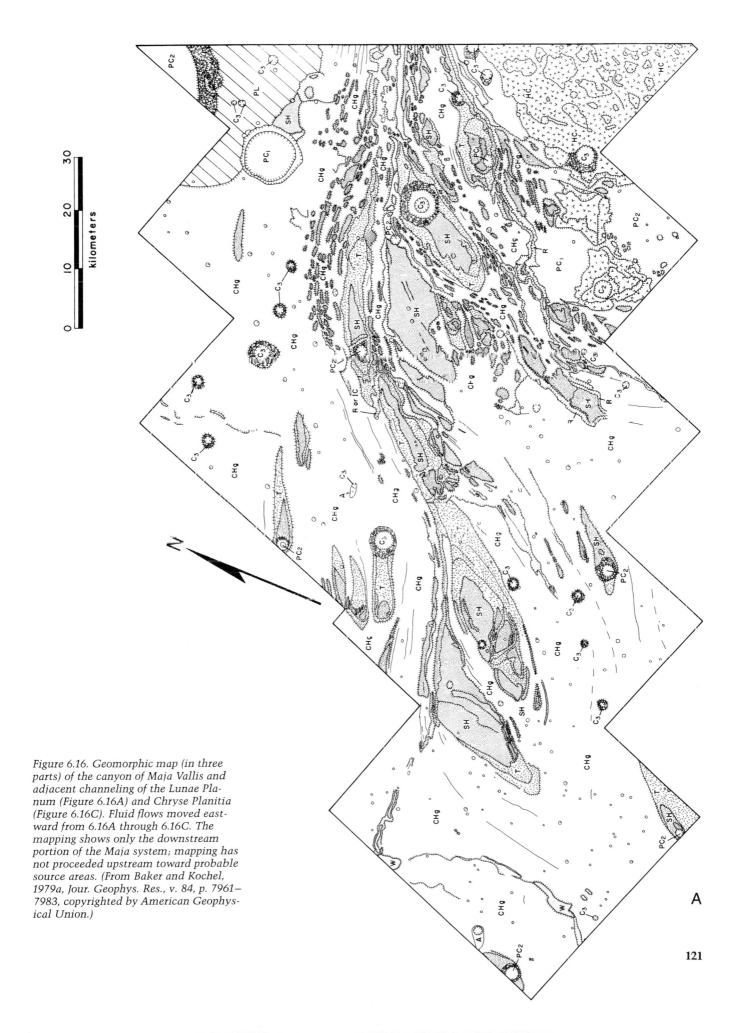

Figure 6.16. Geomorphic map (in three parts) of the canyon of Maja Vallis and adjacent channeling of the Lunae Planum (Figure 6.16A) and Chryse Planitia (Figure 6.16C). Fluid flows moved eastward from 6.16A through 6.16C. The mapping shows only the downstream portion of the Maja system; mapping has not proceeded upstream toward probable source areas. (From Baker and Kochel, 1979a, Jour. Geophys. Res., v. 84, p. 7961–7983, copyrighted by American Geophysical Union.)

A

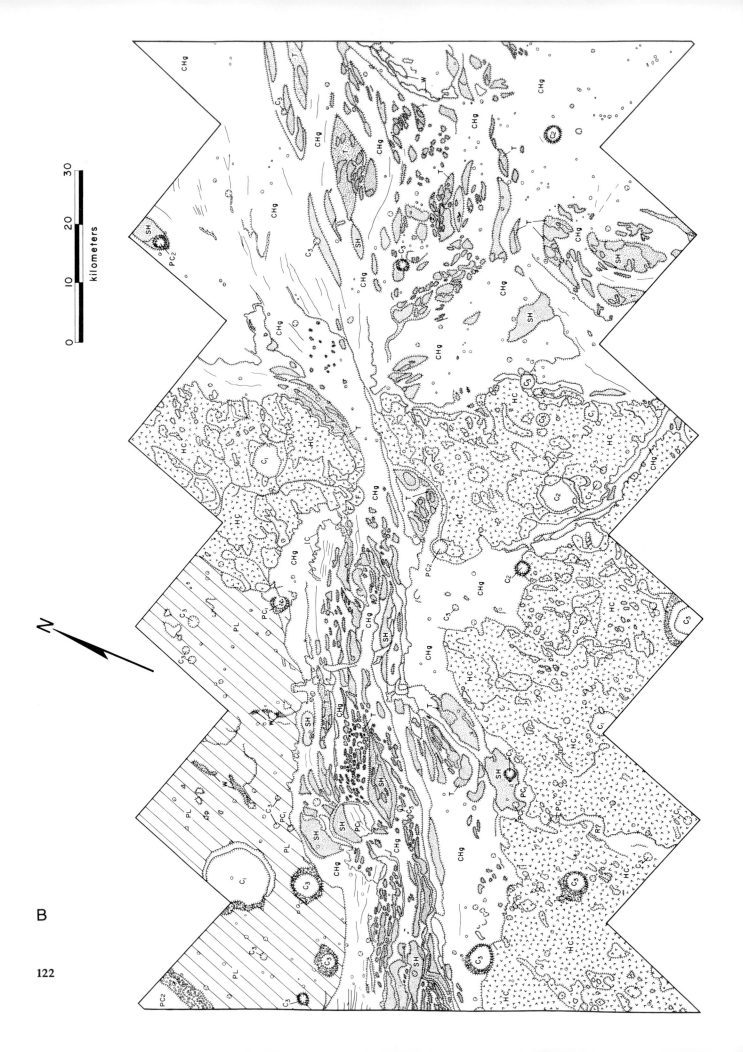

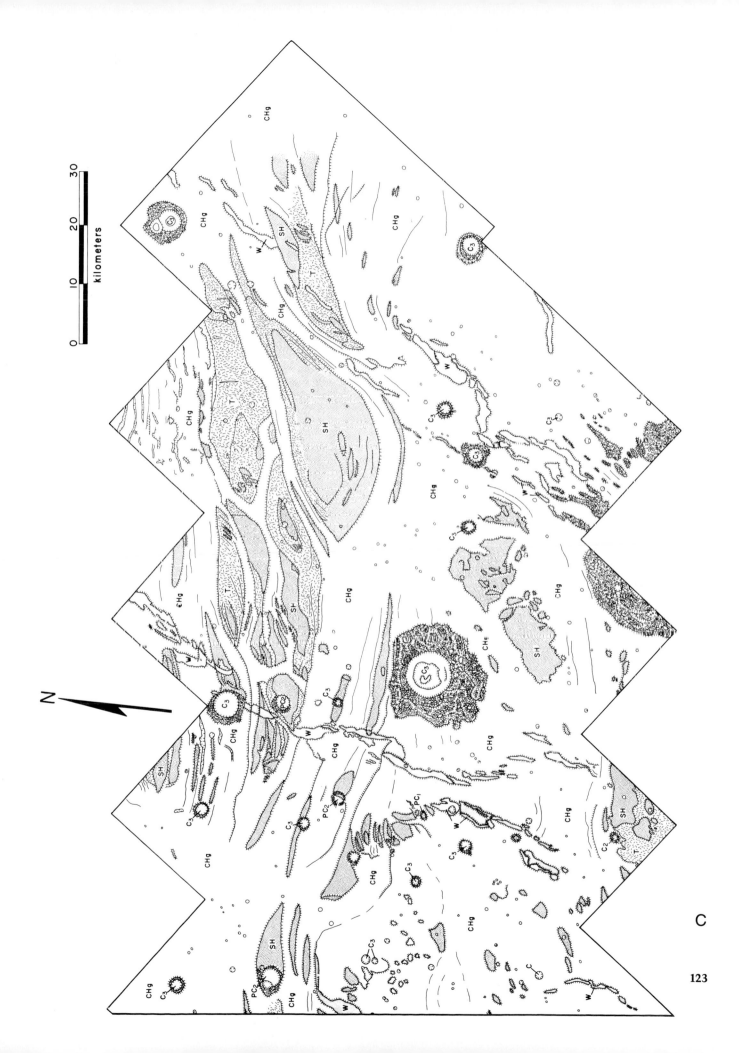

C

123

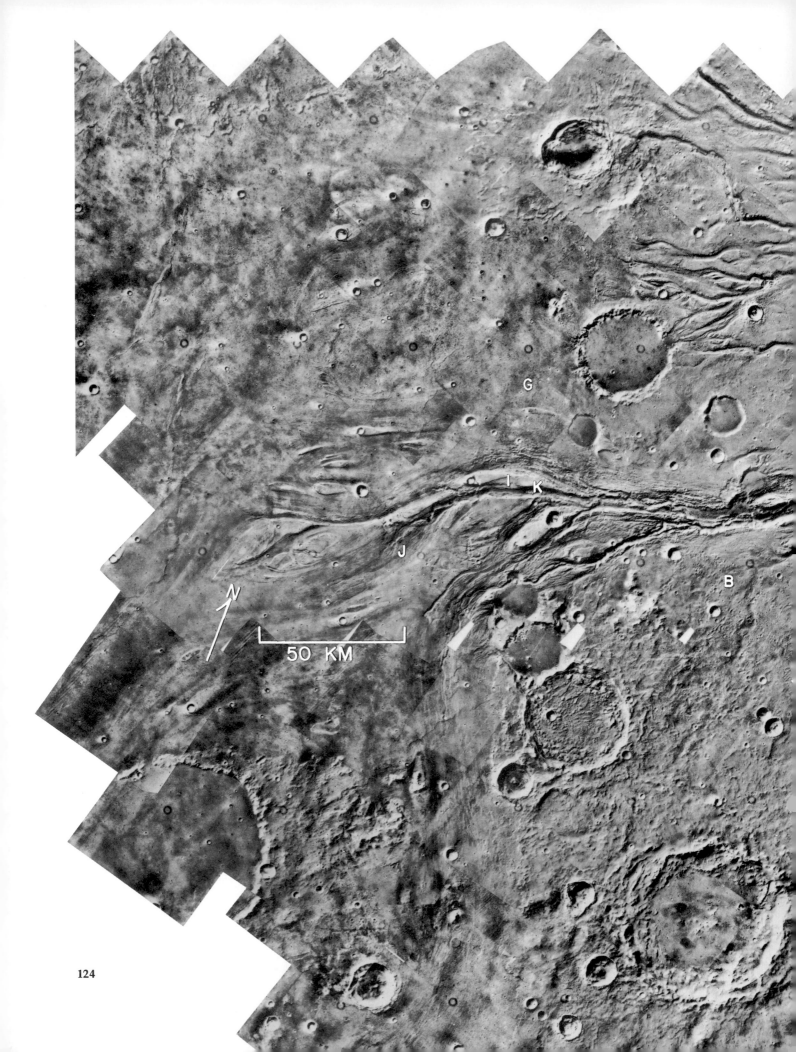

50 KM

124

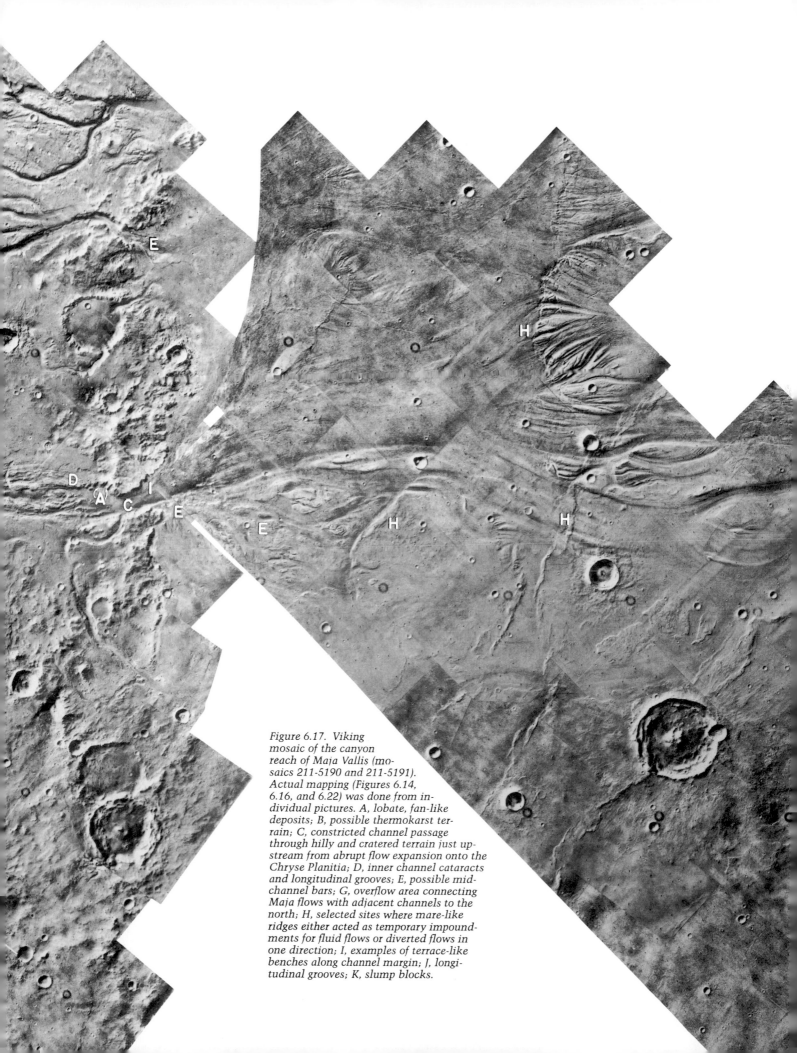

Figure 6.17. Viking
mosaic of the canyon
reach of Maja Vallis (mo-
saics 211-5190 and 211-5191).
Actual mapping (Figures 6.14,
6.16, and 6.22) was done from individual pictures. A, lobate, fan-like
deposits; B, possible thermokarst terrain; C, constricted channel passage
through hilly and cratered terrain just upstream from abrupt flow expansion onto the
Chryse Planitia; D, inner channel cataracts
and longitudinal grooves; E, possible mid-
channel bars; G, overflow area connecting
Maja flows with adjacent channels to the
north; H, selected sites where mare-like
ridges either acted as temporary impoundments for fluid flows or diverted flows in
one direction; I, examples of terrace-like
benches along channel margin; J, longitudinal grooves; K, slump blocks.

vian surface also appear as erosional remnants ("islands") within anastomotic channel reaches.

Plateau Material (PL). This is the material of the Lunae Planum, widespread along the western margin of Chryse Planitia. It comprises relatively smooth plateau or plains surfaces that are extensively developed between Maja and Kasei Valles. Milton (1975) interpreted the Lunae Planum materials as possible basaltic lava flow units underlain by less consolidated materials, perhaps sediments.

Residual hills of plateau materials occur as "islands" within the outflow channel complexes (Figure 6.17). These plateau remnants commonly exhibit a fluid-streamlined shape, and some display horizontal terracing along vertical exposures. The terracing is suggestive of layered basalt flows or sedimentary rocks.

Hilly and Cratered Terrain (HC). This terrain predates all outflow channeling in the Maja and Kasei region (Greeley and others, 1977). The terrain material is characterized by rugged topography formed by abundant hills and overlapping craters. The wide range of impact crater morphologies, including severely degraded varieties, and the high crater density imply a great age for this surface. A relatively narrow zone of heavily cratered terrain separates the Chryse Planitia from the Lunae Planum (Figure 6.14).

Some areas of hilly and cratered terrain appear very mountainous and devoid of craters. It is unclear whether the mountains are remnants of ancient crater rims or were created by other processes. Maja Vallis narrows tremendously and is deeply incised through the hilly and cratered terrain (Figure 6.17). The juxtaposition of hilly and cratered terrain, plateau units, and plains units has resulted in alternating diverging and converging flow patterns (Figure 6.14 and 6.17).

Erosional Landforms

The most striking aspect of the outflow channels is the abundant evidence of fluid erosion on a massive scale. The mapping philosophy for the erosional features has been dictated by the detailed analogy between the outflow channels and the Chan-

neled Scabland of eastern Washington (Baker and Milton, 1974; Baker, 1978a, 1978b). The entire erosional assemblage is well illustrated by the map of the Maja Vallis canyon through the upland of hilly and cratered terrain that separates the Lunae Planum from the Chryse Planitia (Figure 6.16).

Residual Uplands and Streamlined Hills (SH). Erosional remnants of the plateau and hilly, cratered terrains are common within the confines of the larger outflow channels. Residual uplands are irregular in plan view but do show some evidence for modification by the channel-forming fluids. Streamlined hills are remnants which have been so extensively modified that they present fluid-dynamic shapes of minimum resistance to the responsible flows (Baker and Kochel, 1978b).

Streamlined hills have rounded prows on their upstream ends and tapering tails pointing downstream. Crescent-shaped depressions occur on the channel floors upstream from some streamlined forms. These probable scour marks (Baker, 1978b) are best developed in constricted reaches of the channels and absent in broad, shallow reaches. Many of the residual uplands have terraced margins. These indicate differential erosion of lithologic units of varying resistance.

Many small residual hills less than 2 km long occur within the Maja Vallis complex. Where these were too small to determine diagnostic shapes, they were mapped without interpretive connotations (Figure 6.16). For the sake of map clarity, many of the smaller forms were simply deleted. Some of these features may be small remnants of upland surfaces, mass wastage products, fluvial deposits, large boulders on the channel floor, or a combination of several phenomena.

Crater-Channel Interactions. Craters in Maja Vallis exhibit a wide range of morphologies suggesting varying ages and modification by channeling processes. The oldest craters (C_1) may display smooth floors and incomplete rims. Some craters that have smooth rims which are partly incomplete because of probable degradation or burial are designated C_2 if they likewise show no interaction with channel-forming fluids.

Craters showing both fresh-appearing morphology and distinct raised rims, also not interacting with channeling processes, are labeled C_3. The latter are interpreted to be relatively young craters that have not been modified by catastrophic flood flows. Relatively fresh crater interiors commonly show mass wastage features on the crater walls, and many contain raised central peaks.

Craters mapped with the prefix P (PC_1, PC_2) display clear evidence of interaction with the channeling process. Since craters of variable apparent ages have been modified by the channeling process, this designation is necessary to separate those involved with flood erosion from those which have not been affected. In order to reduce clutter on the geomorphic maps, many small craters (less than 1–2 km diameter) have not been labeled. Most unlabeled craters appear to be relatively young (C_3). (The above classification applies only to the maps shown in Figures 6.14, 6.15, 6.16, and 6.22.)

Many residual hills of antediluvian terrain are developed downstream from raised crater rims that apparently served as flow obstructions. Others have no such obstruction at their upstream margins. The latter appear to be most common near the channel margins, while hills behind crater obstructions commonly occupy positions toward the centers of the channels. Many antediluvian crater rims within the constricted channel reaches have rims that were breached by the fluid flows. In contrast, most crater rims in the Chryse Planitia, east of the major constriction of Maja Vallis, do not appear to have been breached (Figure 6.17). A possible explanation is the inability of the expanding fluid flow to develop sufficient turbulence to effectively erode the larger craters. In the constricted zones, deeper fluid flows may have provided the conditions necessary for development of macroturbulence, thus eroding through some of the crater rims (Baker, 1978b).

Mare-like Ridges and Channel Interactions. Greeley and others (1977) identified lunar mare-like ridges in the western Chryse Planitia, and Milton (1975) mapped similar ridges in

Figure 6.18. Mare-like ridges and crater in Maja Vallis which appear to have temporarily ponded fluid flows entering from the left. The flows breached low points in the ridge, eroding at constrictions and expanding downstream. This crater is named Dromore and is located at 20°N, 49°W. (Viking frame 20A62.)

the Lunae Planum. The origin of these ridges, which are also named "wrinkle ridges," is discussed in Chapter 2. The mapping here emphasizes the interaction of the ridges with channel-forming processes.

Antediluvian mare-like ridges may show considerable modification by the fluid flows (Figure 6.18). Some of the higher ridges are breached by narrow channels. In other cases, flow was diverted around the ridges, and erosional processes have streamlined the topography. Breaching probably resulted from temporary ponding of voluminous fluid flows. Various degrees of interaction occur, with some ridges probably totally obliterated by the channeling processes.

Antediluvian mare-like ridges are most common, but other ridges postdate the streamlined forms or lack any definitive contact relationship. Greeley and others (1977) noted that mare-like ridges are most prominent in the western Chryse Planitia and are nearly absent in the central and southeastern parts of the Chryse Planitia. This may reflect burial of the ridges by fluvial sediments. The relationships could also be explained by other burial processes, e.g., eolian or volcanic deposition, or possibly the ridges are simply absent.

Channel Floor Features (Bed Forms). The overall channel pattern and included bedforms were mapped by considering their close analogy to the Channeled Scabland. Channeled areas are designated CHg. In the western Chryse Planitia the channels comprise an overall anastomosing pattern, often with a distinct trimline that separates the channeled region from the unmodified plains. Further upstream, in the hilly, cratered terrain these same channels are narrowly constricted, deeply incised, and exhibit evidence of differential erosive stripping of the rock units.

Maja Vallis narrows from a width of more than 50 km west of the hilly and cratered terrain upland to a width of less than 5 km through a narrow gap in the rugged hilly and cratered terrain (Figures 6.14, 6.16, and 6.17). East of this constricted area the channel rapidly expands and develops a broad anastomosing pattern where deposition may have occurred. The total width of the Maja system east of this probable expansion fan is uncertain, but may be in excess of 75 km. The channel patterns become progressively more diffuse toward the east and eventually become untraceable toward the central Chryse Planitia area. Greeley and others (1977) proposed that the alternation of channel morphology may be due to differential resistance offered by the terrain through which the channels are cut.

Distinctive erosional bed forms developed on the channel floors include longitudinal grooves, inner channels, possible cataracts, scabland, and scour marks around flow obstacles. The longitudinal grooves occur as dark-appearing streaks on the channel floors and display regional conformity to presumed fluid flow directions as dictated by channel boundaries. The Maja Vallis grooves are most common in constricted reaches (Figure 6.17). Grooves are also common along the lateral margins of flow obstructions such as streamlined hills.

Channel Wall Features

Channel wall escarpments exhibit less relief in Maja Vallis than in Kasei Vallis, but similar wall features occur on somewhat different scales. Talus accumulations are best developed along the steepest cliffs (Figure 6.19). Mass wastage probably contributes to postdiluvian backwasting and parallel scarp retreat as described for the Val-

Figure 6.19. Flow expansion of Maja Vallis (center and bottom) onto the Chryse Planitia plains (upper right). Note the terraced channel margin and the fan-like accumulations of material (F) at the base of extensive slopes. These "fans" appear to be connected to gullies which extend upslope. Further upstream are longitudinal grooves (lower left). This frame depicts 45 × 50 km. (Viking frame 044A57.)

les Marineris region by Sharp (1973a) and Lucchitta (1978a). Debris resulting from these processes often exhibits definite lobate patterns in some areas and is mapped as debris fan deposits (F) when the responsible fanhead gullies can be recognized (Figure 6.19). Landslide-like features are scarce in the Maja Vallis area. This may result from lower escarpment relief relative to Kasei Vallis and Valles Marineris canyon walls or from greater postslide modification of slumped deposits in the Maja area.

Depositional Landforms

Some of the streamlined uplands in Maja Vallis may be partly or wholly depositional. In Figure 6.16 some streamlined forms that developed in the lee of flow obstacles are tentatively mapped as pendant bars (B) of fluid-transported sediment. This type of morphology is most common in the Chryse Planitia flow expansion of the Maja Vallis system. In this reach deposition may have been rapid and concurrent with flow expansion, developing a fan-delta at high stage, which was subsequently scoured by waning flows to produce diamond-shaped bar remnants (Figure 6.20).

Longitudinal or midchannel bars (B) would be expected to have a wide distribution following catastrophic water floods. Postdiluvian mantling with eolian, impact, or pyroclastic material may obscure definitive recognition of such forms. Diagnostic bed forms such as giant current ripples if preserved could be below the resolution capabilities of the orbital pictures. Probable longitudinal bars do appear as diamond-shaped light-albedo patches in the constricted canyon of Maumee Vallis (Figure 6.21). Broader light albedo patches appear to mark fan-deltas at the mouths of

both Maumee and Maja Valles (Figures 6.17 and 6.20). These may be areas that escaped significant postdiluvian mantling.

Suggestive evidence for deposition in Martian channels is present, but not overwhelming. It should be recalled that orbital photographs of the Channeled Scabland (Baker and Nummedal, eds., 1978) provide excellent views of erosional features, but clear indicators of flood deposition are often not resolved. This occurs because the flood-depositional surfaces are armored with boulders of the same bedrock found on scoured bedrock floors (Baker, 1973a). In addition, the postflood mantle of eolian sediment, probably fine-grained, loess-like material, accumulates preferentially in the low-relief swales of the depositional terrain. Therefore, the scabland analog suggests that many depositional landforms may not have been recognized

in the geomorphic mapping of the Martian channels.

Adjacent Channel Systems

Vedra and Maumee Valles are developed in the hilly, cratered terrain ridge to the northwest of the Maja Vallis canyon (Figures 6.13 and 6.14). Another channel trends southeast from the Maja canyon at a point 25 km west of the Maja expansion into Chryse Planitia (Figures 6.16 and 6.17). This channel appears to represent overflow from the primary Maja Vallis channel. Evidence for crossing this divide in cratered terrain includes scour within and along the smaller channel and evidence of streamlining and possible deposition at the mouth of this channel.

Masursky and others (1977) suggested that Vedra and Maumee Valles are independent of Maja Vallis and that their morphometric patterns of

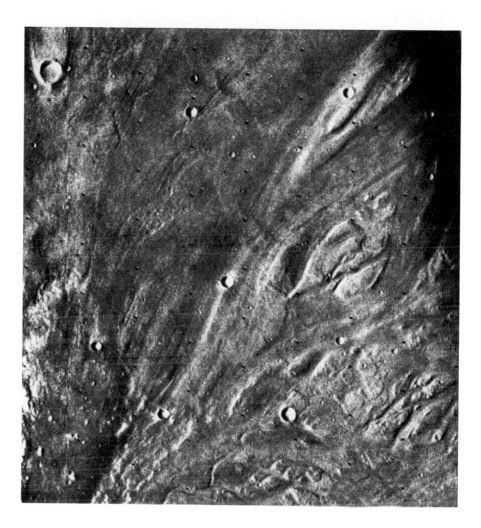

Figure 6.20. Relatively small streamlined hills and small anastomosing channels between apparent depositional features, all of which comprise a fan complex at the mouth of the Maja Vallis canyon. The frame depicts an area 45 × 50 km. (Viking frame 044A57.)

Figure 6.21. Probable mid-channel bars (B) near the flow expansion of Maumee Vallis onto the Chryse Planitia. The probable bars appear as diamond-shaped light albedo patches on the channel floor and resemble terrestrial longitudinal bars found in braided streams. This frame depicts an area approximately 27 × 18 km. (Viking frame 046A59.)

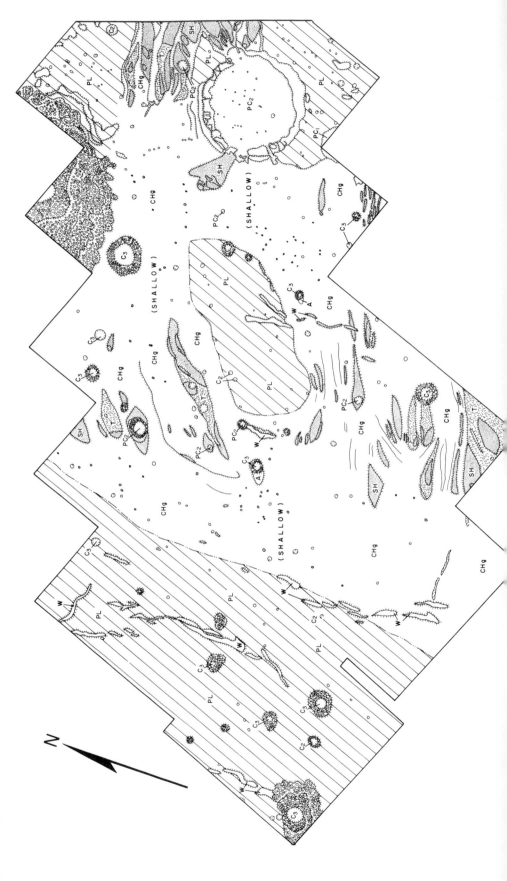

Figure 6.22. Geomorphic map (in two parts) of the Maumee Vallis system immediately north of Maja Vallis (Figure 6.16). The regional relationships of these channels are shown in Figure 6.14 north of the outlined area. (From Baker and Kochel, 1979a, Jour. Geophys. Res., v. 84, p. 7961–7983, copyrighted by American Geophysical Union.)

accordant tributary junctions are similar to terrestrial drainage systems. However, close examination of Viking images in the area between Maja and this system (Figure 6.17) shows streamlined forms and other evidence of shallow flood flow over the low divide area between the Maja and Maumee channel systems. The fluid entered and flowed around the northern end of the large (30 km) flat-floored crater at the head of the Maumee channel system. It appears that overflow from the Maja channel spilled over this divide and was quite shallow, thus leaving little evidence of erosion. Some of the fluid appears to have been funneled into the larger crater at an erosional notch on its western side and then spilled out through a larger outlet on the eastern rim (Figures 6.22 and 6.23). The remainder of the flow appears to have been deflected around the northern edge of the crater, providing a source for the dendritic-appearing channels heading into this region (Figure 6.22). The unusual pattern of Maumee Vallis may have resulted because this flow invaded and/or partially exhumed an antediluvian small valley network.

Flat-Floored Craters

Many of the larger craters in the vicinity of Maja Vallis exhibit dark, smooth, flat floors with relatively few superimposed small craters. They resemble lunar mare regions, but on a much-reduced scale. One possible explanation for this phenomenon could be postdiluvian filling of the crater floors by lava flood basalts. This would explain their partly buried appearance and lack of erosional features produced by channeling. In almost every case, however, these craters also show evidence of rim erosion and breaching associated with channeling processes.

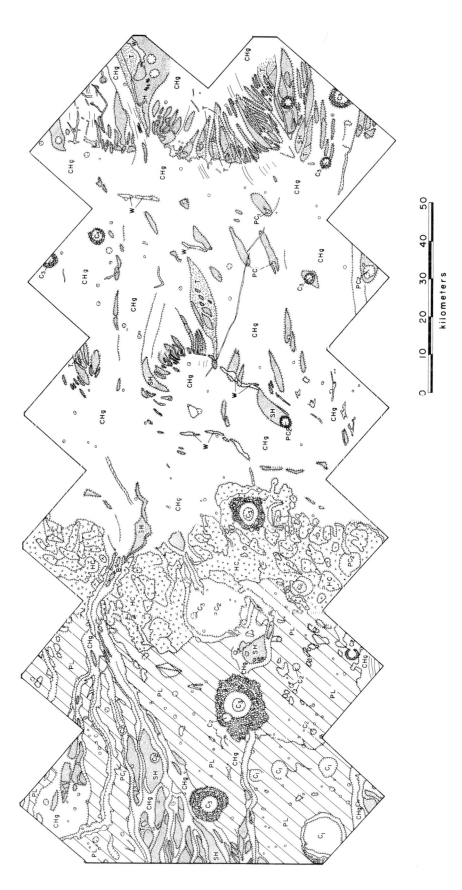

kilometers

50
40
30
20
10
0

They have probable inlets and outlets (Figure 6.23). Therefore, another hypothesis could be that the craters contained large, temporary lakes during the channeling event(s). If this were the case, suspended sediments would likely have been deposited on the crater floors. In this interpretation the flat floors may represent lacustrine deposits from temporary impoundments by the crater rims during the channeling events. A similar process could explain the broad, flat expanses upstream from some mare-like ridges which effectively blocked or slowed flood waters (Figure 6.18).

Temporal Relationships for the Maja Vallis System

Several lines of evidence can be used to investigate the temporal association of channels on the Martian surface, including the ejecta morphology of impact craters, cross-cutting relationships, crater size frequency curves, and morphological preservation. Theilig and Greeley (1979) analyzed the geological history of the Lunae Planum Chryse Planitia region, including Maja Vallis. The oldest terrain in their chronology is the hilly and cratered material (HC in figures 6.14, 6.16, and 6.22). This rugged terrain was partly altered to cratered plateau material, probably by volcanic and/or eolian resurfacing processes. These cratered uplands also display small valley network systems, similar to those described in Chapter 4.

The ancient cratered terrain is embayed by the Lunae Planum from the west and by the lower smooth plains material of Chryse Planitia from the east. The lower smooth plains material of Chryse Planitia is interpreted as overlain by an upper smooth plains unit near the mouths of Maja, Maumee, and Vedra Valles. This upper smooth plains is considered to be fluvial deposition from the outflow channels.

Theilig and Greeley (1979) showed that the major outflow channels were formed after the emplacement of the lower smooth plains. They considered Vedra and Maumee Vallis to have formed first, perhaps by an early flood from Juventae Chasma. The formation of Maja Vallis canyon is consid-

Figure 6.23. Large crater (16 km diameter) in the Maumee Vallis system. The crater apparently interacted with the channeling process and displays an inlet on its upstream side (at A) and a series of outlet channels (B) that drained fluid which was ponded within the crater. (Viking frame 46A54.)

ered to have been the last channeling event. Deposits from the mouth of Maja Vallis canyon appear to overlie the deposits of Vedra and Maumee Vallis in the western Chryse Planitia. Thus the various epochs of channeling are presumably associated with a stratigraphic sequence in the upper plains unit of Chryse Planitia.

KASEI VALLIS

Kasei Vallis (Figure 6.24) is perhaps the largest continuous outflow channel on Mars. It presents an interesting comparison to nearby Maja Vallis because Kasei Vallis appears to have experienced considerable postdiluvian modification. The Kasei Vallis system heads at Echus Chasma and trends northward between a relatively featureless, sparsely cratered plain to the west and the more heavily cratered Lunae Planum to the east. The floor of upper Kasei Vallis has extensive zones of relatively smooth plains (Figure 6.25), perhaps because of mantling with eolian or volcanic materials. Further north the channel floor is characterized by streamlined hills, longitudinal grooves, and inner channels (Figure 6.26). North of latitude 20°N the system splits into two eastward-flowing branches, here named Kasei Vallis Canyon and North Kasei Channel (Figure 6.24). The two channel branches are separated by an upland remnant of the Lunae Planum. The branches recombine at about longitude 63°W (Figure 3.9). The combined system then continues out into the Chryse Planitia. Highly streamlined hills occur at the debouchment of Kasei Vallis into Chryse Planitia (Figure 6.27). The total system has a length of nearly 3,000 km. The width of channeling is locally in excess of 100 km, and portions of Kasei Vallis Canyon are entrenched 2–3 km into the Lunae Planum surface.

Based on the available Mariner 9 imagery, Schumm (1974) concluded that Kasei Vallis originated as a tension fracture that was subsequently modified by wind action and mass wasting. Schumm was especially impressed with the Sacra Fossa region, an obvious fractured valley region on the Lunae Planum immediately east of upper Kasei Vallis (Figures 6.24 and 6.26). He suggested that a similar fracture pattern once existed in the lower Kasei Vallis area and that its angular bends were later smoothed by mass wasting. However, Baker and Milton (1974) described an assemblage of bed forms in lower Kasei Vallis that dictated fluid flow at some time in the valley's history. The bed forms are illustrated by the morphological maps in Figures 6.28 and 6.29. Baker and Milton (1974) concluded that the assemblage was typical of catastrophic flooding, similar to that responsible

for the Channeled Scabland (Chapter 7).

Sharp and Malin (1975) proposed a complex origin for Kasei Vallis. They emphasized a fretting process along structural lineaments, perhaps by a sapping mechanism, as the primary agent of developing the great relief of Kasei Vallis Canyon. This process was then followed by fluidal erosion, either by water or by wind. Baker and Kochel (1979a) examined the Viking imagery of Kasei Vallis and found further evidence for a complex history for Kasei Vallis involving fluidal erosion and extensive postdiluvian modification by mass wasting and probable eolian processes.

In central Kasei Vallis (Figure 6.30) the grooved terrain and the polygonal fracture pattern that cuts it have an "etched" appearance. Part of the fracture network appears to have controlled development of the labyrin-

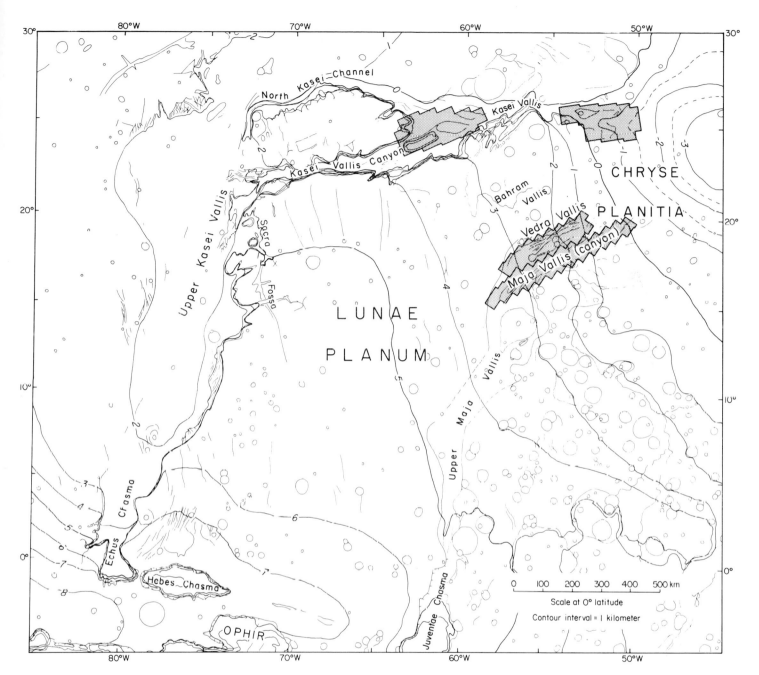

Figure 6.24. Index map of the Lunae Planum and western Chryse Planitia, Mars, showing the general location of Kasei Vallis and related features. The areas outlined are mapped in detail in Figures 6.16, 6.22, 6.28, and 6.29. (From Baker and Kochel, 1979a, Jour. Geophys. Res., v. 84, p. 7961–7983, copyrighted by American Geophysical Union.)

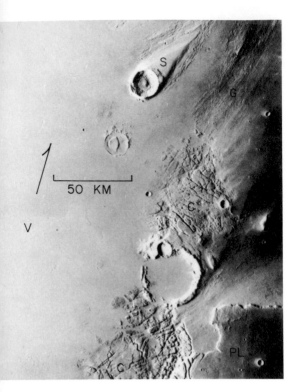

Figure 6.25. A portion of upper Kasei Vallis centered at 14°N, 75°W. Features shown include streamlined upland (S) Lunae Planum surface (PL), chaotic terrain (C), longitudinal grooves (G), and smooth plains (V) which probably mantle part of the channel floor. (Viking frame 519A01.)

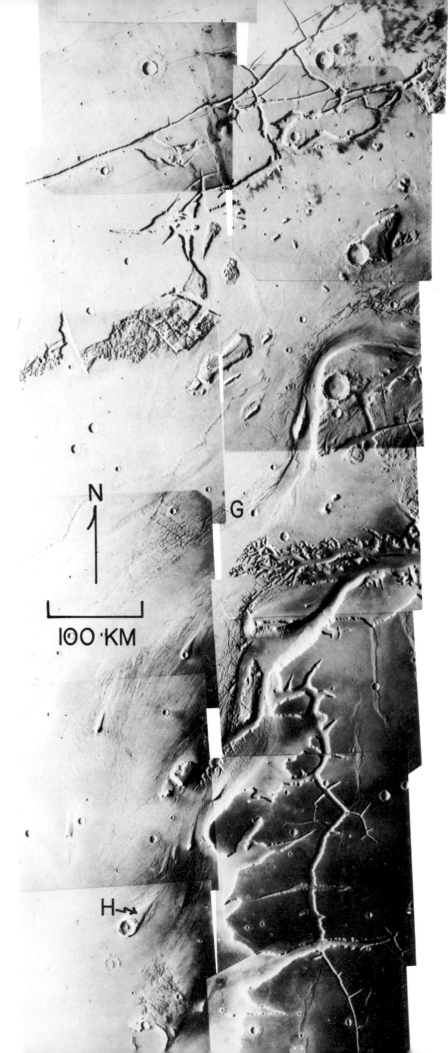

Figure 6.26. Viking mosaic of a portion of Kasei Vallis centered at approximately 20°N, 75°W. This reach is upstream from that mapped in detail in Figure 6.28. Outflow features such as streamlined hills (H) and grooves (G) occupy the channel flow, but valley sides, tributary canyons, and inner channels of the valley floor show pronounced structural control by fracture systems. Antediluvian headward sapping of tributary canyons also appears to be controlled by structure as indicated by the high-angle junctures of tributaries to the main valley. (Viking mosaic 211-5642, rev. 519A.)

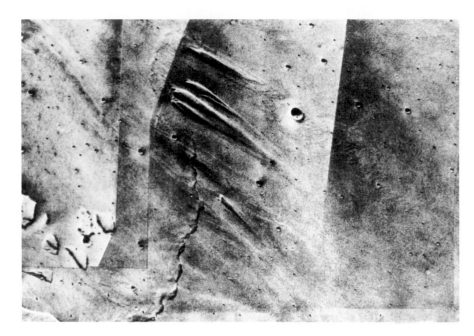

Figure 6.27. Streamlined hills of the western Chryse Planitia near the mouth of Kasei Vallis. These hills display typical long, tapering tails and rounded upstream "prows." The image depicts a scene approximately 100 × 60 km (NSSDC picture 17003.)

Figure 6.28. Geomorphic map of lower Kasei Vallis prepared from Viking images 226A06–226A13 (compare to Figure 3.9). The map shows bed forms indicative of catastrophic flood erosion (Baker, 1978b). Legend for this figure is the same as for Figure 6.15 except that the following crater classification is used: PC_1, craters with raised rims that have interacted with channeling processes; C_1, craters with raised rims that have not interacted with channeling processes; PC_2, partly buried craters that have interacted with channeling processes; C_2, crater with poorly preserved rim morphology; C_3, exhumed craters.

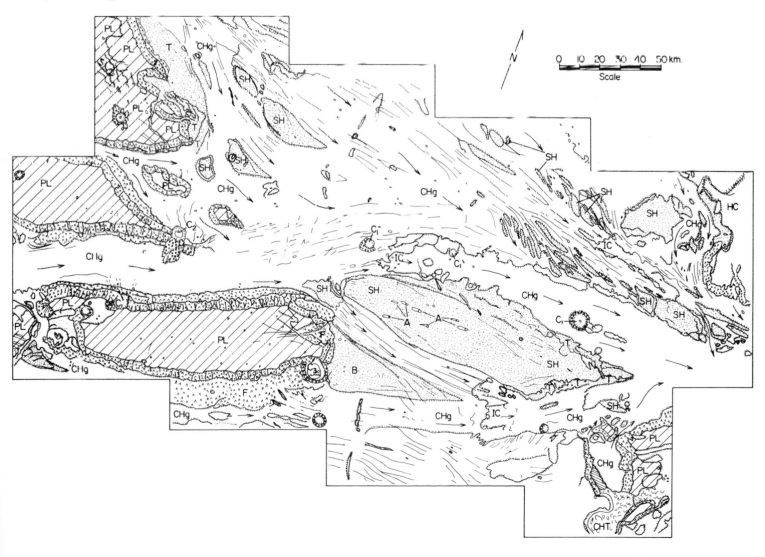

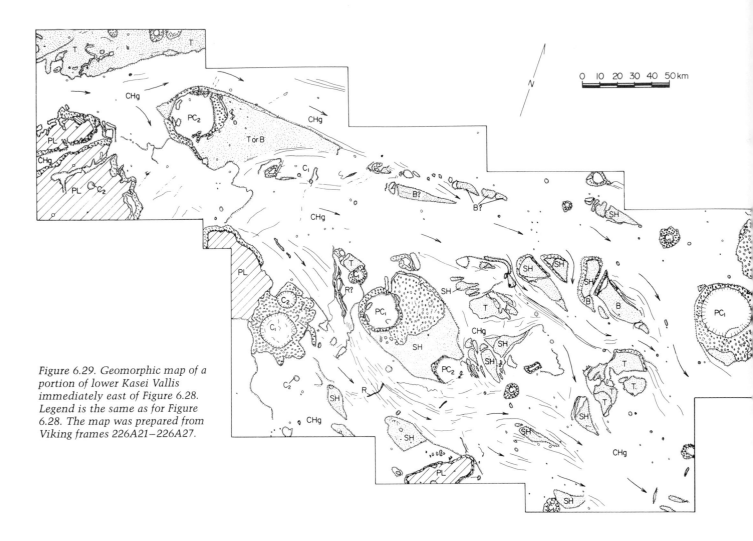

Figure 6.29. Geomorphic map of a portion of lower Kasei Vallis immediately east of Figure 6.28. Legend is the same as for Figure 6.28. The map was prepared from Viking frames 226A21–226A27.

thine inner channel complex shown at the center of Figure 6.30. The apparent erosional enhancement of both grooves and fractures appears to require a postdiluvian process of granular disintegration followed by selective debris removal. Possibly, as suggested by Malin (1974), such "etching" could be achieved by salt weathering and eolian removal of detritus.

In lower Kasei Vallis prominent grooves are well developed upstream from probable inner channel cataracts (Figure 6.28). Except for the larger scale, the grooves occur in similar relationships to those observed in the Channeled Scabland. The scabland grooves are interpreted as the result of action by longitudinal roller vortices in deep, swift flood water.

In the Channeled Scabland, inner channels form by the headward recession of subfluvial cataracts (Chapter 7). Plucking of the resistant rock

layers occurs when flow separation is developed in the water passing over the lip of the cataract. Inner channels and probable erosive headcuts (IC) are well developed in the constricted reach of Kasei Vallis (Figure 6.28). It is probable that the groove topography has been accentuated by an "etching" process, possibly salt weathering and eolian removal of detritus.

Channel walls in Kasei Vallis show prominent modification features that imply significant backwasting since the time of channel formation. High wall relief, approximately 2–3 km (Sharp and Malin, 1975), in the constricted portion of the Kasei Vallis system is associated with prominent mass wastage and sapping features. Cliffs along the channel margins exhibit steep upper slopes which have been dissected into subparallel vertical ribs. These grade into gentler lower talus slopes to produce the di-

agnostic spur-and-gully topography that is also developed on chasma walls in the Valles Marineris (Lucchitta, 1978a). Landslides, debris fans, and debris cones can also be recognized (Figure 6.31). Wall recession by landsliding produces great scars that develop at the expense of slower-operating mass wastage processes.

Portions of upper Kasei Vallis have wall escarpments marked by tributary canyon morphology (Figures 6.26 and 6.32). These canyons, similar to those tributary to the Valles Marineris system (Lucchitta, 1978a), typically have blunt, cirque-like heads and smooth, V-shaped cross-profiles. In most cases the development of tributary canyons appears to be structurally controlled. The headwardly eroding tributaries may have formed through the processes of ground-ice breakup and headward sapping along structurally controlled compartments

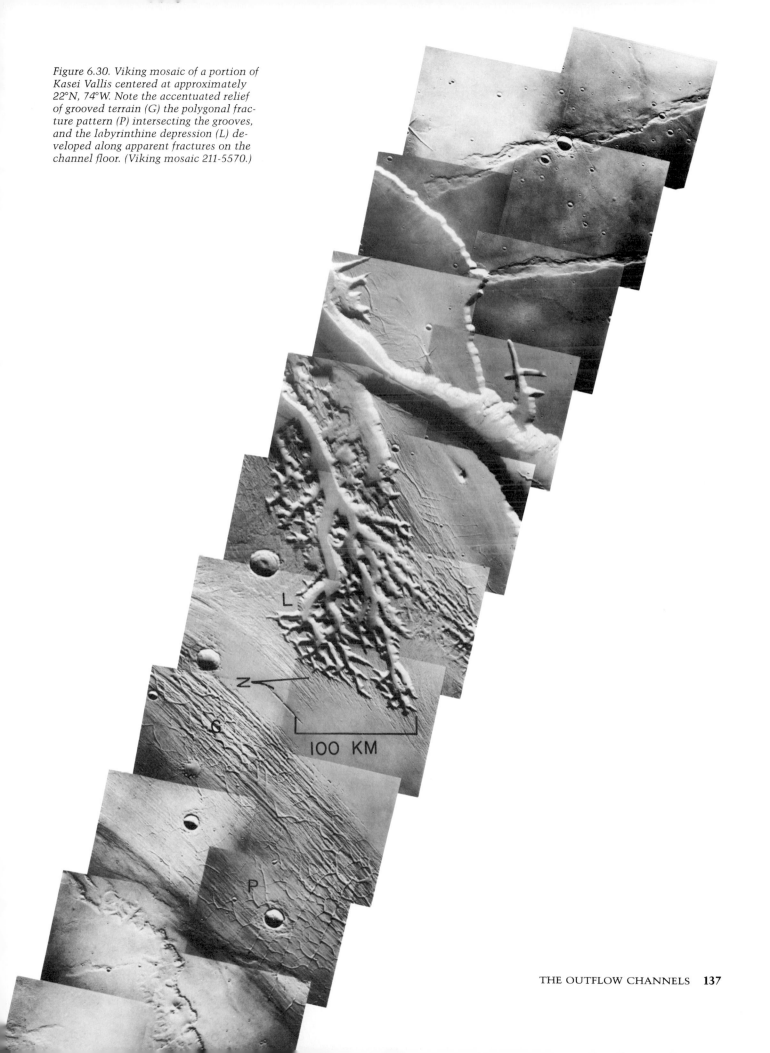

Figure 6.30. Viking mosaic of a portion of Kasei Vallis centered at approximately 22°N, 74°W. Note the accentuated relief of grooved terrain (G) the polygonal fracture pattern (P) intersecting the grooves, and the labyrinthine depression (L) developed along apparent fractures on the channel floor. (Viking mosaic 211-5570.)

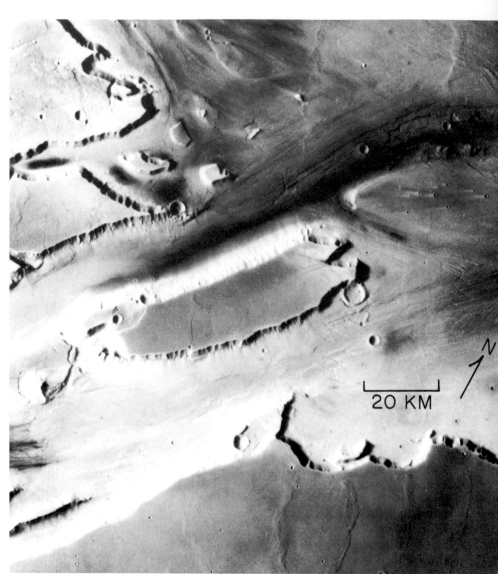

Figure 6.31. Typical spur-and-gully wall morphology in lower Kasei Vallis canyon. The well-developed spurs and ridges suggest considerable backwasting along channel margins since the cessation of channeling processes. At the toes of some slopes are extensive debris aprons and/or talus cones. At the downstream edge of the residual upland is the exhumed crater discussed by Baker and Milton (1974). (Viking frame 520A27.)

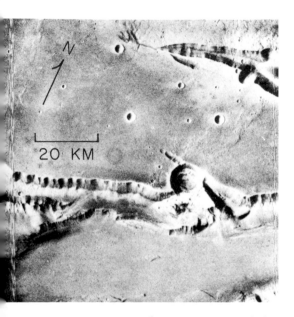

Figure 6.32. Tributary canyon morphology in the eastern canyon section of Kasei Vallis. Typical canyons show upstream terminations in cirque-like canyon heads. Other canyons show smooth V-shaped cross-profiles. (Viking frame 226A04.)

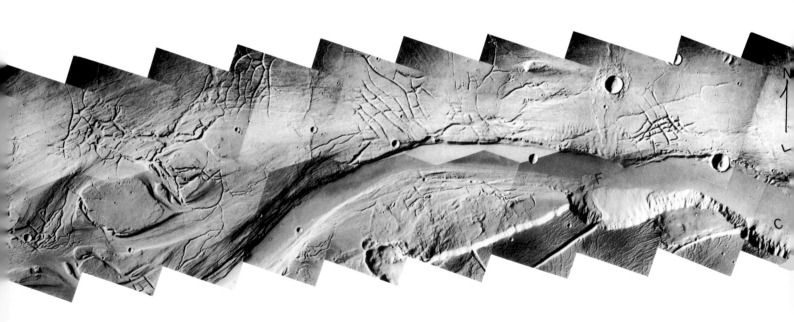

of susceptible rock. Lower Kasei Vallis has been interpreted as having a graben origin (Schumm, 1974). Exposure of permafrost along graben walls would probably initiate the thermokarst backwasting. Another way to expose canyon walls for the initiation of thermokarst backwasting could be by fluvial erosion. Working together or sequentially, both processes could have provided the triggering mechanisms for backwasting by hillslopes and tributary canyon development.

A proposed history for Kasei Vallis can now be summarized. The system began to form after development of the Lunae Planum surface. Steep escarpments were exposed by tectonic processes and subsequent flood incision. The flood fluids probably invaded a fretted margin along the northern Lunae Planum, modifying antediluvian troughs that had developed along zones of prominent structural control. The flooding produced an assemblage of bed forms similar to those that occur in the Channeled Scabland, but both the channel floors and the channel walls have been extensively modified by postdiluvian processes. Upper Kasei Vallis has an accentuated floor relief of grooved ter-

rain. Such an "etching" could be achieved by salt weathering and eolian removal of detritus. Portions of upper Kasei Vallis are also mantled with younger plains materials.

In lower Kasei Vallis, wall relief of 2–3 km is associated with prominent mass wastage and sapping features. Spur-and-gully topography, landslides, debris fans, and debris cones are locally present. The lower depth of incision for North Kasei Channel is associated with much less wall modification (Figure 6.33). Steep escarpments exposed by tectonic influences and subsequent flood incision later became the sites of long-term backwearing by mass wasting and sapping.

Channel wall morphologies in the canyon areas imply a long-term persistence of ground ice which contributed to the modification of the channels since the flood events. Kasei Vallis probably functioned at a time when ice-rich permafrost was widely distributed on Mars. Indeed, the great outflow channels appear to be but one of a great many manifestations of permafrost/ground-ice conditions that existed on a massive scale at least during the diluvian epoch. The channels themselves probably owe their origin to a variety of equilibrium/disequilibrium adjustments in the permafrost. Some suggested fluid-release mechanisms are discussed in Chapter 9.

CONCLUSIONS

Whatever process is invoked to explain the Martian outflow channels, it must also explain a distinctive assemblage of channel landforms that includes regional and local anastomosing patterns, expanding and contracting reaches associated with flow constrictions, streamlined uplands, inner channels with recessional headcuts, pendant forms (bars or erosional residuals) on the downcurrent sides of flow obstacles, longitudinal grooves, irregular "etched" zones on channel floors, and scour marks around obstacles. Some channels show longitudinal grooves converging on constricted reaches cut through ridges oriented transverse to the fluid flow. Elevation-controlled scour features imply an upper flow boundary in the responsible fluid. These features constitute an assemblage of landforms which on Earth is most characteristic of catastrophic flood channeling in bedrock. Chapter 7 will discuss the most spectacular terrestrial example of such channeling: the Channeled Scabland region of the northwestern United States.

Figure 6.33. North Kasei Channel (see Figure 6.24 for location). Note the channel-wall debris flows (F) and the chutes with debris cones (C). (Viking mosaic 211-5882, rev. 665A.)

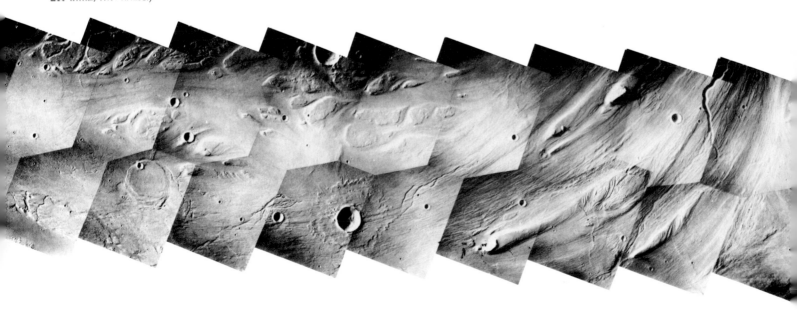

Chapter 7
The Channeled Scabland: An Earth Analog

The Channeled Scabland of the northwestern United States is the best terrestrial landscape·in which to view evidence of catastrophic flooding comparable in scale to that envisioned for the Martian outflow channels. The Channeled Scabland (Figure 7.1) consists of a spectacular complex of anastomosing channels, cataracts, loess "islands," and immense gravel bars created by the catastrophic fluvial erosion of the loess and basalt of the Columbia Plateau (Bretz, Smith, and Neff, 1956). The region is no better described than by the inimitable words of J Harlen Bretz (1928b, p. 446):

No one with an eye for landforms can cross eastern Washington in daylight without encountering and being impressed by the "scabland." Like great scars marring the otherwise fair face of the plateau are these elongated tracts of bare, or nearly bare, black rock carved into mazes of buttes and canyons. Everybody on the plateau knows scabland. It interrupts the wheat lands, parceling them out into hill tracts less than 40 acres to more than 40 square miles in extent. One can neither reach them nor depart from them without crossing some part of the ramifying scabland. Aside from affording a scanty pasturage, scabland is almost without value. The popular name is an expressive metaphor. The scablands are wounds only partially healed—great wounds in the epidermis of soil with which Nature protects the underlying rock.

With eyes only a few feet above the ground the observer today must travel back and forth repeatedly and must record his observations mentally, photographically, by sketch and by map before he can form anything approaching a complete picture. Yet long before the paper bearing these words has yellowed, the average observer, looking down from the air as he crosses the region, will see almost at a glance the picture here drawn by piecing together the ground-level observations of months of work. The region is unique: let the observer take the wings of the morning to the uttermost parts of the earth: he will nowhere find its likeness.

Conceive of a roughly rectangular area of about 12,000 square miles, which has been tilted up along its northern side and eastern end to produce a regional slope approximately 20 feet to the mile. Consider this slope as the warped surface of a thick, resistant formation, over which lies a cover of unconsolidated materials a few feet to 250 feet thick. A slightly irregular dendritic drainage pattern in maturity has been developed in the weaker materials, but only the major streamways have been eroded into the resistant underlying bed rock. Deep canyons bound the rectangle on the north, west, and south, the two master streams which occupy them converging and joining near the southwestern corner where the downwarping of the region is greatest.

Conceive now that this drainage system of the gently tilted region is entered by glacial waters along more than a hundred miles of its northern high border. The volume of the invading water much exceeds the capacity of the existing streamways. The valleys entered become river channels, they brim over into neighboring ones, and minor divides within the system are crossed in hundreds of places. Many of these divides are trenched to the level of the preexisting valley floors, others have the weaker superjacent formations entirely swept off for many miles. All told, 2800 square miles of the region are scoured clean onto the basalt bed rock, and 900 square miles are buried in the debris deposited by these great rivers. The topographic features produced during this episode are wholly river-bottom forms or are compounded of river-bottom modifications of the invaded and overswept drainage network of hills and valleys. Hundreds of cataract ledges, of basins and canyons eroded into bed rock, of isolated buttes of the bed rock, of gravel bars piled high above valley floors, and of island hills of the weaker overlying formations are left at the cessation of this episode. No fluviatile plains are formed, no lacustrine flats are deposited, almost no debris is brought into the region with the invading waters. Everywhere the record is of extraordinarily vigorous subfluvial action. The physiographic expression of the region is without parallel; it is unique, this channeled scabland of the Columbia Plateau.

THE SPOKANE FLOOD CONTROVERSY

In a brilliant series of papers between 1923 and 1932, J Harlen Bretz shocked the geological community with his proposed flood origin of the Channeled Scabland. His ideas were denounced at scientific meetings, and several alternative theories were proposed. Considering the nature and vehemence of the opposition to his hypothesis, which was considered outrageous, its eventual scientific verification constitutes one of the most fascinating episodes in the history of modern science (Baker, 1978a). The controversy was not resolved until the publication in 1956 of a study (Bretz, Smith, and Neff, 1956) which answered with meticulous detail all previous criticisms of the flood hypothesis.

Central to the 1956 investigation was the study of the scabland depositional features. Extensive excavations of the irrigation project and new topographic maps proved that the gravel hills called bars by Bretz (1928a) were indeed that, subfluvial depositional bed forms. Most convincing of all was the presence of giant current ripples on the upper bar surfaces. These showed clearly that bars 30 m high were completely inundated by phenomenal flows of water. Numerous examples of giant current ripples were found on the same bars which other investigators had interpreted as terraces. Such features could only have been produced by the flow velocities associated with truly catastrophic discharges. Bretz, Smith, and Neff (1956) and Bretz (1959) modified Bretz' earlier interpretations to allow for several episodes of flooding. The central theme of their study, however, was that only a hypothesis involving flooding could account for all the features of the Channeled Scabland. More recent studies of the Quaternary geology of eastern Washington have accepted this reasoning (Trim-

Figure 7.1. Regional pattern of the Channeled Scabland as mapped from Landsat imagery (E-1039-1843-5 and E-1004-18201-7). Y, Cheney; R, Ritzville; B, Benge; W, Washtucna; M, Moses Lake; E, Ephrata; S, Soap Lake; C, Coulee City; L, Othello; K, Wilson Creek; A, Odessa.

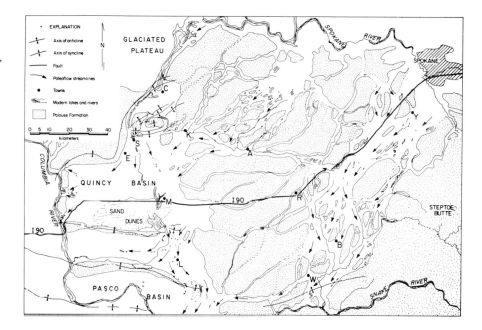

Figure 7.2. Relationship of glacial Lakes Missoula and Bonneville to catastrophic Pleistocene flooding in the northwestern United States.

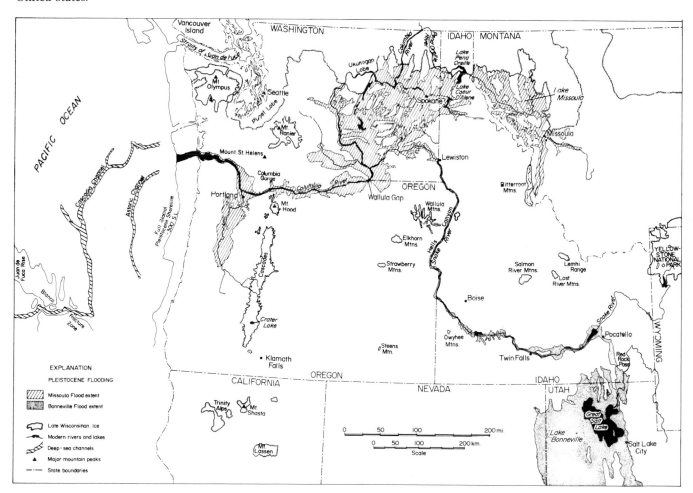

ble, 1963; Fryxell and Cook, 1964; Richmond and others, 1965; Baker, 1973a).

A major difficulty encountered during Bretz' early work on the Channeled Scabland was in establishing a source for the catastrophic flooding. We now know that Pleistocene Lake Missoula was the source for floods released from a glacial ice dam in the vicinity of modern Lake Pend Oreille, Idaho (Figure 7.2). During its maximum outflow, near the end of the last major Pleistocene glaciation, the lake discharged as much as 21.3×10^6 m³/sec into the vicinity of Spokane, Washington (Baker, 1971).

Breakout floods from Lake Missoula occurred repetitively during the Pleistocene (Patton and Baker, 1978a), and another great Pleistocene flood emanated from Lake Bonneville (Malde, 1968). The effects of these floods can be traced downstream, through Portland, Oregon (Trimble, 1963), and out onto the sea floor of the Pacific (Griggs and others, 1970). The recent research on the geomorphology of the Channeled Scabland is summarized by Baker (1978a, 1981) and by Baker and Nummedal (1978).

On November 6, 1979, at the age of 97, J Harlen Bretz was awarded the Penrose Medal, the highest honor of the Geological Society of America. The award cited especially his research in the Channeled Scabland, begun in the summer of 1922. Bretz, in his acceptance, said, "Perhaps I can be credited with reviving and demystifying legendary Catastrophism and challenging a too rigorous Uniformitarianism."

GEOLOGIC SETTING

The many similarities between Martian outflow channels and "scabland-type" channels have been noted in earlier discussions. Because the overall assemblage of landforms is also useful to consider in such comparisons, this chapter will review the pertinent geological history and morphology of the Channeled Scabland. The emphasis will be placed on those large-scale scabland features that would be apparent on orbital photographs, such as would confound an imaginary extraterrestrial intelligence

attempting to explain the Earth's valley systems. Of course, terrestrial scientists have the benefit of extensive field studies in the region. A similar benefit on Mars would greatly aid the quest for the genesis of the Martian channels.

The scabland channels are cut in rock. The Yakima Basalt comprises the bedrock in all but a few parts of the Channeled Scabland. This unit is part of the extensive Neogene eruptions of plateau basalts that cover over 250,000 km² in parts of Washington, Oregon, and Idaho. Most of the lava was erupted during the Miocene. The lava flows are exceptionally thick, and several can be traced over 150 km. Considerable structural and lithologic variation can be found in the basalt sequence, including jointing patterns, pillow-palagonite complexes, sedimentary interbeds (from lakes on the Miocene land surface), and geochemical variation (Swanson and Wright, 1978).

Deformation of the basalt sequence was most extensive during the Pliocene. The entire Columbia Plateau was regionally tilted from an elevation of about 760 m in the northeast to about 120 m in the southwest near Pasco, Washington. Superimposed on the regional structure are east-west fold ridges (Figure 7.1). The upraised northern rim of the plateau is especially significant for the flood history of the Channeled Scabland. Only a truly phenomenal quantity of water could fill the great canyon of the Columbia River between Spokane and Coulee Dam. That filling would be necessary to have water spill over the northern rim of the plateau and flow southwestward, carving the great scabland channels.

QUATERNARY GEOLOGY

The Quaternary history of the Channeled Scabland is characterized by discrete episodes of catastrophic flooding and prolonged periods of loess accumulation and soil formation (Patton and Baker, 1978a). At least one, possibly two phases of flooding predate an extensive blanketing of the region with the Palouse Formation loess. The Palouse Formation apparently accumulated episodically by downwind

accretion, followed by periods of relative stability and soil formation. This loess blanket locally exceeds 75 m in thickness (Ringe, 1970).

The oldest post-Palouse flooding is probably that of Lake Bonneville. Malde (1968) traced the course of this flood through the Snake River Plain of southern Idaho to Hells Canyon. Malde interpreted the date of this event to be about 30,000 years B.P., based on a radiocarbon date for molluscan fossils associated with flood debris and on the relict soil profile developed on the flood gravel (Melon Gravel). The soil has a thick calcic horizon extending to depths greater than 2 m. The soil on the Bonneville Flood deposits is believed to have formed during and since the mid-Wisconsin interglaciation or within the last 30,000 years.

Downstream from Hells Canyon at Lewiston, Idaho, probable Bonneville flood deposits are overlain by slack-water surge deposits from the last major episode of flooding from glacial Lake Missoula. Because Bonneville flooding was confined to the Snake River Canyon, it skirted to the south of the Channeled Scabland (Figure 7.2).

The last major episode of Lake Missoula flooding extended from about 18,000 to 13,000 years ago (Baker, 1978d). It probably involved several lake outbursts, and some evidence has been interpreted as representing forty independent flood events (Waitt, 1980). The dating of this phase of flooding has been established by studies of the relatively low-energy slackwater facies of the flood deposits. These deposits are intercalated with volcanic ash that was erupted from Mount St. Helens, one of the Cascade volcanoes.

Eruptions of volcanoes in the Cascade Mountains showered ash over wide areas of the northwestern United States during the late Quaternary. Where these ashes can be distinguished, they provide a valuable series of marker beds for dating the associated sedimentary deposits and for tying together the geologic history of widely separated provinces.

Most prolific of the ash sources in the Northwest was Mount St. Helens. The individual eruptions of this vol-

Figure 7.3. Exposure of the Mount St. Helens ash "triplet" in fine-grained "slackwater" sediment that immediately overlies scabland gravel at the mouth of Lynch Coulee, near its junction with the Columbia River at West Bar. A coarse white ash layer occurs at B and a fine-grained thinner layer occurs at A. The lowest of the three layers occurs discontinuously in the granule gravel C. The sediment containing the ash (Mount St. Helens set S) is considered to be time equivalent to the last major scabland flood (Mullineaux and others, 1977).

cano have been dated by using carbonaceous materials intercalated with the coarse airfall-erupted deposits on the flanks of the volcano. The tephra occur in distinct groups, called "sets." The oldest is set S, erupted between 18,000 and 12,000 years B.P. Mullineaux, Hyde, and Rubin (1975) presented arguments for the most widespread members of the set S eruption (layers Sg and So) being dated at about 13,000 years B.P.

The Mount St. Helens set S ash occurs as "couplets" and "triplets" in fine-grained clastic sediments, often designated as "slackwater facies" of Missoula flooding. A typical exposure is shown in Figure 7.3. Mullineaux and others (1977) concluded that tephra set S dates the last major scabland flood. They support the age estimate of 13,000 years B.P. for the flood with a radiocarbon date of 13,080 ± 350 on peat directly overlying the "Portland delta," a Missoula Flood deposit at Portland, Oregon.

After the above words were written and during editorial work on this book, Mount St. Helens provided a rare glimpse of catastrophic geology. On May 18, 1980, the volcano experienced a major eruption, sending ash upward to 19 km above sea level. The ash fell over a wide area of Washington, northern Idaho, and western and central Montana. A laterally directed blast devastated a zone on the north

flank of the volcano extending over 20 km outward (Decker and Decker, 1981). The eruption left a crater 1.6 km across and 900–1,500 m deep. The spectacle was a modern analog to the set S eruptions of 13,000 years earlier.

PALEOHYDROLOGY

Geomorphic features result from forces acting on resistant materials at the interface between the lithosphere and the atmosphere or hydrosphere. Until the last decade the dynamics of the forces involved in making the Channeled Scabland were generally ignored in the controversy that surrounded the origin of those forces. I first used quantitative procedures in relating the pattern of scour and deposition to the regimen of scabland floods in 1973 (Baker, 1973a). Because the Missoula floods involved the largest discharges of fresh water that can be documented in the geologic record, continued study of scabland processes establish some upper limits to our knowledge of the short-term erosive and transport capabilities of running water.

Several fortuitous circumstances have combined to allow an approximate reconstruction of maximum flood flows through the Channeled Scabland: (1) the last flooding episode was the last major event of the Pleis-

Figure 7.4. Loess scarps, scabland, and a major divide crossing in the Cheney-Palouse Scabland tract. Topography is from the U.S. Geological Survey Texas Lake Quadrangle (7.5 minute, 10 foot contour interval).

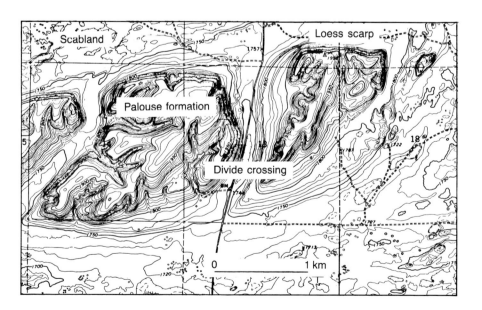

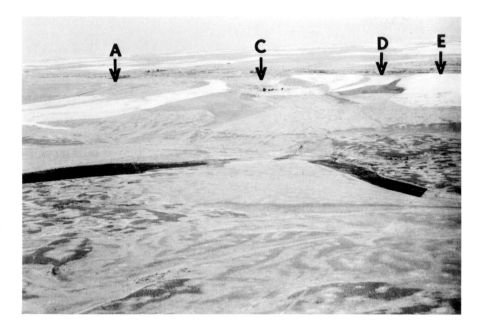

Figure 7.5. Oblique photograph of minor divide crossings (arrows) cut through a loess divide between Crab Creek and South Fork (Sections 26, 27, and 28, T. 21N, R. 36E). The floors of the divide channels A, C, and D are at 600, 588, and 594 m (1,970, 1,932, and 1,950 ft) elevation respectively. The uncrossed divide at E is at 606 m (1,990 ft).

tocene, occurring perhaps as recently as 13,000 years B.P.; (2) postflood drainage on the Columbia Plateau was isolated by the paths of major rivers around the plateau; (3) the postglacial dry climate produced only intermittent streamflow; (4) flood deposits and erosive effects contrast sharply with the loess sediments and processes immediately adjacent to scabland channels; (5) exotic lithologies transported by the flooding are easily recognized; and (6) earlier flood events were followed by the massive loess deposition over the plateau.

For the last major Missoula flood to cross the Channeled Scabland, a variety of features have been preserved which serve as high-water marks. These features may be studied by geomorphic field work and by the interpretation of topographic maps and aerial photographs. The most abundant type of high-water mark is a divide crossing, which occurs where flood water filled valleys and spilled over the loess-mantled interfluves (Figure 7.4).

Divide crossings always establish lower limits for the flood high-water surface elevation and, in some cases, may fix upper limits as well. Obviously there is an expected range of error on these elevations. Elevation estimates are most accurate when divides were crossed by shallow water. These marginally crossed divides are characterized by poorly developed troughs. In areas where divides were crossed by deep flows, the approximation for the water-surface elevation is less precise. Thus, the most useful divide crossings are also the most difficult to recognize. Further refinement of the high-water surface elevation is possible through the study of aerial photographs and field observations of highest flood gravels, erosion of interfluves, and highest erratics.

The magnitude of the range of elevations obtained for the high-water surface depends on the contour interval of the topographic map and upon the nature of the crossing. Estimates are most accurate when uncrossed as well as crossed divides are present in the local study area (Figure 7.5). A substantial portion of the topographic map coverage of the Channeled Scabland is at a 3 m (10 ft) contour interval. Thus the error range is rather small for maximum water depths of 100 m.

The plotting of the flood high-water surface begins with locating the obvious evidence, such as highest eroded scabland and major divide crossings. These data provide a rough approximation to the lower elevation limit. Refinement of the high-water surface elevation may follow through the location of the lowest divide not crossed (an upper limit) and the highest divide of marginal nature that was crossed (lower limit). It is important to select those divides just barely crossed when establishing lower limits. There remains a nebulous zone between these two limits where it is uncertain whether divides have been crossed. The midpoint of this range can be taken to represent the high-water surface elevation for a local reach of the profile. Figure 7.6 shows the high-water surface profile along a portion

of the Cheney-Palouse Scabland tract (the eastern portion of the Channeled Scabland). The water surface slopes show a marked steepening at points where the flood waters encountered constricted channel sections. Water was ponded upstream from these constrictions and flowed at high velocity through the constrictions.

The high-water surface slope can be used as an approximation of the energy slope in the slope-area method of indirect discharge calculation (see Benson and Dalrymple, 1968, and Baker, 1973a). Paleoflow depths are obtained by measuring the difference in elevation between the channel bottom and the high-water surface. Channel cross sections can be derived from the large-scale topographic maps. I followed standard hydraulic engineering procedures to combine these data into estimates of probable flood discharges and mean flow velocities at the maximum flood stage (Baker, 1973a).

The distribution of mean flow velocities in the Channeled Scabland (Figure 7.7) is important in understanding flood erosion processes. Constricted channels in the western scablands, such as Lenore Canyon, lower Grand Coulee, and Othello and Drumheller Channels, achieved flood flow velocities as high as 30 m/sec. Such high velocities were reached only because of the unique combination of great flow depth (60–120 m) and very steep water surface gradients (2–12 m/km) that characterized the Missoula Flood. The constrictions contain the best-developed erosional topography. Adjacent basins such as the Hartline, Quincy, and Pasco Basins produced ponded water and the accumulation of sediment eroded from the constrictions (see Baker, 1973a, p. 15).

In my 1973 paper (Baker, 1973a), I made some preliminary limiting calculations on the draining of Lake Missoula and the routing of flood flows through the Channeled Scabland. The peak discharge of flood flows in the Rathdrum Prairie, close to Lake Missoula's breakout point, was 21.3×10^6 m³/sec. Even at this phenomenal discharge, the lake volume of 2.0×10^{12} m³ would have maintained the flow for about a day. Of course, the declining head in the lake would gradually reduce the discharge in time, spreading the water release over a broader time scale. It seems more likely that the flood peaks recorded by the high-water surface persisted no more than a few hours. Lower stages could have been maintained for a week or more.

The ultimate control on outflow from the entire region of eastern Washington flooded by Lake Missoula

was Wallula Gap, an antecedent canyon eroded by the Columbia River through the Horse Heaven Hills anticline (Bretz, 1925). Wallula Gap was inadequate to carry all the flood water supplied from upstream scabland channels. This constriction hydraulically dammed the flood, causing ponded water to reach 350 m elevation behind the "dam." I calculated the maximum discharge achieved by this control point as 9×10^6 m³/sec (Baker, 1973a). Thus, the total volume of Missoula Flood water entering from upstream would have required 2½ days to pass this point, even at the maximum discharge. More likely the high flows persisted for at least a week.

These discharge calculations can only be considered a first approximation to flood hydraulics in the Channeled Scabland. The calculations were based on hydraulic formulae which were derived for use in streams whose discharges are three or four orders of magnitude less than those of the Missoula flooding through the Channeled Scabland. The data input into these formulae was not based on a dynamic record of the flooding, but rather on the time-transgressive high-water surface. Time variant aspects of the flood surges cannot be quantitatively deduced from the existing field evidence. The complex geometry of Lake Missoula itself probably exerted an unknown dynamic influence on its draining. As discussed above, the

flows were routed through an extremely complex set of anastomosing channel ways.

The exact significance of the high-water surface itself is open to some question. If the flooding burst on to the Columbia Plateau at near its peak discharge, then one might expect the high loess divides to have been eroded almost immediately. A subsequent long recession of the flood hydrograph might then have greatly deepened the channels, enlarging their capacity beyond what is implied by the high-water surface. Two lines of evidence argue against this view. First, the hydrographs of *jökullhlaups* (Icelandic: "glacier bursts") are precisely opposite to the above hypothesis. Flood flows rise slowly to a peak and then drop abruptly (Meier, 1964). The second line of evidence involves the field relationships. The extraordinary preservation of high-stage bed forms throughout the Channeled Scabland implies that the *jökullhlaup* hydrograph applies. Waning flood stages lasted so short a time that bed forms such as giant current ripples were not scoured away.

A more difficult problem arises in the estimation of roughness for the scabland channels. I followed Malde (1968) in assuming a value of Manning's *n* despite the many difficulties in scaling from the empirical Manning equation (derived for "normal" rivers) to the immense flow depths

Figure 7.6. High-water surface profile through a portion of the Cheney-Palouse Scabland tract. Note the pronounced ponding of water at Rattlesnake Flats and the steepening through Staircase Rapids.

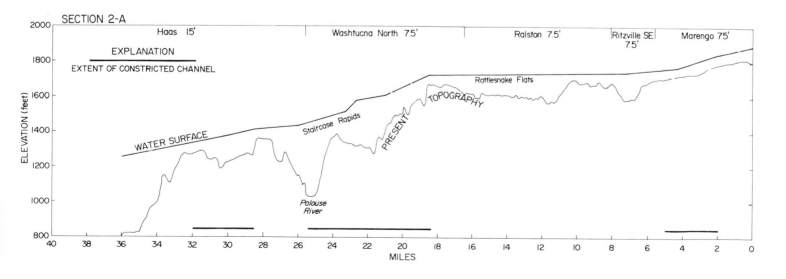

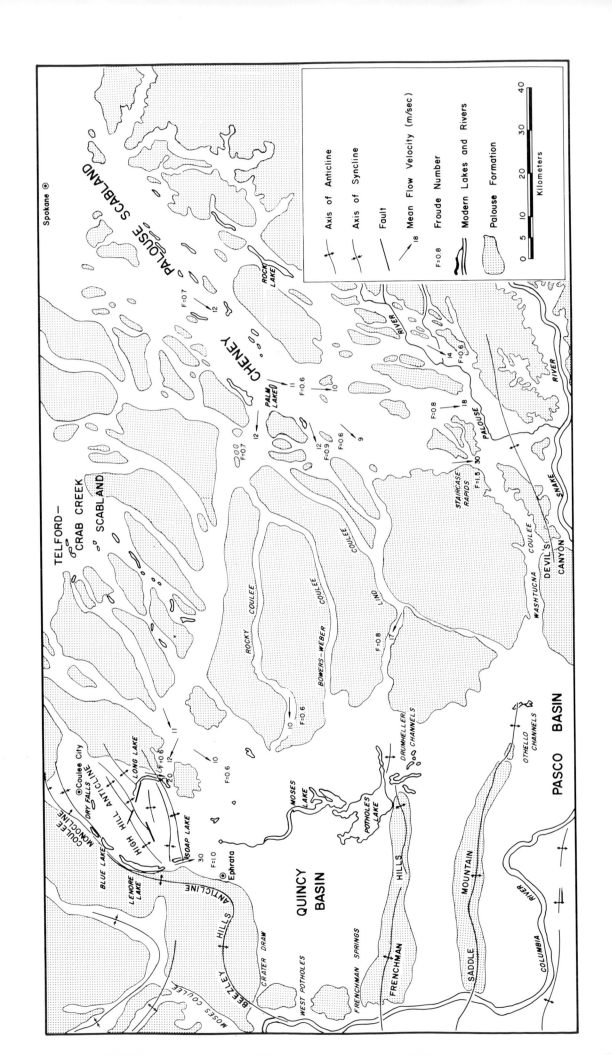

of the scabland flood (Baker, 1973a, p. 19). The value chosen from Chow's (1959) empirical tables was $n = 0.040$ for rock-bounded channels.

Despite the various limitations on the paleohydraulic reconstruction of flooding in the Channeled Scabland, a variety of sediment transport phenomena have proven to be generally consistent with the quantitative reconstruction provided by my preliminary analysis (Baker, 1973a). These include the bottom shear stresses for particle transport, the boulders carried by the flow, and especially the giant current ripples. These latter forms are almost certainly confined to the upper part of the lower flow regime of Simons and Richardson (1966). As expected, the calculated Froude numbers of the reaches containing those bed forms were in the range 0.5–0.9. There is no doubt that the future development of a better hydraulic theory for large floods may modify the absolute magnitudes of the events, but the relative pattern will probably remain unchanged.

LARGE-SCALE EROSIONAL AND DEPOSITIONAL FEATURES

J. H. Mackin has been quoted as saying, "To understand the scabland, one must throw away textbook treatments of river work" (Bretz, Smith, and Neff, 1956, p. 960). Certainly a failing of Bretz' critics in the Spokane Flood debate was their insistence that the Channeled Scabland conform to "established" geomorphic processes. The scale of the problem was the key, and a completely new frame of reference was required. Bretz (1932, p. 28) provided the required viewpoint: "Chan-

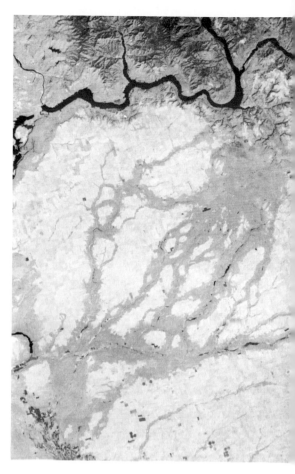

Figure 7.8. Anastomosing channel pattern in the Telford–Crab Creek Scabland complex. This Landsat image depicts a scene 70 × 150 km. (Landsat E-1039-18143-5, 31 August 1972.)

neled Scabland is river bottom topography magnified to the proportion of river-valley topography."

The Spokane Flood debate might have been resolved more easily if the participants could have viewed modern orbital photography of the region (Figure 7.8). At a glance one can appreciate Bretz' observation. Clearly the Channeled Scabland is a plexus of channels rather than a network of valleys.

The general pattern of the Channeled Scabland is large channels eroded in loess and underlying basalt. The channels form locally anastomosing complexes with individual channels that have relatively low sinuosity. The term "anastomosis" should not be confused with "braiding." "Braiding" refers to branching and rejoining around alluvial islands or bars. Braided streams are part of a continuous series of fluvial forms that develop in quasi-equilibrium with external controls on the river systems. "Anastomosis" has no genetic connotation. It refers to channel morphology whether in alluvial streams ("braided") or bedrock streams. The scabland anastomosis is deeply cut into rock. Anastomosis occurs in the Channeled Scabland because preflood valleys did not have the capacity to convey the Missoula Flood discharges without spilling over preflood divides into adjacent valleys. This crossing of di-

Figure 7.7. Regional paleohydraulic features of the Channeled Scabland. Mean flow velocities were determined from high-water-mark evidence and channel geometry (Baker, 1973a). Also shown is the regional structural pattern on the Columbia Plateau. The Palouse Formation is not shown on High Hill and Pinto Ridge (between Soap Lake and Long Lake) in order to emphasize the structural detail in that area.

TABLE 7.1. Important Bed Forms in the Channeled Scabland

	Scoured in Rock	Scoured in Sediment	Depositional
Macroforms (Scale controlled by channel width)	Channel anastomosis and residual uplands between channels	Large-scale streamlined uplands	Bars
Mesoforms (Scale controlled by channel depth)	Longitudinal grooves Potholes Inner channels Cataracts	Scour marks	Large-scale transverse ripples (giant current ripples)
Microforms (Not discussed here)	Scallop pits	Not preserved	Small-scale ripple stratification

Figure 7.9. Excellent association of depositional macroform (bar) with superimposed mesoforms (giant current ripples). The bar occurs along the Snake River just downstream from its junction with the Palouse River (top center). The Snake River is impounded by a dam just downstream from the reach depicted in this oblique aerial photograph.

vides produces the effect of channels dividing and rejoining.

The erosional and depositional features of the Channeled Scabland can be classified in a hierarchical arrangement (Table 7.1). The classification is analogous to one developed for rivers by Jackson (1975). It treats the channel features as bed forms and it has genetic implications. The macroforms, including various kinds of eroded, streamlined hills and depositional bars, do not respond to local flow conditions. They rather respond to long-term hydrologic and geomorphic factors. Mesoforms include large-scale ripples (dunes), antidunes, and large-scale lineation. The spacing of mesoforms depends on the outer zone of the turbulent boundary layer as the flow varies through a dynamic event such as a flood. In rivers the boundary layer control is approximated by flow depth. Microforms include current lineation and small-scale ripples. Microforms respond to flow structure in the inner part of the turbulent boundary layer, and their lifetime is much shorter than the periodicity of dynamic events.

The Channeled Scabland is remarkable for its excellent preservation of both macroforms and their superimposed mesoforms (Figure 7.9). This association probably results from the nature of the hydrograph for flood flows in the Channeled Scabland. In most rivers the hydrograph shows a long recession. Depositional bed forms that are stable at high stage (mesoforms) are washed out, and post-flood surfaces show only the highly stable macroforms such as alternate bars. In the Channeled Scabland an abrupt cessation of flood discharge resulted in preservation of many of the mesoforms, especially those located on the higher bar surfaces.

STREAMLINED ERODED HILLS AND DEPOSITIONAL BARS

The eastern part of the Channeled Scabland is called the Cheney-Palouse Scabland tract. Its overall pattern is a complex of channel ways and interchannel divides. The pattern is especially pronounced on orbital photography (Figure 7.10) because the loess on interchannel divides contrasts sharply with the eroded basalt scabland on the channel floors.

In overall pattern the Cheney-Palouse Scabland tract resembles that of a braided stream. Instead of bars of sediment laid down by high discharges, however, the "islands" in the Cheney-Palouse are erosional residuals of basalt and loess. The largest of these have a crude quadrilateral shape in plan, often forming diamonds or parallelogram shapes. Morphologically these are similar in appearance to eroded bar remnants that are common in braided gravel streams. The flow split around these residual elements, eroding their upstream ends (Figure 7.11). The flow then reconverged on the downstream ends of the residual forms. Many of the residuals were further modified by relatively shallow flood water flowing obliquely across their surfaces.

Figure 7.12 is a detailed geomorphic map of features in the central Cheney-Palouse Scabland tract. The residual loess hills are the most prominent features in an assemblage that also includes various types of bars, scabland, and giant current ripples. The loess hills have three genetic varieties: (1) subaerially exposed, (2) par-tially submerged, and (3) submerged.

The subaerially exposed loess hills were not overtopped during the Missoula Flood and are characterized by the well-developed steep flanking scarps which truncate and behead the preflood drainage systems still evident on their crests. Many of these islands were crossed by small distinct channels during the last flood, but these channels had little effect on the overall morphology and sedimentation. The islands have well-developed quadrilateral shapes, many very similar to the rhombic or diamond shape of longitudinal bars typical of braided streams.

The partially submerged loess hills were transected by one or more major channels still shallow enough not to have eroded into the underlying basalt. Many such channels eroded large re-entrants on the downstream margins of the islands which were later partially filled with sediment. Consequently, these loess islands have the thickest accumulation of flood sediment.

The submerged loess hills were completely covered by the flood. Flanking scarps are not as well developed, and all preflood drainage topography on the tops of the islands was obliterated. These islands have gravel bars attached to their tails in much the same manner that wind-shadow dunes form in the lee of flow obstructions. In this case, the gravel bars drape over the tail of the eroded loess form, and the shape of the resulting streamlined hill is partly influenced by the loess island and partly by the gravel bar. Well-developed submerged hills (Figure 7.13) often show several or all of the following characteristics: (1) flow obstacles that localized the resistant landform, (2) upstream crescent-like scour marks, (3) downstream tapering streamlines on the adjacent channel floor, and (4) oblique channels cutting through small divides at the crest of the streamlined form.

The Cheney-Palouse tract contains groupings of erosional residuals (loess islands) that have been modified by three or four levels of stream erosion as indicated by channel size and degree of scabland topographic development. The characteristic arrangement

Figure 7.10. Landsat photograph of the
southern part of the Cheney-Palouse
Scabland tract. The outlined region is
mapped in Figure 7.12. Reference to that
figure will explain the depicted landforms. This computer enhancement of
the original Landsat image was supplied
by Dr. J. C. Boothroyd, University of
Rhode Island.

Figure 7.11. Oblique aerial photograph of
the upstream end of a quadrilateral loess
residual on the Palouse-Snake divide.
Note the steep scarps (40 m high) that
were eroded by flood water streaming
over the divide. Washtucna Coulee is visible in the background.

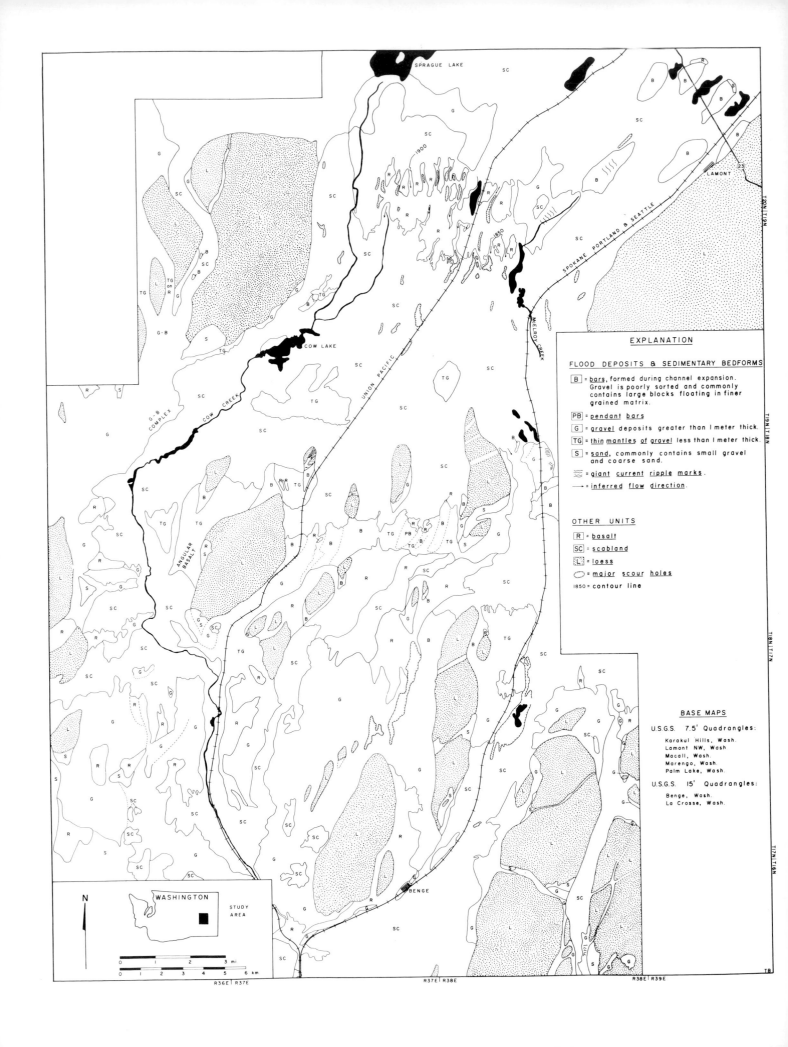

SPRAGUE LAKE

LAMONT

SPOKANE PORTLAND & SEATTLE

McELROY CREEK

COW LAKE

UNION PACIFIC

COW CREEK

G-B COMPLEX

ANGULAR BASALT

BENGE

EXPLANATION

FLOOD DEPOSITS & SEDIMENTARY BEDFORMS

B = <u>bars</u>, formed during channel expansion. Gravel is poorly sorted and commonly contains large blocks floating in finer grained matrix.

PB = <u>pendant bars</u>

G = <u>gravel</u> deposits greater than I meter thick.

TG = <u>thin mantles of gravel</u> less than I meter thick.

S = <u>sand</u>, commonly contains small gravel and coarse sand.

= <u>giant current ripple marks</u>.

→ = <u>inferred flow direction</u>.

OTHER UNITS

R = <u>basalt</u>

SC = <u>scabland</u>

L = <u>loess</u>

= <u>major scour holes</u>

1850 = contour line

BASE MAPS

U.S.G.S. 7.5' Quadrangles:

Karakul Hills, Wash.
Lamont NW, Wash.
Macall, Wash.
Marengo, Wash.
Palm Lake, Wash.

U.S.G.S. 15' Quadrangles:

Benge, Wash.
La Crosse, Wash.

N

WASHINGTON

STUDY AREA

0   1   2   3 mi
0 1 2 3 4 5 6 km

R36E | R37E   R37E | R38E   R38E | R39E

of erosional elements probably represents variations in flow velocities, rates of erosion, and rates of deposition for various elevations in the flood channel-way. In modern braided stream environments, Williams and Rust (1969) noted a decrease in the flow regime, water discharge, rate and mode of sediment transport, and period of activity with increasing elevation or ranking of the channel.

The same conditions may have existed during the creation of the Cheney-Palouse. The lowermost channels probably carried the greatest discharges for the longest durations generally with the greatest velocity and turbulence. These channels would logically be expected to exhibit the highest degree of scabland development and the least amount of associated sediment deposition. Sedimentation was greatest in the secondary channels probably because sediment concentration was still high, although velocities were somewhat reduced. These factors, when combined with the numerous channel expansions and large flow obstructions, created numerous zones conducive to deposition.

The analogy between the subaerially exposed loess island complexes and the bars of braided streams has already been noted. As in braid bars, the greatest potential for gravel deposition was downstream from the largest channels dissecting individual loess islands. The loess islands created low velocity zones in their lees which localized the deposition of secondary gravel bars. The loess islands also behaved as major elements within a braided stream channel as they forced the main flow against the channel margins. Therefore, the loess islands must be, at least in part, responsible for the width of the Cheney-Palouse tract.

The term "bar" is used for all large-scale depositional forms in streams and rivers. Bars originally denoted impediments to navigation. Here the term is applied to all depositional

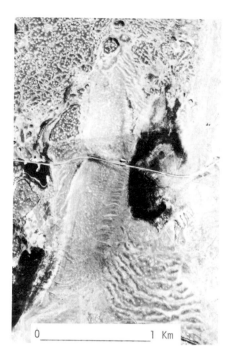

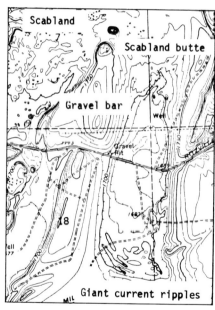

Figure 7.12. Geomorphic map of the central portion of the Cheney-Palouse Scabland tract. (Geomorphic mapping by P. C. Patton and V. R. Baker.)

Figure 7.13. Residual loess hill streamlined by flood erosion. A small cataract, heading an inner channel, has worked its way around the blunt upstream end of the hill (on map). Longitudinal grooves and butte-and-basin topography can be seen in the marginal scablands. Water depths and velocities averaged 12 m/sec for depths of 30–40 m in this area during the flood maximum. Map contour interval is 10 ft. This hill is located in the Cheney-Palouse Scabland tract in sections 1 and 12, T. 18N, R. 37E.

Figure 7.14. Vertical aerial photograph and topographic map of a small pendant bar in the Cheney-Palouse Scabland near Macall, Washington. Contour interval is approximately 3 m (10 ft). The bar accumulated downstream from a residual butte of basalt. The bar occurs in sections 7 and 18, T. 18N., R. 38E.

Figure 7.15. Oblique aerial photograph of Bar 2 near Wilson Creek, Washington. This is a relatively small bar about 1 km in length.

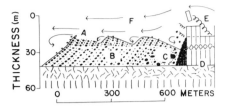

Figure 7.16. Schematic development of a hypothetical pendant bar. Bedload is transported across the surface of the bar by giant ripples and deposited at the downstream end of the bar. A, foreset bedding associated with giant current ripples; B, foreset bedding within the bar; C, chaotically deposited flood gravel; D, basalt bedrock; E, erosion of basalt columns by kolks; F, flow directions.

macroforms in scabland channels. Unfortunately there is no generally accepted classification of fluvial bars. Indeed, a hard and fast classification of bars is probably impossible for the following reasons: (1) no classification can satisfy all three major purposes of bar studies, morphologic, hydrodynamic, and paleohydraulic; and (2) many bar forms are ephemeral members of evolutionary sequences, complexly related to initial conditions and transitory flow conditions.

Longitudinal bars, which are elongated parallel to the flow direction, tend to occur in those scabland reaches which lack abrupt expansions and constrictions. In braided gravel rivers, longitudinal bars tend to be broad, low forms with massive bedding or crude horizontal structure within. In the Channeled Scabland, however, the longitudinal bars are mounded, streamlined forms tens of meters thick. Stratification is dominated by foreset beds that were accreted to avalanche faces on the downstream margin of the bar.

Malde (1968) introduced the term "pendant bar" to refer to streamlined mounds of Bonneville Flood gravel that occur downstream from bedrock projections on scabland channel floors. I found that this was the most common type of bar in the Channeled

Scabland (Baker, 1973a). The locus for bar initiation may be a knob of basalt (Figure 7.14) or the bend of a preflood meandering valley (Figure 7.15). Bar deposition was apparently initiated by gravel deposition in flow separations that developed downstream from a variety of flow obstructions (Figure 7.16). Additional material was then added as huge foresets on the downstream margins of the bars. By this downstream accretion, scabland bars maintained a zone of flow separation that induced deposition from flows that otherwise would be competent to transport even coarse boulders.

Rather than being purely depositional forms, pendant bars are best viewed as the consequences of special flow conditions that locally reduced the fantastic competency of the peak flood discharges. The transport rate of flood gravel into the separation zone downstream from the bar was simply greater than the transport rate out of that zone. The relationship of the bars to the flow conditions is evinced by the fact that they are streamlined to present minimum resistance to the flood water. Moreover, in curving reaches pendant bars never abut on the channel walls, but are separated from the walls by depressions which Bretz, Smith, and Neff (1956) termed "fosses."

BEDROCK MESOFORMS

The bed forms eroded by the macroturbulent flood flows on bare basalt surfaces are of immense variety and unique size in the Channeled Scabland. Indeed these are the features most characteristic of "scabland." Bretz' studies of these bizarre erosional forms were largely descriptive. It is now apparent, however, that the rock bedforms exist in an evolutionary sequence that is related both to the flood flow hydrodynamics and to the resistant characteristics of the jointed basalts. Here the forms will first be described, and then their evolutionary sequence will be discussed.

The most impressive erosional forms created by the Missoula floods are probably the abandoned cataracts (Figure 7.17). Most famous of these is Dry Falls (Figure 7.18), 5.5 km wide and 120 m high. Bretz, Smith, and Neff (1956, p. 1029) recognized that many of the cataracts were formed subfluvially rather than by the plunge-pool undercutting classically illustrated by Niagara Falls. This contention is supported by the high-water mark evidence left by the early Pinedale (late Wisconsin) flood (Baker, 1973a, Fig. 7). Bretz (1932) described in detail the initiation of a 250 m cataract near Coulee City, Washington,

Figure 7.17. Oblique aerial photographs of scabland cataracts and inner channels. A, West Potholes cataract on the western rim of the Quincy Basin; B, Frenchman Springs cataract, also on the western rim of the Quincy Basin; C, Hudson cataract at the head of Hudson Coulee in the Hartline Basin; D, Palouse Falls, a small cataract in the inner channel eroded through the Palouse-Snake divide crossing (the Palouse River now occupies this flood channel).

and its 32 km upstream recession to create the upper Grand Coulee (Figure 7.19). The unique capacity of vertically jointed basalt to maintain the lip of a recessional cataract was held to be the primary consideration in this type of erosion.

All scabland cataracts show multiple horseshoe-shaped headcuts. Potholes and Frenchman Springs have two parallel headcuts or alcoves (Figure 7.20). At the base of each alcove is a large closed depression. Even with postflood modification, the closure in these depressions is as much as 35 m. Probably the cataract headcuts acted as efficient funnels during the maximum flood flows. The water surface was sharply drawn down over the cataract, producing intense macroturbulent scour beneath the locally steep water-surface gradient. Plucking erosion was concentrated in the columnar-jointed zones, and large blocks of

entablature were undermined at the cataract lip.

The headward recession of scabland cataracts produced distinct inner channels. Field mapping shows that the margins of these channels and the cataract lips are held up by relatively resistant basalt entablature (Figure 7.21A and B). Bretz, Smith, and Neff (1956) envisioned powerful kolks at the plunge pool locations, undermining the cataract lip. Sediment, usually basalt columns and large blocks of entablature, was transported away from the cataract lip in a state of quasi-suspension, buoyed by the intense macroturbulence (Baker, 1973a, p. 26–29). The competency of the kolks is difficult to estimate theoretically. Boulders up to 30 m in diameter (estimated from aerial photography) were transported from the lip of West Potholes cataract. Certainly the vertical vortices (kolks) were exceedingly pow-

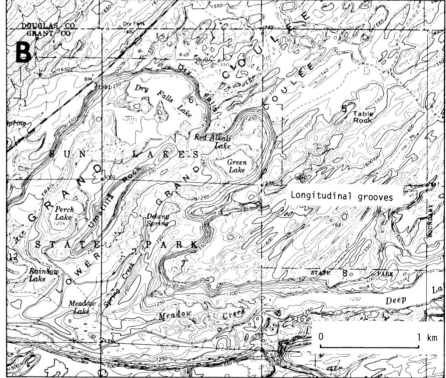

Figure 7.18. A: Oblique aerial photograph of the Dry Falls cataract group. The cataract is 120 m high and 5.5 km wide. Longitudinal grooves are visible just upstream from the cataract head, and the upper Grand Coulee (containing Banks Lake) extends to the horizon. B: Topographic detail of the Dry Falls area showing the inner channel development (contour interval 10 ft; elevations in feet). Note topographic expression of longitudinal grooves.

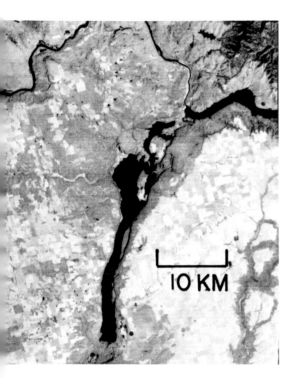

Figure 7.19. Orbital image of the upper Grand Coulee showing the immense inner channel left by headward retreat of the 250 m deep cataract. (Landsat image E-2 936-17451-5.)

Figure 7.20. Topographic map of the Potholes Cataract. Topography is from the Babcock Ridge 7.5 minute quadrangle.

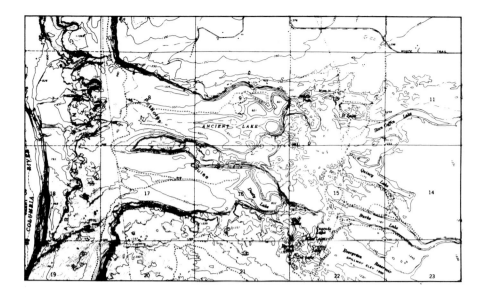

Figure 7.21. Representative cross sections of scabland channels showing structural characteristics of basalt, erosional features, and flood stage plus maximum mean flow velocity (V̄). The sections at Soap Lake (A) and Long Lake (B) show inner channels with flanking butte-and-basin scabland topography. The section at Palm Lake (C) shows longitudinal groove topography.

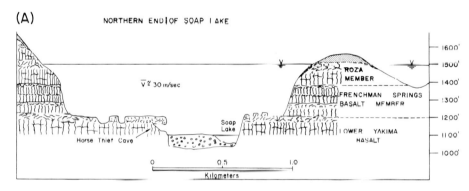

(A) NORTHERN END OF SOAP LAKE

V̄ ≅ 30 m/sec

ROZA MEMBER

FRENCHMAN SPRINGS BASALT MEMBER

LOWER YAKIMA BASALT

Horse Thief Cave

Soap Lake

0   0.5   1.0
Kilometers

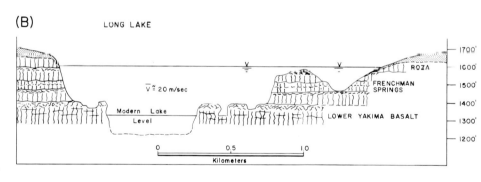

(B) LONG LAKE

V̄ ≅ 20 m/sec

ROZA

FRENCHMAN SPRINGS

LOWER YAKIMA BASALT

Modern Lake Level

0   0.5   1.0
Kilometers

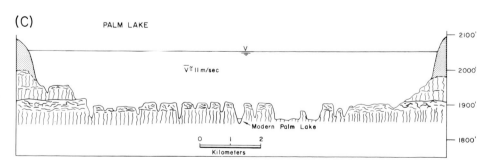

(C) PALM LAKE

V̄ ≅ 11 m/sec

Modern Palm Lake

0   1   2
Kilometers

Basalt entablature

Basalt colonnade

Flood Gravel

Palouse Formation

V̄
Maximum height of flooding

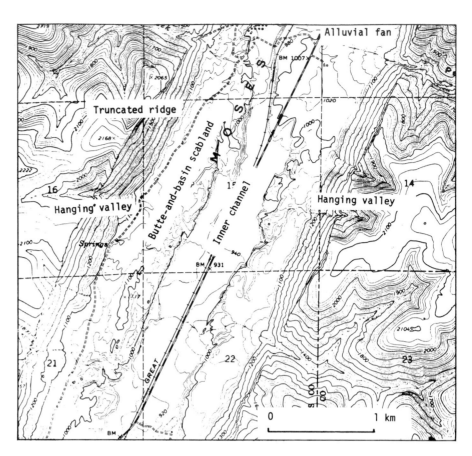

Figure 7.22. Inner channel development in Moses Coulee. This oblique aerial photograph shows a distinct inner channel (l) on the coulee floor with marginal butte-and-basin scabland (S). Giant current ripples (R) occur on the surface of a flood gravel bar at the mouth of the inner channel. Preflood tributaries to the drainage were truncated by flood erosion to form hanging valleys (V) separated by truncated spurs.

Figure 7.23. Topographic map of lower Moses Coulee showing the relationships discussed in the caption to Figure 7.22. The contour interval is 10 ft (elevations in feet).

Figure 7.24. Butte-and-basin topography near Long Lake (A) and Blue Lake (B). Potholes show undercutting of their entablature rims particularly on the downstream side.

Figure 7.25. *Longitudinal grooves devel-*
oped on the scabland surface near Palm
Lake. The small closed contours are
mostly mounds of postflood silt. The
larger closed depressions are scab-
land basins. Topography is from the
Palm Lake, Washington, 7.5 minute
quadrangle.

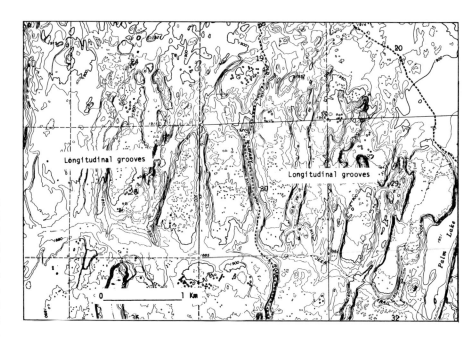

erful agents of lift and transport.

In the upper Grand Coulee (Figure 7.19), cataract recession was initiated at the structural step provided by the Coulee Monocline. The undermining and recession process proceeded rapidly enough for the cataract to recede 32 km to the gorge of the Columbia River on the northern margin of the Columbia Plateau. On the western rim of the Quincy Basin (Figure 7.20), cataract recession was initiated at the canyon walls carved by the Columbia River.

Many scabland channels were excavated from preflood stream valleys. At Lenore Canyon and Moses Coulee (Figure 7.22) the preflood tributaries now form hanging valleys that drain Palouse loess topography unmodified by catastrophic flooding. The floors of these valleys were deepened by flooding, often with the production of distinct inner channels (Figure 7.23).

Perhaps the most prevalent topographic form in eroded rock of the scabland tracts is butte-and-basin topography (Figure 7.24). The usual development is small anastomosing channels and rock basins surrounding buttes and mesas, with a total relief of 30–100 m. The rock basins range in size from shallow saucers to the scale of Rock Lake, 11 km long and 30 m deep. Bretz (1932, p. 26–28) described this combination of features as follows: "The channels run uphill and downhill, they unite and they divide, they deepen and they shallow, they cross the summit, they head on the back-slopes and cut through the summit; they could not be more erratically and impossibly designed."

Where broad expanses of a single basalt surface were eroded by catastrophic flood flows, the characteristic form of the erosion is a series of elongate grooves. These features generally have their long axes parallel to the prevailing flood flow streamlines. The grooves are common on basalt entablature surfaces near Dry Falls (Figure 7.18). They average 5 m in depth and 50 m in width. Relatively large grooves are developed near Palm Lake in the Cheney-Palouse Scabland tract (Figure 7.25).

Experimental studies of fluvial erosion utilizing simulated bedrock (Shepherd, 1972; Shepherd and Schumm, 1974) indicated that a sequence of erosional bed forms may develop in bedrock as a function of time. First to appear in these experiments were the faint streaks of longitudinal lineations associated with potholes and transverse erosional ripples. The lineations then became enlarged into prominent longitudinal grooves. Eventually the grooves decreased in number, and finally one narrow, deep inner channel formed. In the experiments the inner channels were incised below base level, and nickpoints migrated headward upstream (see Schumm and Shepherd, 1973, p. 7, for a longitudinal profile). Deposition of bedload occurred downstream from the headcut, culminating in a low-gradient inner channel with a sand bed and simulated bedrock banks. The lower part of this channel was actually an elongate basin, incised below base level.

The progressive erosion of a typical scabland divide crossing (Figure 7.26) is envisioned as follows. The first flood water to overtop a divide encountered soft Palouse loess and Ringold Formation (Phase I). The high-velocity water quickly exposed the underlying basalt, leaving an occasional streamlined loess hill as a remnant of the former cover (Phase II). The entablature of the uppermost basalt flow was then encountered. This probably yielded to groove development, possibly associated with longitudinal roller vortices. The first exposure of well developed columnar jointing, perhaps at the top of a flaring colonnade along the irregular cooling surface, introduced a very different style of erosion (Phase III). Large sections of columns could now be removed at this site with the simultaneous development of vertical vortices (kolks). With the enlargement and coalescence of the resultant potholes, the surface assumed the bizarre butte-and-basin topography that characterizes much of the Channeled Scabland (Phase IV). The eventual topographic form was the development of a prominent inner channel (Phase V). Such inner channels may have been initiated at downstream structural steps in the basalt, and then migrated headward by cataract recession. The lateral enlargement of inner channels probably proceeded by the undercutting of re-

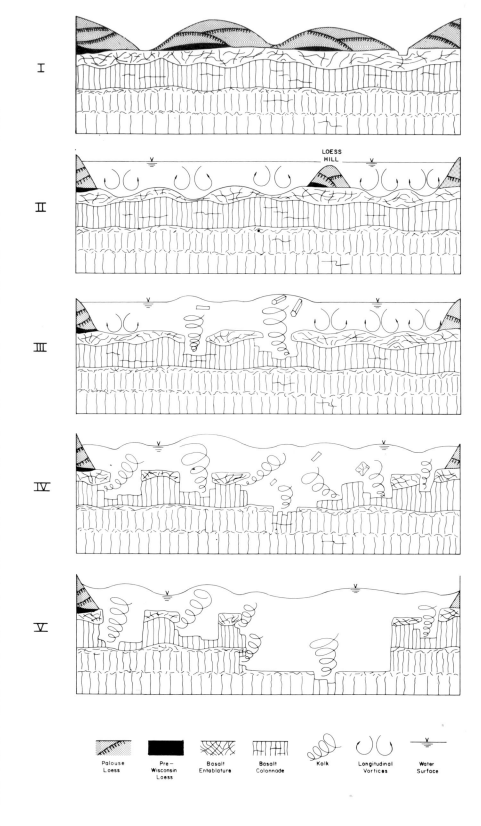

sistant entablature as columns were plucked out by kolks. Horsethief Cave, north of Soap Lake, is an excellent example of this type of erosion.

GIANT CURRENT RIPPLES

Aerial photographs of some scabland gravel bars reveal patterns of parallel ridges and swales which Bretz, Smith, and Neff (1956) identified as "giant current ripples." These constitute the most important bed forms used in the paleohydraulic reconstruction of the last major scabland flood (Baker, 1973a). Over 100 sets of these bed forms have been identified in various Missoula Flood channel ways. Figure 7.27 illustrates some of the variety that exists in ripple morphology.

Measured heights and chords of the scabland ripple fields show a remarkable symmetry of form. Mean ripple heights for a ripple field are closely related to mean ripple chords (Figure 7.28). The chords generally range from 20 to 200 m. Ripple heights, ranging from 1 to 15 m, have probably been somewhat reduced by waning flood stages and by the modification of postflood processes.

Bretz, Smith, and Neff (1956, p. 980) suggested, "An interesting sidelight on the hydraulics of these glacial rivers will appear when the giant current ripples are given careful detailed study." I analyzed forty-three sets of giant current ripples in Missoula Flood reaches (Baker, 1973a). Statistical correlation and regression analyses were used, treating the values of mean ripple height (H) and chord (B) as the dependent variables. The independent variables were the various hydraulic parameters calculated for each reach. A typical result is presented in Figure 7.29, which gives the relationship between ripple chord (B) and stream power (ω).

Stream power is generally measured as the product of a stream's mean flow velocity ($\bar{V}$) and the shear on its bed (T) (Bagnold, 1966). Bed shear is obtained from

Figure 7.27. Typical sets of giant current ripples in the Channeled Scabland: A, Lind Coulee; B, Marlin; C, Artesian Lake; D, West Bar. Locations for these ripple trains are given in Baker (1973a).

Figure 7.28. Logarithmic relationship of height as a function of chord for forty sets of giant current ripples. The general relationship for large-scale current ripples determined by J. Allen (1968) is shown by a dashed line. Five additional measurements by various authors are also plotted. Standard errors on the regressions are indicated by the letter σ.

SOURCES OF DATA

• CHANNELED SCABLAND
× PARDEE (1942)
+ BRETZ (1959)
✳ FAHNESTOCK AND OTHERS (1969)
■ PRETIOUS AND BLENCH (1951)
⊙ WHETTEN AND FULLAM (1967)
▲ WHETTEN AND OTHERS (1969)

$\bar{H} = 0.074\bar{B}^{0.77}$ (ALLEN, 1968)

$\bar{H} = 0.0029\bar{B}^{1.50}$

±1σ

±2σ

RIPPLE HEIGHT $\bar{H}$ (FT)

RIPPLE CHORD $\bar{B}$ (FT)

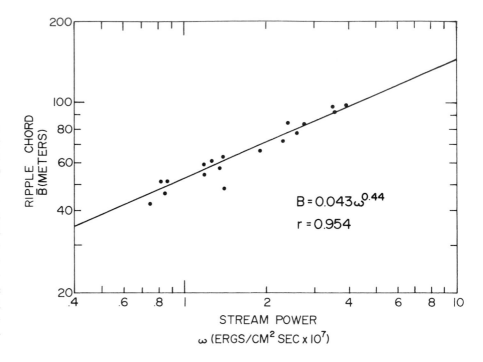

$$T = \gamma DS,$$

where D is the flow depth, S is the energy slope (here approximated by the flood high-water surface), and γ is the specific weight of the fluid (1000 kg/m³ for water).

Figure 7.29 illustrates the remarkable consistency in scale and genesis between the scabland bed forms and the probable hydraulics of the scabland flooding. Although this consistency is best illustrated quantitatively, one is led to an appreciation of the qualitative insights made by J Harlen Bretz over fifty years ago as he defended his catastrophic flood hypothesis (Bretz, 1928b, p. 475–476):

All scabland channels possess discontinuous mounds, hillocks or hills of stream gravel. . . . They are unlike any other detrital accumulations except the much smaller features of river channels commonly called bars. With these there is exact parallelism except for size. When considered in their setting in the scabland system, with all its other evidence for great volume and great erosion, they are seen to be an integral part. They should exist! And if they are bars, the great scoured channel ways should exist! Again this assemblage of unique land forms in the Pacific Northwest is seen to be a genetic group. A lively imagination is required for the acceptance of the hypothesis, but a scientific imagination withal.

SUMMARY

The Channeled Scabland region coincides with the portion of the Columbia Plateau that was subjected to periodic catastrophic flooding during the late Pleistocene. J Harlen Bretz pioneered in the geomorphic study of this remarkable region and introduced the term "scabland" to refer to the chaotically eroded tracts of bare basalt that occupy the floors of channels cut through the thick loessial cover of the plateau. Bretz (1923, 1928a, 1928b, 1932) described the amazing assemblage of erosional and depositional forms in the scablands, including

rock basins, anastomosing channels, cataracts, gravel bars, and coulees. Numerous features, especially the huge gravel bars, multiple high-level divide crossings, and rock basins as much as 40 m deep led him to postulate that a catastrophic flood of enormous proportions had swept across the Columbia Plateau. The source of the flood was eventually shown to be glacial Lake Missoula, which was dammed by a glacier in the vicinity of modern Lake Pend Oreille.

The Channeled Scabland is the closest terrestrial analog to the outflow channels of Mars. This is appreciated by comparing tables 7.1 and 3.4. The bedrock erosional forms are the most comparable and include anastomosing channels eroded in rock, streamlined uplands within and between channels, cataracts, hanging valleys, inner channels, scour depressions, and grooves.

The same detailed assemblage of landforms characterizes the Martian outflow channels and the Channeled Scabland, leading to the conclusion that the channels on Mars were caused by catastrophic flooding. Nevertheless, significant differences exist between the two regions. The outflow channels have experienced considerable postdiluvian modification (Chap-

ter 6). Moreover, the relative abundance of various bed form types differs on Mars and Earth. The dominance of erosional grooves on Mars contrasts with the predominance of butte-and-basin scabland in the terrestrial channels. The scabland analog is qualitatively complete but quantitatively skewed. The reason awaits the results of further research.

Following the considerable criticism of his ideas in the late 1920's, Bretz (1932, p. 82) agonized over his flood hypothesis for the Channeled Scabland. He stated, "Somewhere must lurk an unrecognized weakness. Where is it? If it exists, it probably lies in the hydraulics of the concept." Bretz asserted that the turbulence of the glacial flood and the jointing of the basalt may be important, an idea gained from his earlier studies of the Columbia River (Bretz, 1924). He added (Bretz, 1932, p. 83), "We do not know enough about great flood mechanics to make any conclusions valid . . . Hydraulic competency must be allowed the glacial streams however much it may differ from that of stream floods under observation."

The above quotations apply equally well to the hypothesized floods of Mars. The problem will be discussed in Chapter 8.

Chapter 8
Catastrophic
Flood Processes

The detailed assemblage of erosional forms on the floors of the Martian outflow channels has been described in earlier chapters. The analogy of these forms to those in the Channeled Scabland is apparent from Chapter 7. This chapter will consider the process implications of that analogy. It will essentially elaborate the hypothesis that the outflow channels experienced deep, high-velocity flows of water (or dynamically similar fluid) (Baker and Milton, 1974; Baker, 1978e). These processes are inferred from bed forms on the channel floors. They are not necessarily responsible for eroding the entire valley depth, much of which may have been created by collapse and headward growth as fluids were released from subsurface reservoirs (Baker 1977, 1978b).

Because the science of geomorphology is plagued by problems of equifinality, i.e., the generation of similar landform assemblages by different combinations of processes, this chapter is necessarily speculative. It may also appear presumptuous to speculate on the details of fluid-erosional processes before the source of the responsible fluid has been clearly identified. Whether or not future research verifies the physical reality of the Channeled Scabland analog to the Martian outflow channels, the history of scabland investigations demonstrates that source relationships need not be clearly established to discern the response of landscapes to physical processes (Baker, 1978a).

This chapter will focus on processes that appear capable of eroding rock. It will attempt to set some limits on the operation of such processes in large-scale water flows under terrestrial conditions as well as under present and (assumed) past Martian environmental controls. The first consideration will be the probable fluid properties and hydraulics of water flows on Mars. Next will come a discussion of resistance factors in the material being eroded. For scabland-type erosion this is largely a function of the structural attributes of the rock at the scale of jointing, bedding planes, etc. (Baker, 1973b). Finally, each of the major genetic processes (macroturbulence, streamlining, river ice processes, and cavitation) will be discussed.

FLUID PROPERTIES AND FLOW HYDRAULICS

The calculation of flow velocities, discharges, and other hydraulic parameters is a useful exercise in comparing the scale of fluid flows. However, accurate calculations are difficult even for terrestrial examples of known fluid properties in ancient flood channels (Baker, 1973a, 1974). The problem on Mars is immensely compounded by the unknown nature of the fluid properties and by the less precise definition of the channel geometries involved. For these reasons it seems best only to make very simple hydraulic estimates for the Martian channels and to recognize that these estimates are subject to considerable error.

With a surface temperature in the range of 150–300°K (Leighton and Murray, 1966) and a surface pressure in the range of 7–10 mb, liquid water would either evaporate or freeze on Mars under the present climatic conditions. However, hydraulic calculations may be appropriate for probable ancient conditions when Mars had a denser atmosphere. Liquid water flows could also be maintained if an overburden of ice formed over the channels, effectively insulating the water flow from the atmosphere (Lingenfelter, Peale, and Schubert, 1968). Another hypothesis is that the water emanating onto the Martian surface from subsurface sources could have had a high dissolved salt content. Ground-water systems in the Antarctic "dry valleys" are known to function with concentrated brines (Cartwright and Harris, 1976). The lowered freezing point and raised boiling point favor maintenance of the liquid state in the otherwise adverse thermodynamic regimen.

The indirect calculation of flow hydraulics from channel dimensions is usually accomplished by using the semi-empirical Manning equation:

$$\bar{V} = \frac{1}{n} D^{2/3} S^{1/2}, \qquad (8\text{-}1)$$

where $\bar{V}$ is the mean flow velocity in m/sec, D is the flow depth in meters, and S is the energy slope. The term n is the Manning roughness coefficient, which requires estimation for any application of the equation (e.g., Chow, 1959). Other considerations applying to the use of this equation include the substitution of depth for hydraulic radius and channel bottom slope for energy slope. These complications are discussed more fully elsewhere (Baker, 1973a). Remembering that only crude comparisons are required, however, the present discussion will be limited to the simplest possible form of the Manning equation.

The difference in gravitational accelerations (g) on Mars and Earth is one factor that can be estimated in the hydraulic equations. For Earth, g is 981 cm/sec², but on Mars it is only 372 cm/sec². The appropriate modification to equation 8-1 must be made for the resistance factor, Manning's n. Komar (1979) presented a rigorous discussion of the influence of gravity on the hydraulic calculations. The effect does not permit a simple adjustment of equation 8-1, since this equation depends on the scale of the flow (depth). Nevertheless, for the probable range of peak flow depths in Martian channels, perhaps 10–100 m (Carr, 1979b), it is possible to use an approximate formula suggested by Carr (1979b):

$$V = \frac{0.5}{n} D^{2/3} S^{1/2}, \qquad (8\text{-}2)$$

where $\bar{V}$ and D are measured in m/sec and m respectively, and the multiplier 0.5 represents an approximate adjustment of equation 8-1 for Martian gravity.

Since $\bar{V}$ is the mean flow velocity, the discharge along a channel is simply the product of $\bar{V}$ and the cross-sectional area of the fluid-filled portion of the channel. For wide, shallow channels this can be approximated by

Figure 8.1. Hydraulic data for various scabland channels as estimated in Baker (1973a). Note that most of the flood flows were subcritical (F < 1.0) and were also less than the critical depth-velocity combination necessary to induce cavitation according to the criterion of Barnes (1956).

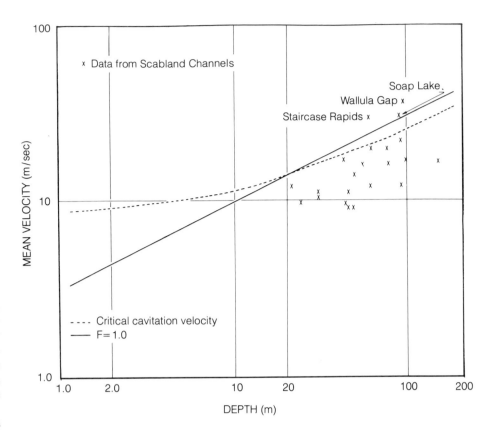

$$Q = \bar{V} D W, \qquad (8\text{-}3)$$

where Q is the total discharge, D is the average depth of flow, and W is the channel width. Table 8.1 summarizes the results of several hydraulic calculations for Martian outflow channels. Comparisons are illustrated for the Mississippi River and the Missoula Floods. The calculations show a general similarity in flow characteristics between the various outflow channels and Missoula Flood channels. However, both systems are markedly different from common large terrestrial rivers such as the Mississippi.

A dimensionless parameter that is useful in comparing flows of different scales is the Froude number F,

$$F = \frac{\bar{V}}{\sqrt{gD}}, \qquad (8\text{-}4)$$

where g is the acceleration of gravity and $\bar{V}$ and D are defined as before. The critical value $F = 1.0$ separates the two flow regimes of tranquil flow (subcritical, $F < 1.0$) and shooting flow (supercritical, $F > 1.0$). Shooting flow tends to occur either where gradients are very steep or where flow is abruptly constricted.

The calculated maximum values of mean flow velocities for Lake Missoula flooding mostly reveal subcritical flow (Figure 8.1). Shooting flow occurred only at several very abrupt flow constrictions, such as Staircase Rapids, Wallula Gap, and Soap Lake (Figure 8.1). Erosion was at its maximum at those constrictions (Baker, 1973a, 1973b).

The preliminary Mars calculations (Table 8.1) show that Mars channels also have variable Froude numbers. The Ares and Maja reaches were supercritical, in the range $F = 1.0-2.0$, depending on the chosen depth. Mangala was subcritical, $F = 0.5-0.8$. The high Froude number reaches display

TABLE 8.1. Estimated Flow Parameters for Martian Outflow Channels and Possible Earth Analogs

Channel	Depth D (m)	Slope S	Channel Width W (m)	Mean Velocity $\bar{V}$ (m/sec)	Discharge (m³/sec)
Maja Vallis	100	0.02	8×10^4	38	3×10^8
	10	0.02	5×10^4	8	4×10^6
Ares Vallis	100	0.01	2.5×10^4	27	7×10^7
	10	0.01	2.5×10^4	6	1.5×10^6
Mangala Vallis	100	0.003	1.4×10^4	15	2×10^7
	10	0.003	1.4×10^4	3	4×10^5
Mississippi River	12	0.00005	8.3×10^2	3	3×10^4
Missoula Floods	150	0.001	7×10^3	16	2×10^7
	60	0.002	2.4×10^3	17	2.6×10^6

Note: The assumed roughness for all calculations was 0.04 except the Mississippi ($n = 0.012$). Equation 8-1 was used for terrestrial flows and equation 8-2 for Martian flows. Results for Maja and Ares are comparable to estimates made by Carr (1979b). Results for Mangala and Mississippi are comparable to those of Komar (1979). Missoula Flood values are from Baker (1973a). Recent analysis of radar data for Ares Vallis indicates that the slope may be 0.001 (H. Masursky, oral communication, 1981). If this is true, the calculated discharges for Ares should be multiplied by a factor of 0.3.

Figure 8.2. Oblique aerial photograph of the head of Lenore Canyon near Park Lake. The dipping beds of the Coulee Monocline are exposed at top center. Downstream from this point, the dipping portion of the flexure has been completely eroded away to form Lenore Canyon. Longitudinal grooves mark the bare basalt surface in the foreground.

Figure 8.3. Erosion by scabland flooding at the crest of High Hill anticline in the southern Hartline Basin. Note the dip of Yakima Basalt units to the right and left of the eroded anticlinal crest (center).

steep gradients and constricted flow as in the Channeled Scabland.

The immense scale of the proposed Martian flood flows requires an immense scale of sediment transport. The reduced gravity on Mars means that mean flow velocities for similarly scaled flows are less on Mars than on Earth (compare equations 8-1 and 8-2). However, the lower weight of otherwise similar sediment particles on Mars means that those particles are more easily transported than on Earth. Moreover, lower settling velocities would prevail on Mars, enhancing the transport of suspended load (Komar, 1980). At the flow scales postulated above, Komar (1980) estimated that gravel and cobbles would have been transported in suspension through the Martian outflow channels. Boulders several meters in diameter would have moved as bedload (Baker and Ritter, 1975).

RESISTANCE

The Martian outflow channels are carved into a variety of probable rock types. Layered sequences, probably of basaltic lava flows, are especially common. These have been described in Kasei and Maja Valles. Other channeling probably occurs in impact breccia and in previously deposited flood sediments. Some consideration of resistance factors in the Channeled Scabland will illustrate the influence of the basaltic rock types.

Large-scale flow resistance is provided by the regional form of the land surface. The Channeled Scabland was carved from a basalt surface that was warped and deformed (Figure 7.1). The catastrophic flooding followed the troughs and gradients of this preflood topography.

The Lower Grand Coulee illustrates structural control of scabland flood erosion to a remarkable degree. Flood water spilling over a preflood divide to the north encountered the Coulee monoclinal flexure near present-day Dry Falls (Bretz, 1932). This created a huge cataract, 250 m high, that receded northward to form the Upper Grand Coulee. Below Dry Falls the flood water preferentially eroded the fractured rock on the steep eastward-dipping limb of the monocline.

Figure 8.4. Joint-controlled rock basins and channels on the Palouse-Snake divide crossing southeast of Washtucna, Washington. The Palouse River occupies the inner channel of the main canyon.

This process excavated Lenore Canyon (Figure 8.2) from the bent and broken basalt units.

Scabland divide crossings across anticlinal ridges also show the preferential erosion of fractured zones. Plucking erosion was concentrated at the tension-jointed anticlinal crests (Figure 8.3). Joint control of rock basins and scabland channels is perhaps most spectacular on the Palouse-Snake divide crossing (Figure 8.4). There it appears that the flood erosion simply etched out the regional joint pattern from the exposed basalt surface (Trimble, 1950). Large-scale structural control of channel erosion is also common in the Martian outflow channels. Examples have been described in Kasei Vallis and in the western Chryse Planitia, where flooding eroded the prominent wrinkle ridges. Another important observation on Mars is the abundant evidence of erosive stripping of channel floors. This stripping was layer by layer, exposing resistant ledges and terraces in channels such as Maja and Ares Valles. Similar erosive stripping was achieved in the Channeled Scabland.

Planes of weakness within the basalt bedrock were an important influence on fluvial erosional forms produced by Missoula flooding. The individual basalt flows average 25–60 m in thickness and are characterized by a variety of depositional and cooling features which allowed variable resistance to flood erosion (Figure 8.5). Most flows have a scoriaceous upper portion grading into the rubbly top typical of aa lava. Some flows exhibit the classic Tomkeieff (1940) sequence of two-tier columnar jointing. Columns in the lower colonnade may be 1–2 m in diameter. An entablature of long slender columns and hackly fragments is present in all flows. In some flows an upper colonnade of much smaller columns occurs above the entablature. The cooling history of the flows controls the nature of columnar jointing (Spry, 1962). Irregular cooling surfaces developed because of erosion during the intervals between basalt emplacement. This often resulted in flaring columns in the next higher Yakima basalt flow. Rapid cooling produced hackly fragmented lava.

Between the outpourings of basalt, enough time elapsed to permit weathering, growth of forest cover, and the formation of lakes. Local sedimentary intercalations in the basalts include conglomerate beds, clay layers, and fresh-water diatomite. The lavas which overrode these lake beds formed pillow-palagonite complexes and spiracles (gas chimneys) as described by Waters (1960).

Observations of basalt boulders transported by Pleistocene flooding (Baker, 1973a) revealed that the largest boulders were always portions of hackly jointed entablature (Figure 8.6). The phenomenal size of the flood-transported boulders also indicates that unusual hydrodynamics prevailed in the eroding water flows.

MACROTURBULENCE

Conventional fluid mechanics theory recognizes the generation and degeneration of turbulent eddies that are superimposed on the average motion of deep, fast-moving boundary-layer flows. These turbulent eddies are unsteady, strongly interactive, and not well understood. The smallest eddies, termed "microturbulence," are almost totally independent of the size of the flow, vary with flow Reynolds number, and are generally treated as stochastic processes. There is another scale of turbulent phenomena in which the size of the eddies corresponds to the scale of the boundary-layer flow (flow depth in a river). These phenomena, sometimes termed "macroturbulence," may assume a structured arrangement in the flow field that depends strongly on the scale of the flow. The relevant flow scale might be measured, for example, as the height and diameter of an obstacle on a stream bed or as the flow depth. Although large- and small-scale turbulent processes are closely coupled, it may be possible to discuss macroturbulence qualitatively as an end member in a continuum of phenomena. The following speculative discussion will treat macroturbulence as a more-or-less deterministic secondary flow that is superimposed on (indeed, results from) the stochastic microturbulence that dictates turbulent boundary-layer flow. The

reader is warned that a conventional scientific treatment of turbulence would begin with a complete discussion of microturbulence, emphasizing concepts of diffusion and temporal fluctuations of shear stress and lift on bed materials.

Although the physical complexity of turbulence may appear overwhelming to nonspecialist readers, the concept should be intuitively obvious to anyone who has observed the surface of a deep, swift river. Jackson (1976) cited Mark Twain's (Clemens, 1896, p.44–48) description of distinctive patterns of turbulent fluid motions on the water surface of the Mississippi. Such observations contributed to the original eddy concept of turbulence, a concept perhaps best summarized in the following rhyme, attributed to L. F. Richardson (1920):

Big whorls have little whorls,
    Which feed on their velocity;
Little whorls have smaller whorls
    And so on unto viscosity.

Thus eddies of a given size, or order, develop from larger eddies by borrowing energy from their "parents." This division process continues to such a small scale (microturbulence) that the eddies can no longer borrow sufficient energy to further divide.

The first qualitative classification of macroturbulent phenomena in rivers was that of Matthes (1947). The most important erosive form recognized by Matthes was termed a "kolk."

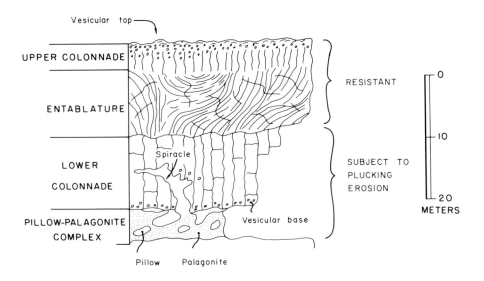

Figure 8.5. Cross section of an idealized Yakima basalt flow showing structural features important to flood erosion processes. The upper colonnade and the pillow-palagonite complex are present only in some basalt flows. Diagram is modified from Swanson (1967) and Schmincke (1967) and uses the terminology of Tomkeieff (1940), Waters (1960), and Spry (1962).

Figure 8.6. Boulder of basalt entablature measuring 18 × 11 × 8 m. This boulder is located on Ephrata Fan, 2.5 km west of the Rocky Ford Creek Fish Hatchery, Ephrata, Washington.

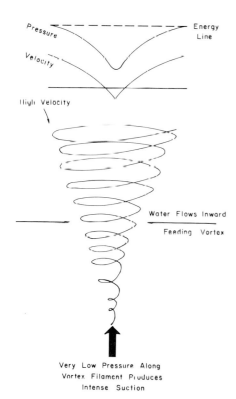

Figure 8.7. Characteristics of a macroturbulent kolk. The intense pressure and velocity gradients of the vortex produce a hydraulic lift. This concept is highly simplified, since kolks interact with other turbulent phenomena. Moreover, they are unsteady, growing and dissipating in time and space. (From Baker, 1979a, Jour. Geophys. Res., v. 84, p. 7985–7993, copyrighted by American Geophysical Union.)

Kolks are transient eddies possessing near-vertical axes and associated with "boils" and other large-scale disturbances of the river surface. Matthes envisioned a kolk as a site of intense fluvial energy dissipation by upward vortex action. The intense pressure and velocity gradients of the vortex produce a phenomenal hydraulic lift force along the filament of the vortex (Figure 8.7). The precise magnitude of this suction effect is unknown, but it certainly is greatly in excess of normal hydraulic lift forces.

The conditions necessary for the generation of kolks according to Matthes (1947) include (1) a steep energy gradient (steep water surface slope), (2) a low ratio of actual sediment transported to potential sediment transport, and (3) an irregular, rough boundary capable of generating flow separation.

Jackson (1976) presented an alternative model in which he attributed kolk generation not to simple vortex action, but rather to the oscillatory growth and breakup stages of the turbulent bursting phenomenon. Bursting, as described by Offen and Cline (1975), characterizes the turbulent structure of the outer part of the turbulent boundary layer. It is not yet clear, however, that this mechanism, as observed in smooth-walled flumes, will apply to the extremely irregular flow boundaries of scabland channels.

The concept of kolk action was applied by Bretz, Smith, and Neff (1956) and by Bretz (1969) to the large-scale erosion of bedrock in the Channeled Scabland. Bretz (1924) had earlier demonstrated that scabland-type erosion was a result of plucking rather than abrasion. He observed that eddies generated by the irregular channel walls at the Dalles area of the Columbia River are responsible for the plucking of basalt columns. He stated that inner channels of the "Dalles" type, potholes, and other scabland features could be produced only by a river of high discharge and steep gradient eroding close and vertically jointed rock. The analogy between the erosional bed forms in the Channeled Scabland and in the Martian outflow channels suggests a similar mechanism. However, the kolk process remains poorly documented by

both physical theory and observation.

Two other classes of macroturbulence are better understood than kolks. The first of these occurs downstream from points of flow separation, such as cavities or downward steps, which appear on the stream bed. The region of separated flow developed downstream from a step and at right angles to the flow is characterized by a recirculation, which J. R. L. Allen (1971) termed a "roller." The second class of macroturbulence is the "vortex," distinguished from the roller by not being isolated from the free external stream. Unlike rollers, which are separated from the external flow field, vortices are fed by fluid from that field. The fluid enters and provides motion along the vortex axis in a spiral pattern.

The irregular rock steps of both the Channeled Scabland and the Martian outflow channels provided numerous points for possible flow separation. The rollers which could form at such localities must conform to the physics of such flow patterns. Separated flow fields reattach at a range of positions on the bed, depending on the Reynolds number and the relative roughness (ratio of flow thickness to step height). For a relative roughness of about 1 and for large Reynolds numbers the reattachment point is located a distance downstream from the flow step equal to four to eight times the step height (J. R. L. Allen, 1971).

The maximum time-averaged fluid pressure measured on the stream bed downstream from a flow step reaches a maximum at the point of reattachment and gradually decays downstream. Throughout the separated region and downstream there are large fluctuations of instantaneous pressure at any given observation point. The bed experiences rapidly varying normal forces that alternatively pull and push at it. Turbulent shear stresses developed in and downstream from separated regions are much greater than in the boundary layer. Experiments by Allen (1971) confirm that the greatest erosion by turbulent separated flows occurs at the points of flow reattachment. This conclusion applies to any of several erosional mechanisms, including cavitation, corrosion, fluid stressing, and solution. Thus, erosion by secondary flow is easily induced around various kinds of obstacles or obstructions on a river bed, and can produce many of the scour marks described in Martian outflow channels.

The hydrodynamics of scour mark formation has been especially well studied by engineers in order to protect bridge piers during floods (Laursen, 1960). The scour is generated by two basic systems of vortices (Shen, 1971): the horseshoe-vortex system and the wake-vortex (Figure 8.8A).

The prominent crescentic scour hole on the northwest (upstream) side of the boulder in Figure 8.8B was probably caused by the hydrodynamic stretching and accumulation of vortex filaments in the front of the boulder. P. D. Richardson (1968) described this process as a characteristic effect of a blunt-nosed obstacle on an approaching two-dimensional velocity field. The strong pressure field produced by the blunt obstacle causes a separation of the boundary layer, which then rolls up ahead of the obstacle to form a horseshoe vortex. Karcz (1968) suggested that this mechanism is responsible for the current crescents that commonly occur upstream from obstacles on the bed of ephemeral streams.

The large elliptical scour hole that formed on the downstream side of the boulder in Figure 8.8B formed as a result of a wake-vortex system generated by flow separation in the rear of the boulder. Engineering experiments have shown that the wake vortex system is a function of Reynolds number (velocity of approach × obstacle diameter × fluid density, ÷ dynamic viscosity of the fluid). With other variables in the Reynolds number held constant, increasing velocity results first in a pair of vertical vortices. When the flow changes from laminar to turbulent, vortices form and migrate downstream. Karcz (1968) noted that the average velocity in the wake region is quite low at this stage, and deposition is likely in the lee of the obstacle. Indeed both the scablands and the Martian outflow channels contain many examples of pendant bars that were deposited as elliptical forms in the lee of obstacles. Malde (1968) described similar features that occurred during the Bonneville Flood in the Snake River Plain of Idaho.

At very high flow velocities, vortex intensity grows and the sucking action of kolks dominates. Shen (1971, p. 23–25) observed that the wake-vortex system then acts like a vacuum cleaner in removing bed material at this stage. Thus, the scour hole in the lee of the boulder may indicate high flow velocities.

An important class of Martian outflow channel bed forms is the elongate grooves that parallel the presumed flow directions. Similar erosional grooves are developed on basalt channel beds in the Channeled Scabland (Figure 8.9). The macroturbulent structure consistent with these features is that of longitudinal vorticity, which is believed to form as a result of transverse instability in straight stream channels (Einstein and Li,

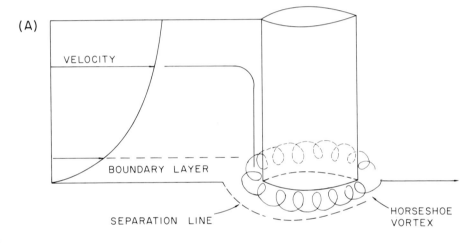

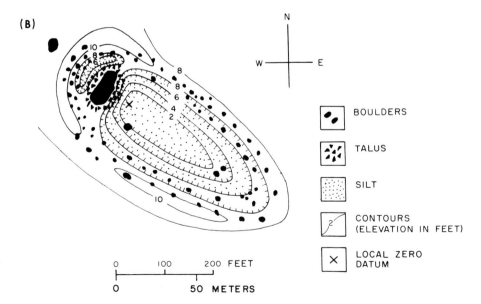

Figure 8.8. A: Formation of a horseshoe-vortex system at the front of a vertical cylinder mounted on an experimental flume bed (after Moore and Masch, 1963). B: Scour hole development near an 18 × 11 × 8 m boulder (Figure 8.6) 2.5 km west of Rocky Ford Creek Fish Hatchery, Ephrata, Washington (Baker, 1973a).

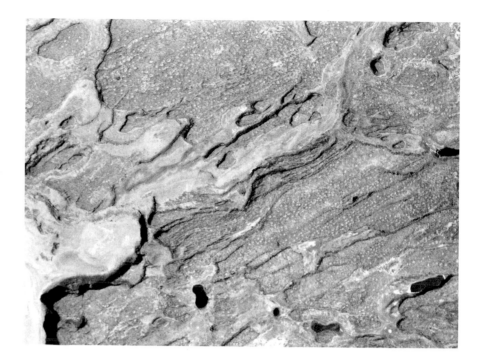

Figure 8.9. Longitudinal groove development on a basalt scabland surface near Jasper Canyon in the Grand Coulee portion of the Channeled Scabland. The small light-colored patches are mounds of eolian silt that settled into swales along the linear groove trend.

1958). The result of superposing transverse movements on the main flow in a stream is a helical array of vortices aligned parallel to the flow direction and showing alternating senses of rotation. These longitudinal vortices are distinctive in having their axes in the streamwise direction. The theory initially yields two helical flows near the two banks of a relatively wide shallow channel. These two helical flows will then induce similar flows throughout the flow section until the entire fluid mass is split into secondary rotating cells (Karcz, 1967).

The Einstein and Li (1958) analysis applied to fully turbulent flow in mathematically straight channels. Of course, any irregularity will create additional vorticity to interact with the longitudinal flow field. If the longitudinal vortices produce longitudinal bed forms, the bed forms will further enhance the flow field, leading to stronger longitudinal erosion. This will eventually produce morphological forms that mimic certain hydraulic attributes of the responsible fluid. If defects occur in the bed, they may produce transverse flow separations that break up the longitudinal pattern of vorticity. Self-enhancement would then create transverse bed forms such as flow steps and cataracts. Cataract retreat would generate inner channels. Thus, eroded bed forms produced by a variety of macroturbulent flow structures can easily coexist, as implied by detailed mapping of the Martian outflow channel floors.

STREAMLINING

Streamlined landforms, such as drumlins, yardangs, longitudinal river bars, and the scoured loess hills of the Channeled Scabland, develop in erodible materials by a process of minimizing resistance in the responsible fluid flows. The Martian outflow channels contain striking examples of streamlined uplands (Chapter 6), but it is emphasized that the ideal "teardrop" shape of such uplands (Figure 8.10) is an end member. The outflow channels and the Channeled Scabland also contain many partially streamlined uplands that do not display this ideal form. These larger elements are best analyzed as components of braided channel systems (see Chapter 7). Morphometric studies of Martian channel braiding (Trevena and Picard, 1978) yield results most consistent with the hypothesis that those channels are erosional features of fluvial origin.

Leonardo da Vinci presented some of the first drawings of what we now call "streamlined" bodies. In more modern applications, thousands of aircraft wing sections, struts, airship models, boat hulls, etc., have been tested in wind tunnels and water flumes (Hoerner, 1958). These studies show that streamlining is the reduction of drag (resistance) of an object in a moving fluid to a minimum. The total drag F_T on the object is composed of several components:

$$F_T = F_D + F_f + F_w, \qquad (8\text{-}5)$$

where F_D is the pressure drag (or form resistance) of the object, F_f is the shear drag (or skin resistance) of the object, and F_w is wave drag produced by a free fluid surface. For a totally immersed object, F_w can be ignored.

Pressure drag arises from the flow separation that is generated by obstacles, and it is proportional to the area A of the obstacle section normal to the flow direction (Prandtl, 1949, Prandtl and Tietjens, 1934).

$$F_D = C_D A \frac{\rho V^2}{2}, \qquad (8\text{-}6)$$

where C_D is a dimensionless drag coefficient (for pressure drag), A is the area of the section of the immersed body perpendicular to the direction of fluid motion, and ρ and V are the fluid density and velocity respectively.

Shear drag is proportional to the surface area of the obstacle S over which the fluid passes (Prandtl and Tietjens, 1934):

$$F_f = C_f S \frac{\rho V^2}{2}, \qquad (8\text{-}7)$$

where C_f is a dimensionless drag coefficient (skin resistance), S is the surface area of the object exposed to the fluid, and ρ and V are as defined above.

In a given fluid flow field the critical factors in drag minimization will be the drag coefficients, C_D and C_f,

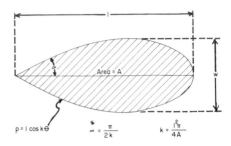

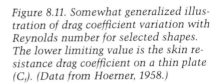

Figure 8.10. Morphometry of a stream-lined form. The figure shown is one loop of a lemniscate derived from the depicted equation in polar coordinates. (From Baker, 1979a, Jour. Geophys. Res., v. 84, p. 7985–7993, copyrighted by American Geophysical Union.)

Figure 8.11. Somewhat generalized illustration of drag coefficient variation with Reynolds number for selected shapes. The lower limiting value is the skin resistance drag coefficient on a thin plate (C_f). (Data from Hoerner, 1958.)

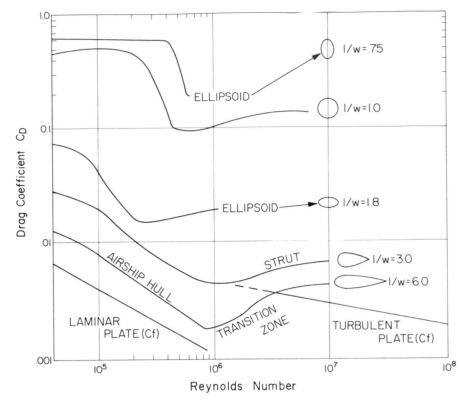

and the two measures of submerged object area, A and S. The latter terms are complex functions of streamlined body geometry, beyond the scope of this brief discussion. In qualitative terms, one can envision the influence of these variables in terms of the elongation of a body already possessing a streamlined shape. As the object elongates, it presents less cross-sectional area perpendicular to the flow, lowering its pressure drag. However, the shear on the added total surface area will increase with increasing elongation.

Both drag coefficients are also functions of the flow Reynolds number R calculated from

$$R = \frac{\rho VD}{\mu}, \qquad (8\text{-}8)$$

where D is the flow depth, μ is the dynamic viscosity, and P and V are as defined above. The relationship between drag and Reynolds number has been thoroughly investigated by design engineers concerned with stream-lining problems. Note in Figure 8.11 the effect on the drag coefficient of object elongation, where elongation is

measured as the length-to-width ratio (l/w).

Many equations are used to describe streamlined shapes (Hoerner, 1958). A simple one, approximating the two-dimensional form, is the lemniscate shown in Figure 8.10. More complex expressions are advocated by Roberts and Cutts (1978), who found that Martian streamlined forms have somewhat longer "tails" than can be approximated with a lemniscate. However, when dealing with shapes that are already highly streamlined, it may be expedient to compare the lengths, widths, and planimetric form areas of the objects. Kochel and I used this procedure (Baker and Kochel, 1978b, 1979b), with results that will be summarized here.

The three physical parameters that proved easiest to measure from Viking orbital images were length L (km), measured parallel to the inferred flow direction; width W (km), measured as the maximum width of the stream-lined form perpendicular to the inferred flow direction; and area, A (km²) measured with a polar planimeter. We measured the best-developed

streamlined shapes, i.e., shapes for which fluid dynamic considerations dictate minimum flow separation in the responsible erosive fluid.

The morphometric data for the Martian streamlined forms compare closely to the data for hills and bars in the Channeled Scabland of eastern Washington (Figures 8.12, 8.13, and 8.14). In a very generalized manner all the data conform to the following simple model:

$$L = 2\sqrt{A} \qquad (8\text{-}9)$$
$$W = \tfrac{2}{3}\sqrt{A} \qquad (8\text{-}10)$$
$$L = 3W \qquad (8\text{-}11)$$

There is, however, a tendency for the Martian forms, especially in Maja Vallis, to be slightly more elongate than their scabland counterparts ($L = 4W$). This is confirmed statistically by the "best-fit" regressions summarized in Table 8.2.

Kochel and I (Baker and Kochel, 1978b) presented an empirical study of the Channeled Scabland stream-lined forms that showed an increase in streamlined form elongations with Reynolds number, R, in the range of

Figure 8.12. Length of streamlined forms versus planimetric form area for the Channeled Scabland, Kasei Vallis, and Maja Vallis. The plotted equation is a general model, not a regression line. (From Baker, 1979a, Jour. Geophys. Res., v. 84, p. 7985–7993, copyrighted by American Geophysical Union.)

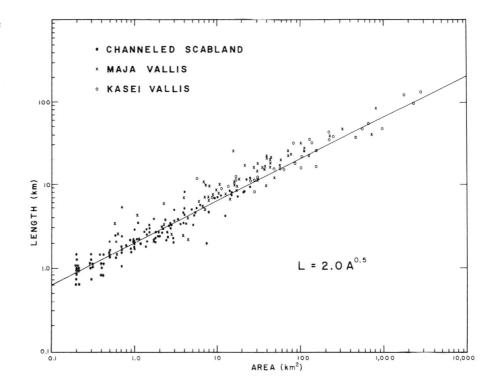

Figure 8.13. Maximum width versus planimetric form area for the same streamlined forms analyzed in Figure 8.12. The plotted equation is a general model. (From Baker, 1979a, Jour. Geophys. Res., v. 84, p. 7985–7993, copyrighted by American Geophysical Union.)

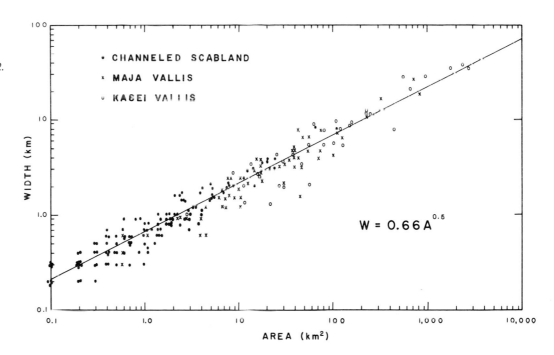

TABLE 8.2. Regression Equations for Shapes of Streamlined Forms on Earth and Mars

Scabland Loess Hills and Bars	Maja and Kasei Valles
$N = 137$	$N = 95$
$L = 1.93A^{0.479}$	$L = 2.96A^{0.438}$
$r = 0.94$	$r = 0.91$
$W = 0.66A^{0.496}$	$W = 0.50A^{0.56}$
$r = 0.92$	$r = 0.95$
$W = 0.34L^{0.98}$	$W = 0.23L^{1.05}$
$r = 0.87$	$r = 0.85$

Note: r is the product-moment correlation coefficient for the least-squares regressions.

10^8 to 10^9 (Figure 8.15). If this dimensionless relationship applies to Mars as well, and if the slightly more elongate Martian forms developed in deep, high-velocity water flows, then this relationship indicates that Reynolds numbers for the Martian flows exceeded the scabland values by a factor of about 0.5. Other factors being equal, this result would imply that the Martian streamlined forms probably developed at slightly higher flow velocities and/or greater flow depths than the scabland forms, a result that is consistent with the greater scale of channel erosion on Mars than in the Channeled Scabland.

RIVER ICE PROCESSES

Present environmental conditions on Mars do not favor the maintenance of liquid water in contact with the atmosphere. Surface temperatures at the Viking 1 lander site typically range from 190°K to 250°K (Hess and others, 1977). Atmospheric pressure at the lander sites varies with seasonal carbon dioxide condensation and sublimation at the south polar cap and falls in the range of 7–10 mb (Leovy, 1979). In such an environment water would quickly evaporate and freeze. However, as pointed out for the lunar environment of even more extreme conditions, rapid evaporative cooling of water produces a surface layer of ice that inhibits subsequent evaporation and permits the underlying water to retain its liquid state for a considerable period of time (Lingenfelter, Peale, and Schubert, 1968).

Wallace and Sagan (1979) performed the relevant calculations for the present Martian conditions. Although liquid water will evaporate very rapidly, evaporative cooling will cause a layer of floating ice to form at the water surface. This will reduce the evaporation rate above the ice to very low values, on the order of 10^{-6} g/cm²/sec. Wallace and Sagan (1979) speculated that even modest discharges of water could be maintained as ice-choked rivers flowing for hundreds of kilometers. In addition, the ice cover could contribute to a subfluvial environment of relatively high pressure, semi-isolated from the low-pressure atmospheric environment. Very deep floods, as envisioned for the large outflow channels, could be maintained easily under these theoretical conditions.

Wallace and Sagan (1979) suggested that ice-covered rivers on Mars would resemble terrestrial high-latitude rivers in winter, except that the cooling of the Martian stream would be accomplished mainly by evaporation in the thin atmosphere rather than by temperature changes. They observed that such rivers would be "smooth enough, if not thick enough, to skate on." The persistence through time of a perfectly ice-mantled river depends on its depth. For the depth of a catastrophic flood, about 100 m, Wallace and Sagan calculated a persistence time of 30,000 years. Smaller depths would persist for much shorter time periods, but it is obvious that the theory allows a broad range of fluvial phenomena.

The Wallace and Sagan analysis might be questioned in its simple application to the highly turbulent flood flows envisioned in the genesis of the large Martian outflow channels. Turbulence would tend to break up an ice cover that formed by the evaporative cooling mechanism (Masursky and others, 1977). The result would be a mixing process that would shorten the persistence time of the water flow. However, in addition to fostering evaporative cooling throughout the water column, the turbulence and broken ice cover would probably contribute to a variety of river ice processes. The constrictions and expansions of the Martian channels would

be ideal sites for the development of ice jams and attendant phenomena.

In terrestrial environments, the formation of ice in turbulent supercooled river water begins with the development of small crystals, called "frazil" (Ashton, 1979). The next stage is the development of frazil slush and ice pans. The slush masses and pans collide, become pressed together, and form continuous surface ice sheets (Michel, 1966). Reduction of the area of open water then inhibits the production of frazil. In river reaches of alternating flow depth the ice cover tends to break up in shallow, high-velocity reaches and reform in deep, low-velocity reaches.

Ice drives can occur when surface ice is flushed from a channel by a pulse of high discharge. Such drives are well known in very large rivers flowing northward to the Arctic Ocean and in the rapid spring snowmelt runoff rivers of Alberta and Montana (Smith, 1979). Ice drives can result in significant channel enlargement by bank scour and incision (Figure 8.16).

Ice jams occur when ice drives are obstructed at channel constrictions, sharp meander bends, and islands (Smith, 1979). Jamming leads to ice scour at these locations and to back-flooding by water impounded above the jam. Spectacular floods can be achieved when hydrostatic pressure builds up above a firmly grounded ice jam. When the water pressure behind such a dam reaches a critical value, the jam breaks up and releases the ponded water.

CAVITATION

Cavitation is the erosive effect of collapsing vapor bubbles in a fluid caused by dynamic-pressure variations. In general terms, the condition of vapor bubble growth in a liquid is termed "boiling" if caused by a temperature rise and "cavitation" if caused by dynamic-pressure reduction. The phenomenon of cavitation is physically complex, involving numerous interacting variables in the discrete states of cavitation inception, cavitation maintenance and development, and bubble collapse (with resultant erosive effects at the flow boundary). De-

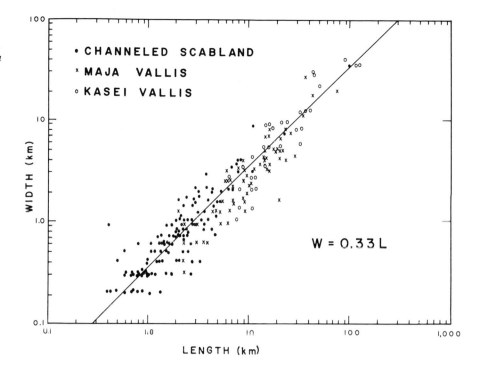

Figure 8.14. Maximum width versus length for the same streamlined forms analyzed in Figures 8.12 and 8.13. The plotted equation is a general model, not a regression line. (From Baker, 1979a, Jour. Geophys. Res., v. 84, p. 7985–7993, copyrighted by American Geophysical Union.)

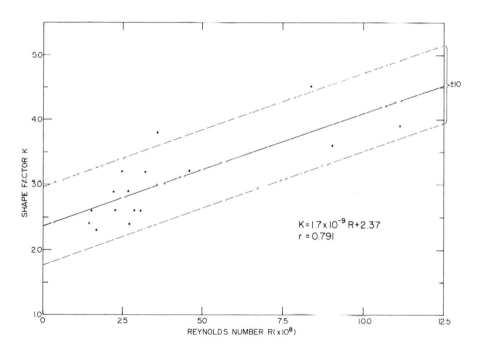

Figure 8.15. Streamlined shape factor k versus maximum flow Reynolds number for peak discharges associated with streamlined forms in the Cheney-Palouse Scabland tract. The dashed lines indicate one standard error.

Figure 8.16. Ice drive on the Red Deer
River, Alberta, Canada (1976). Note the
bank erosion along the flow margin. The
ice blocks are about 1 m thick, and they
were observed to tumble as they dragged
and sheared past the stream bank. (Photograph by Dr. Derald Smith, University
of Calgary.)

spite the concentrated efforts of phys-
icists, hydraulic engineers, and other
investigators, the intricacies of cavitation phenomena remain incom-
pletely understood.

As a first approximation, cavitation
occurs when absolute pressure P_a at
some point in a liquid is reduced below the liquid vapor pressure P_v, generally by the increase of flow velocity
V according to Bernoulli's theorem
(Eisenberg and Tulin, 1961):

$$\sigma = \frac{P_a - P_v}{\frac{1}{2}\rho V^2}, \qquad (8\text{-}12)$$

where σ is the critical cavitation
number and ρ is the fluid density. At
points in the flow field exhibiting
critically low σ values, vapor will
form from the liquid as growing bubbles that are carried downstream.
However, hydraulic experience has
shown that the cavitation parameter
σ is only a rough guide to the dynamics of cavitation phenomena (Knapp,
Daily, and Hammitt, 1970). Equation
8-12 applies only to fluid-dynamic pressure effects. Moreover, the
precise σ value at cavitation inception varies considerably with surface
tensions, undissolved gas particles,
boundary effects, and other controls.
A more generalized statement of the
critical cavitation number is that it
represents the ratio of the pressure
tending to collapse a cavity to the
pressure tending to induce cavity formation and growth (related to the
fluid velocity head).

Cavitation inception tends to occur
in zones of local pressure drop, such
as venturi-like flow constrictions. It
also appears at low pressure points
along solid boundaries. The transient
bubbles are then carried downstream
in this "traveling cavitation" condition. "Vortex cavitation" occurs when
cavities develop in the cores of vortices that develop in zones of high
fluid shear. The third variety of impor-

tance is "fixed cavitation," in which
liquid flow detaches from the rigid
boundary of an immersed body. This
cavity may then grow downstream
from the immersed body, a condition
known as "supercavitating."

The actual stress required to create
a cavity is the tensile strength of the
liquid at various temperatures. Experiments summarized by Knapp, Daily,
and Hammitt (1970, p. 51–58) showed
that appreciable tensions develop at
zones of weakness in the liquid caused
by various contaminants. For exam-

ple, suspended solids, bubbles of undissolved air, and other "defects" in
the fluid probably serve to nucleate
cavitation bubbles. Thermodynamic
and hydrodynamic characteristics of
the flow are also important. These
considerations limit the quantitative
precision of the estimates presented
here.

For terrestrial river flow, Barnes
(1956) began his estimate of the critical conditions for cavitation inception by assuming that fluvial bed
obstructions will increase local ve-

locities to about twice the mean flow velocity V. This allows calculation of the critical mean stream velocity permitting terrestrial cavitation, V_e (m/sec), from Bernoulli's theorem:

$$V_e = \sqrt{\frac{2g}{3}} \sqrt{\frac{(P_a - P_v)}{\gamma} + d}, \quad (8\text{-}13)$$

where $g = 9.81$ m/sec^2, γ is the specific weight of water (9.8×10^3 N/m^3), and d is stream depth (in meters). For $P_a = 10^5$ N/m^2 (1 atmosphere) and $P_v = 0.024$ P_a (21°C), equation 8-13 reduces to

$$V_e = 2.6 \sqrt{10 + d}. \quad (8\text{-}14)$$

The extrapolation of cavitation behavior from one set of conditions to another is a complex problem, even in normal terrestrial applications. The scaling to possible water flow on Mars is thus tentative, but several simplifications may be possible (Baker, 1979b). For relatively cold water, thermodynamic properties of the fluid become minor. Adjusting for the Martian gravity, equation 8-13 can be rewritten:

$$V_m = 1.6 \sqrt{\frac{(P_a - P_v)}{4,000} + d}, \quad (8\text{-}15)$$

where V_m is the critical mean flow velocity for cavitating water flow on Mars measured in m/sec, P_a and P_v are measured in N/m^2, and d in meters.

For the present Martian atmospheric pressure of 700–1,000 N/m^2 (7–10 mb), the $(P_a - P_v)$ term in equation 8-15 becomes negligible:

$$V_m = 1.6 \sqrt{d}, \quad (8\text{-}16)$$

where the same units apply as in equation 8-15. For the higher atmospheric pressures of various postulated ancient atmospheric conditions,

the $(P_a - P_v)$ term will become more important, especially at flow depths less than about 100 m. A set of representative curves of critical cavitation velocities on Mars at various atmospheric pressure values is presented in Figure 8.17. These curves assume that vapor pressures for water were always small in comparison to the absolute pressures of cavitation inception. The important point is that the combination of lower gravity and lower atmospheric pressure allows Martian fluvial cavitation to occur at much lower flow velocities than are required in terrestrial rivers.

Clearly the inception of cavitation in Martian water flows poses no problems. Indeed, the very low atmospheric pressure values might be thought to pose difficulties for the maintenance of a coherent liquid flow because of cavitation throughout the entire flow depth. High-velocity water flows in direct contact with the present Martian atmosphere would not be able to achieve the pressures necessary for bubble collapse, thereby maintaining the liquid flow. Pieri and Sagan (1980) calculated that Martian cavitation bubble pressures become important for bedrock erosion only at flow depths of approximately 30 m or more. However, this picture is too simple for the indicated dynamics of

fluid flow in the outflow channels. The alternating constrictions and expansions of the channel cross sections (Baker, 1978b) require alternating changes in flow depth and velocity. In the deep, slow-moving water of an expanding reach, the cavitation parameter will be drastically increased because of lower flow velocities and higher absolute pressures beneath the thick water column (equation 8-12). Thus foaming water flows initiated at the throats of constrictions could revert to coherent liquid in the adjacent expansion.

It should also be noted that the very high velocities envisioned for the Martian floods are associated with extremely great flow depths (Baker and Milton, 1974). The relevant portion of Figure 8.17 is not the left side but the right portion, where the critical conditions become independent of atmospheric pressure and equation 8-16 applies. The deep Martian floods could have had very high velocities and merely been near the critical point for cavitation inception. In addition, the probability of an ice cover over the flows would lead to a microenvironment of relatively high absolute pressure that would also locally increase the cavitation parameter. The suppressive action of ice on cavitation would probably be localized at zones of rel-

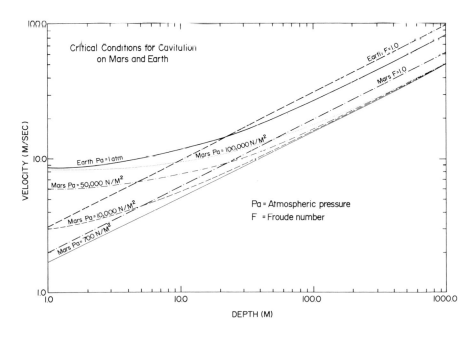

Figure 8.17. Critical conditions for cavitation in flood flows on Mars and Earth for various atmospheric pressures. The critical cavitation velocities are compared to the critical Froude numbers for both planets. (From Baker, 1979a, Jour. Geophys. Res., v. 84, p. 7985–7993, copyrighted by American Geophysical Union.)

atively great ice thickness, such as jams at the entrances to constrictions. These considerations might explain why the observed population of outflow channels is restricted to very large features. Water flows of smaller scale would simply cavitate out of existence, while exceptionally deep flows could be maintained at high velocity.

The processes of cavity collapse and resulting erosion are commonly viewed as results of pressure shock waves that radiate from the collapse centers of the bubbles. The classic analysis of the problem for a spherical bubble was performed by Lord Rayleigh (1917). However, modern high-speed photography of bubble behavior near flow boundaries (Benjamin and Ellis, 1966) shows that exceedingly violent jets of liquid shoot from the high- to low-pressure sides of collapsing cavities. These microjets are responsible for the transmission of large transient forces to the boundary. Theoretical studies indicate that the intensities of these transmitted stresses are at least 10^8 N/m^2, and some experimental work indicates local pressures of at least $2 \times 10^9 \text{ N/m}^2$ (Knapp, Daily, and Hammitt, 1970). Barnes (1956) cited a figure of $3 \times 10^{10} \text{ N/m}^2$. Abundant empirical evidence shows that the process can rapidly shatter the surface layers of steel, rock, or nearly any other boundary material. Fractured surface layers are quickly carried away by turbulent flow, and the pitting-type erosion continues.

Since it is the $(P_a - P_v)$ term in equation 8-12 that fosters cavity collapse, Barnes (1956) argued that very low vapor pressures favor faster and more violent collapse. Conversely, since the vapor pressure of water increases dramatically with temperature, cavity formation is favored at high temperatures. It is interesting to speculate that relatively warm water emanating from subsurface reservoirs to the present Martian surface environment would initially have a low cavitation parameter that would subsequently increase with the decreasing vapor pressure induced by cooling. The collapsing cavities in the recently released flows could then contribute to erosion of the channel walls. The tre-

mendous variety of thermodynamic states, flow dynamics, and local pressure environments on Mars seems to be ideal for repeated changes through the thresholds for cavitation inception and collapse.

A considerable body of experimental data shows that cavitation erosion in channelized flows is localized at the following points: (1) venturi-like flow expansions, where adverse pressure gradients favor the collapse of bubbles that formed at upstream constrictions (Batchelor, 1967), (2) zones of flow separation generated by discontinuities in the flow boundary (Kenn, 1966), and (3) zones of vorticity generated within the boundary layer (Kenn and Minton, 1968). Because situations 2 and 3 constitute macroturbulent flow structure in the responsible fluid, the detailed action of cavitation erosion on the stream bed should mimic the macroturbulence of the generating fluid flow.

DISCUSSION

Fluvial processes involving ice covers, macroturbulence, streamlining, and cavitation appear likely in the large-scale flood flows thought responsible for shaping the Martian outflow channels. The four processes can be mutually interacting.

When jointed, brittle rocks such as basalt are eroded, they form numerous flow steps that favor flow separation (Baker, 1973b). The rollers and kolks that might form along such irregular stream beds would further contribute to plucking or cavitation erosion, maintaining the irregular channel floor of scabland, inner channels, and cataracts. Streamlining is the physical antithesis of this process. Forms of minimum resistance develop in such a way that there is minimal separation of flow. Because of the association between cavitation erosion and macroturbulence, the erosion of channel floors can be extremely intense even though the adjacent streamlined uplands are preferentially preserved.

A feedback mechanism can be envisioned that links cavitation and macroturbulence as the erosive agents for the beds of the Martian outflow chan-

nels. Catastrophic cavitating water flows moving over the irregular Martian surface rapidly eroded constrictions and expansions with the aid of venturi pressure effects. Residual areas were preserved as streamlined uplands that subsequently minimized cavitation erosion. Longitudinal vorticity in the remaining channelized flow became the dominant macroturbulent flow structure. Critical cavitation zones were initiated along the low-pressure vortex filaments, and the vapor bubbles subsequently produced destructive forces as they were carried outward and to the bed by the vortex rotation. The consequent development of longitudinal-bed erosional forms then served to further enhance the flow field, leading to stronger longitudinal erosion. This hypothetical process would have fixed the longitudinal vorticity of the flowing liquid to the form of the developing grooves, resulting in a series of grooves paralleling the paleoflow streamlines with a spacing dependent upon the depth of the responsible liquid flow.

While the above processes can explain many of the features found in the Martian outflow channels, we should remember that explanation in science is not synonymous with truth. It is the lack of any alternative explanation that better explains the unknown phenomena which allows us comfort in our chosen hypothesis. Thus, the above theory has been presented in some detail, so that it too can be probed, questioned, and perhaps rejected when a better theory becomes available.

Now that the channels of Mars have been described, terrestrial analogs suggested, and the probable processes outlined, the remaining question is the largest. What do the channels mean? Percival Lowell knew exactly what his faintly seen "canals" meant—an intelligent civilization. The channels of Mars, by contrast, have been imaged, mapped, and scrutinized, but their meaning remains a question. The implications of their existence will be discussed in Chapter 9.

Chapter 9
Mars: A Water Planet?

When Mariner 9 began to orbit Mars in 1971, the planet was assumed to be impoverished in water. The discovery of the channels by Mariner 9 became the major anomaly in the general picture of a volatile-poor planet. The Viking mission has now generated an impressive list of Martian phenomena associated with water liquid or water ice (Table 9.1). Despite the various models and theories to the contrary, these observations dictate that substantial quantities of water were involved in surficial processes on Mars. Water is simply the only plausible, cosmically abundant substance that meets the morphological constraints imposed by all the various channels of Mars (Pollack, 1979). This chapter will attempt to reconcile theory and observations, but, as will become apparent, substantial unknowns remain. The riddle is not unlike many that have been posed during the history of Mars studies (Chapter 1).

WAS MARS VOLATILE-POOR OR VOLATILE-RICH?

Various arguments for the previous existence of a water-rich permafrost on Mars have already been summarized in Chapter 5. The thickness of the zone containing abundant ground ice was about 1–3 km. This figure is supported by abundant topographic data on chaotic terrain, fretted escarpments, channel walls, and other landforms (Soderblom and Wenner, 1978). It is also consistent with theoretical limits on Martian permafrost development that are imposed by surface temperature regimens and probable thermal gradients (Fanale, 1976). For reasonable porosities in an impact-generated megaregolith, the indicated

permafrost thickness could accommodate up to the equivalent of about 400 m of water spread over the planetary surface (Fanale, 1976). The quantity of water in this hard-frozen megaregolith would therefore be several orders of magnitude greater than is present today in Mars' atmosphere and polar caps. Other volatiles could also be trapped in the Martian regolith, making the present tenuous atmosphere of Mars a mere hint of the planet's richness in volatiles.

Mars could have achieved this immense permafrost if it had a volatile inventory similar to Earth's but scaled to Mars' lower mass and surface area. Fanale (1976) performed a trial calculation for these conditions and found that as much as 1 km of water (averaged over the planet's surface) would be achieved. The problem for proving this result is that all other volatile substances must be scaled to this high water content. Many of these other volatiles are not so easily hidden from measurement as is the frozen water.

Anders and Owen (1977) presented the most detailed model for the abundance of Martian volatiles. Their argument begins with the Viking atmospheric analyses that show noble gas abundances on Mars to be two orders of magnitude lower than on Earth. They also believe that Mars, like Earth, achieved its initial volatile inventory late in the process of planetary accretion. Both planets presumably received a late accreting veneer rich in carbonaceous chondrite materials (a class of meteorite presumed to represent primordial compositions). However, Mars, perhaps because of its smaller size, was less well endowed in volatiles than Earth. This impoverishment is thought to explain the argon abundances on Mars.

Argon is a noble gas that is volatile enough to outgas rather easily during a planet's history, but it is too heavy to escape the Martian atmosphere and too unreactive to be trapped in sediments. Argon has two isotopes: argon 40 and argon 36. Argon 40 is fairly abundant on Mars (Table 5.2), and taken alone this would indicate huge outgassing. Scaling other volatiles to argon 40, Pollack (1979) calculates that Mars might have outgassed 500

mb of carbon dioxide (nearly 100 times the amount in the modern atmosphere) and 2,500 mb of water (about 75 m of water spread over the planet's surface). However, argon 40 can accumulate as a daughter of the radioactive isotope potassium 40. Anders and Owen (1977) therefore tie their analysis to argon 36, whose abundance is anomalously low on Mars. This assumption leads Anders and Owen to scale downward the global abundances of volatiles on Mars from those on Earth by as much as a factor of 35.

The Anders and Owen (1977) model predicts that Mars had an initial water content corresponding to a layer of liquid water 9.4 m deep over the entire planet. The initial carbon abundance was seventy times the present atmospheric carbon dioxide value. If Mars had outgassed to the same degree as Earth, this could have given it an ancient atmospheric pressure of about 500 mb, or nearly half that of Earth. However, because outgassing was about four times less complete than on Earth, the ancient dense atmosphere was probably 140 mb, or about twenty times as dense as the present atmosphere. Anders and Owen (1977) noted that these factors could be increased two- to threefold, if Mars' original volatile endowment was higher than they assumed. However, they claimed that much higher factors are ruled out because of the very high argon 40/argon 36 ratio in the Martian atmosphere and because of the need to account for huge amounts of lost or hidden nitrogen, hydrogen (in water), and carbon (from carbon dioxide).

Despite their very restrictive allotment of volatiles to Mars and the restrained outgassing in their model, the ancient atmosphere and the amount of water predicted by Anders and Owen (1977) do provide the wherewithal for flowing water on Mars. However, the result is not compatible with the immense extent of ice-rich permafrost implied by the photographic observations. At least one order of magnitude more water would be probably involved in a planet-wide permafrost 1–2 km thick. Even more water could occur as liquid below the ground-ice layer. Does current planetary atmosphere theory rule out this geological observation?

Mutch and others (1976) calculated how much water Mars would receive by the crystallization of its observed volcanic rocks. Assuming that these rocks expel about 3 percent water by volume on crystallization, they calculate that about 2×10^6 km^3 of water is yielded, or about 20 m of water averaged over the planet's surface. If the Anders and Owen (1977) calculation is correct, then the extensive volcanic rocks of Mars must have erupted very impoverished in water. This conclusion runs contrary to the extensive nature of the many lava flow units on Mars. Morphologically the Martian lavas most closely resemble terrestrial low-viscosity basaltic flows. Such lavas are commonly rich in iron oxide and water.

Anders and Owen (1977) noted that their model is inconsistent with the common cosmological view that primordial planetary volatile content decreases with decreasing distance from the sun. They offer the alternative that small differentiated bodies in the solar system, like Mercury, Mars, and the Moon, are volatile-poor, while large bodies, Venus and Earth, are volatile-rich. However, their own suggestion is inconsistent with the fact that small differentiated bodies further from the sun are volatile-rich. Examples are the parent bodies for carbonaceous chondrites and other meteorites of the asteroid belt between Mars and Jupiter. What is missing from the Anders and Owen model is a history of planetary development and atmospheric evolution. The relationship of such a history to planetary water was summarized by Ringwood (1978). In Ringwood's view the mystery about water on Mars is not where it came from but rather what has happened to the immense quantities of Mars' original water endowment.

Ringwood (1978) traced the origin of Martian and terrestrial water to the process of solar system formation. The sun and planets formed by contraction of a cool interstellar cloud of dust and gas. Much of this cloud had to be water-rich to account for the immense quantities of water associated with the outer planets of the solar system and their satellites. As the pri-

TABLE 9.1. Indicators of Present or Former Water-Related Processes on Mars

Category	Observation	References
Atmosphere	Formation of atmospheric condensate (probably water ice)	Jones and others (1979)
	Atmospheric water in equilibrium with permafrost ground ice	Farmer, Davies, and LaPorte (1976)
	Atmosphere locally saturated with water	Farmer and Doms (1979)
Channels	Outflow channels	Baker and Milton (1974); Baker (1977, 1978b, 1979c); Baker and Kochel (1979a)
	Small valley networks	Pieri (1976, 1980a, 1980b); Pieri and Sagan (1979)
	Fretted channels	Sharp and Malin (1975)
	Runoff channels	Sharp and Malin (1975)
Craters	Craters as channel sources	Maxwell and others (1973)
	Floor-fractured craters	Schultz (1978)
	Crater ejecta morphology	Carr and others (1977a)
	Phase of intense degradation in planetary cratering history	Jones (1974); Chapman (1974); Mutch and others (1976)
	Crater degradation	Sharp (1968)
Glaciers	Moraines on Arsia and Olympus Montes	Lucchitta (1979)
Hillslopes and mass movement	Landslide morphology	Sharp (1973a, 1973d); Lucchitta (1978d, 1979)
	Spur-and-gully topography	Luchitta (1978a)
	Debris fans and cones	Baker and Kochel (1979a)
	Seepage gullies	Sharp (1973a)
	Debris lobes and aprons	Carr and Schaber (1977)
	Rock glaciers	Squyres (1978, 1979)
	Recurring escarpment heights (ice-water interface)	Soderblom and Wenner (1978)
Patterned ground	Polygonal patterns	Carr and Schaber (1977)
	Stripes	Carr and Schaber (1977)
Polar regions	Layered polar deposits	Cutts, Blasius, and Roberts (1979); Howard (1978)

TABLE 9.1. (continued)

Category	Observation	References
	Residual north polar cap composed of water ice	Farmer, Davies, and LaPorte (1976); Kieffer and others (1976)
Special terrains	Chaotic terrain	Sharp (1973b)
	Fretted terrain	Sharp (1973b); Gatto and Anderson (1975)
	Layered valley floor sediments in Valles Marineris	Blasius and others (1977)
	Tributary canyons, Valles Marineris	Sharp and Malin (1975); Blasius and others (1977)
Thermokarst	Alases	Theilig and Greeley (1978)
	Alas valleys	Carr and Schaber (1977)
	Ground-ice collapse features	Sharp (1973b)
Volcanism	Table mountains	Hodges and Moore (1978); C. C. Allen (1979b, 1979c)
	Olympus Mons and its aureole	Hodges and Moore (1979)
	Crater-controlled igneous processes	Schultz and Glicken (1979)
	Evidence for explosive volcanism associated with permafrost terrains	Reimers and Komar (1979)
	Amphitheater-headed valleys	Baker (1980b)
	Palagonite formation	Toulmin and others (1977); Soderblom and Wenner (1978)
	Pseudocraters	Frey, Lowry, and Chase (1979)
Weathering processes	Salt weathering	Malin (1974)
	Probable formation of iron-rich smectite clays (nontronite), carbonates, sulfates, iron oxides, and hydrated salts by the interaction of iron-rich magma and ground ice	Toulmin and others (1977); Clark (1978)
	Duricrust (probably of hydrated sulfate salts)	Toulmin and others (1977)

mordial cloud condensed to form planetesimals, the more distal regions of the cloud were rich in water and other low-temperature condensates. In contrast, those regions proximal to the cloud nucleus, the protosun, accumulated high-temperature condensates. Ringwood (1978) argued from terrestrial and meteorite geochemistry that the primordial high-temperature condensate was rich in iron-nickel and silica. The low-temperature condensate was similar to type 1 carbonaceous chondrites, which contain about 20 percent water, largely bound in hydrous magnesium silicates. The composition of the earth is satisfactorily explained by the homogeneous accretion from a mixture of 15 percent low-temperature solar nebula condensate and 85 percent high-temperature, devolatilized, metal-rich condensate (Ringwood, 1979).

After planetesimals formed by condensation, gravitational effects led to the accretion process of planets. In planetary growth the energy of accretion increases approximately with the square of the planetary radius. On large planets, like Earth, the low-temperature condensate is easily evaporated to the atmosphere. Ringwood (1978) envisioned that the early Earth, with its water-rich primordial atmosphere, was bombarded by dominantly high-temperature condensate. The chemistry of this process was such as to destroy water. The iron of the high-temperature condensate was oxidized by the water, leaving hydrogen to easily escape from the planet. The Moon apparently experienced this process to an extreme. Its rocks contain metallic iron throughout, indicating that all the original, relatively small amount of lunar water has been reduced by iron to leave a totally dry body. Mercury also suffered this fate, having even less low-temperature condensate to begin with because of its proximity to the protosun.

How the Earth was able to retain some of its volatiles is explained by the formation of its iron-rich core. Ringwood (1979) demonstrated the solubility of ferrous oxide in molten iron under the pressure and temperature conditions prevailing at the core, and he explains the anomalously low density of the core as a result of the presence of about 10 percent oxygen. Formation of the core removed iron from the mantle, thus allowing the preservation of water. Because of its high internal heat, the Earth has been rather effective at transferring these volatiles to the surface. Plate tectonics is the surface expression of Earth's great heat engine.

In Ringwood's model Mars differed from Earth in two regards: (1) it had a much lower accretion energy, about 25 percent of the terrestrial value, and (2) it had a much greater preponderance of water-rich low-temperature condensate in its primordial composition. Factor 1 led to inhibition of the process that differentiates an iron-rich core, as evinced by Mars' extremely weak magnetic field. However, unlike Mercury and the Moon, Mars was rich in volatiles (factor 2). Ringwood (1978) predicted that Mars will be found to be a totally oxidized planet, a complete contrast to the Moon. If Mars' iron is all oxidized, then abundant water is likely in its mantle. The question emerging from this scenario then is not why we see an abundance of water-related surficial features on Mars. Rather the question is why there is not even more evidence for water on the planet's surface. The answer could come from several processes: (1) incomplete outgassing of the original volatile inventory, (2) development and loss of an early dense atmosphere, and (3) trapping of water and other volatiles in subsurface reservoirs.

INCOMPLETE OUTGASSING

The Viking atmospheric analyses show that Mars was not as efficient in outgassing its volatiles as was Earth. Anders and Owen (1977) tried to estimate this inefficiency by calculating the "relative release factor" for various volatiles, defined as the ratio of observed atmospheric abundance to predicted crustal abundance. For argon 36 this value is 0.27, and it probably corresponds to the total fraction of volatile release from the Martian crust. Since hydrogen release is identical to argon 36 release on Earth (Anders and Owen, 1977), Mars can be assumed to also have outgassed only about one-quarter of its original wa-

ter. (Note that the global abundances of argon and water are only slightly higher than the crustal abundances.)

The lack of large-scale crustal movements over Martian geologic history is consistent with the concept of incomplete outgassing. On Earth such movements, known as plate tectonics, have led to relatively efficient outgassing. Plate tectonics is important to planetary outgassing because it provides a mechanism for large movements of liquid and solid material from a planet's interior to its surface. Thus, the mantle chemistry comes in contact with the atmosphere, reacts with it, and subsequently is taken back into the planet's mantle (Ringwood, 1978). This has certainly happened in the later history of Earth, but it has not happened on Mars.

The concept of incomplete outgassing is very important in the later phases of atmospheric history. But the earliest phases, especially those associated with accretion, may be the most important. These are also the phases about which the least is known.

AN EARLY DENSE ATMOSPHERE

The great age of most channels on Mars (Table 3.2) suggests that the responsible fluids may have flowed at a time when Mars had a dense residual volatile-rich atmosphere (Sharp and Malin, 1975). The Viking measurements of the isotopes nitrogen 14 and nitrogen 15 prove that the present atmosphere of Mars does not contain all the nitrogen that was degassed from the planet. Nitrogen was probably lost by diffusive separation in the upper atmosphere of Mars followed by photochemical escape. McElroy, Yung and Nier (1976) made the appropriate calculations to show that an amount of nitrogen was lost equal to 10 times the present value. If nitrogen compounds are deposited in the Martian soil, this factor would have to be even higher. Since Earth has outgassed about 60 times as much carbon dioxide and about 300 times as much water as nitrogen, it is likely that Mars also outgassed much more of these two gases than is implied by the present atmospheric composition. Al-

though today these gases are probably in sinks, such as the polar deposits and the permafrost, they may once have formed a denser atmosphere.

The concept of hot accretion of Earth (Hanks and Anderson, 1969; Anderson, 1972) leads to the probability of rapid early degassing of volatiles. This contrasts with earlier theories that the earth degassed at a uniform rate over geologic time (Rubey, 1955; Holland, 1962). The hot accretion model has Earth melting and differentiating rapidly into core, mantle, and crust. Atmospheric gases would nearly all be released from their volatile sources at this time (Fanale, 1971). Mars may have experienced a similar history.

Atmospheric gases, especially water, carbon dioxide, and ozone, are important in absorbing and scattering sunlight and absorbing thermal radiation. Mars, like Earth, now has an oxidized atmosphere in which the gases do not lead to appreciable self-enhanced warming effects. However, if a primordial reducing atmosphere occurred, then substantial "greenhouse effects" could be generated (Sagan and Mullen, 1972; Sagan, 1977). Such an atmosphere could have evolved if sufficient water was available in the primordial source materials (Pollack, 1979). Pollack (1979) performed calculations showing that this type of atmosphere could be expected if Mars experienced a short-lived period of intense outgassing early in its history. He calculated that as this atmosphere evolved it would achieve an intermediate state containing both oxidized gas species (carbon dioxide) and reduced species (ammonia). This atmosphere would produce an optimum high atmospheric pressure and high surface temperature on the planet. The optimum might have come at a period of especially intense outgassing, such as the time when the northern volcanic plains of Mars were emplaced. The optimum might explain both the planet-wide distribution and the age of the small valley networks.

Cess, Ramanathan, and Owen (1980) objected to the concept of an early reducing atmosphere on Mars. They showed that, assuming a past abundance of carbon dioxide of about one atmosphere, a carbon dioxide–

water greenhouse effect could be achieved. Their model demonstrated a large enough increase in the global temperatures to permit the existence of liquid water on the planet's surface.

Anders and Owen (1977) noted that the key observation for the volatile-poor scenario of Mars is the very high argon 40/argon 36 ratio in Mars' atmosphere. However, another explanation for this ratio is that Mars was originally volatile-rich and extensively outgassed. This primitive atmosphere would have contained the bulk of the argon 36 and a major fraction of the planet's carbon and nitrogen. The atmosphere would have contained considerable liquid water only if it was relatively warm. Most likely the planet's water was retained as ice even at very early points in its history. If the primordial massive atmosphere was lost in some catastrophic event during the early history of Mars, then the argon 40/argon 36 ratio is explained by the continued slow outgassing of argon 40 after the catastrophe. Anders and Owen dismissed this hypothesis on two grounds: (1) it is catastrophic and therefore suspect; and (2) it has not yet been developed into a detailed quantitative model to explain various other noble gas abundances.

Despite the above arguments, the concept of catastrophic loss of Mars' primordial atmosphere is not disproven. A speculative scenario follows. The primitive atmosphere formed on both Earth and Mars by degassing during accretion. (Indeed, if hot accretion prevailed, such an atmosphere must have formed because of the extremely high temperatures developed during the impacting of volatile-rich accretionary materials.) This atmosphere would have contained the bulk of Mars' argon 36, especially if many of the volatile materials were introduced as a late-accreting veneer. As the planets cooled, Mars, because of its further heliocentric distance, could have formed a permafrost cap to trap its water and slow the escape of degassing volatiles. The atmosphere-loss event could have come from an active phase of the sun. This phase, called the "T-Tauri phase" of solar-type stars, might have produced an intense solar wind that could strip the atmosphere of a planet unprotected by a magnetic field (Anders and Owen, 1977). Core formation on Earth gave it a protective magnetic field, but Mars was unprotected. Whatever the cause of the atmospheric catastrophe, it can explain the argon values. Mars would have been relatively enriched in argon 40 because of its argon 36 loss, but Earth would have retained its argon 36 in the observed proportion. Plate tectonics on Earth and its lack on Mars would have been important in the efficiency and nature of subsequent degassing over billions of years. Exospheric escape of some gases, such as nitrogen, and losses of volatiles to major planetary sinks, such as Earth's oceans and Mars' permafrost, would then explain the remaining atmospheric differences.

The loss of a dense early atmosphere explains the problem recognized in comparisons of Martian and lunar cratering. Early in its history Mars experienced a phase of intense crater degradation, presumably induced by exogenic processes from Mars' early atmosphere (Jones, 1974). However, subsequent to this ancient phase of degradation, Martian cratering, channeling, and volcanism have all produced land forms that are preserved with a remarkably pristine appearance. This difference cannot be attributed to age alone because the degradation event probably occurred at least 3.5 billion years ago, assuming equivalence of the Martian and lunar cratering curves (Mutch and others, 1976). This is when many of the planet's surface features were developing (Table 2.2). Except for local eolian modifications, which probably involve relatively soft materials, there is little evidence for extensive atmosphere-controlled degradation on Mars over the billions of years following its ancient phase of very pronounced degradation. The cratering record is thus most in accord with the development and loss of a dense, early atmosphere on Mars. Although the origin and development of such an atmosphere are highly speculative, its existence cannot be denied merely because models are not yet available to fully explain it.

Results from Pioneer Venus, in combination with data for Earth and Mars, suggest that the primordial noble gas abundances may vary appreciably from planet to planet (Hoffman and others, 1979). Thus, the usual assumption of scaling volatile abundances to noble gases has recently come into serious question.

Attempting to reconcile the Mars noble gas abundance with the surprising Pioneer Venus results, Pollack and Black (1979) proposed a two-stage degassing model for Mars. The first phase was an early period of extensive degassing. This was followed by later, localized degassing through volcanism. Pollack and Black predicted a Martian water inventory of 80–160 m water averaged over the planetary surface. If Mars had the same proportions of volatiles as Earth, then this amount would be increased by a factor of five. The inefficient degassing of Mars was attributed to a smaller accretional heating and a smaller energy release per unit mass during core formation (Pollack and Yung, 1980).

AN OCEAN OF PERMAFROST

Today nearly all of Mars' degassed water must reside in the permanent polar caps, regolith adsorption, hydrated minerals, and subsurface permafrost. The amount of water in the polar caps is not known precisely, but it is clearly much less than the total water predicted by even conservative outgassing models. Fanale and Cannon (1974) estimated that the polar water total is equivalent to about 0.7 m of water spread over the planet's surface. They showed that considerable adsorbed water and carbon dioxide are theoretically possible in the Martian regolith, but frozen interstitial water is much more likely from terrestrial experience. Hydrated minerals are undoubtedly present in the near-surface Martian regolith (Toulmin and others, 1977). Considerable degassed water could be tied up in such minerals (Clark, 1978). But this would require an inordinately thick layer of regolith rich in these hydrated minerals. From terrestrial experience, and given the abundant geomorphic evidence for its existence, the thick ice-rich permafrost is probably the major sink for Mars' degassed water.

The Martian permafrost can act as the major volatile sink in any of the schemes proposed thus far. If Mars experienced massive degassing, then permafrost formation was probable as the planet's atmosphere cooled and its water precipitated (Mutch and others, 1976; Carr, 1979b). In that scenario, the permafrost ice would form, as in the terrestrial case, by the downward percolation of water through porous surface materials. Alternatively, the planet may have experienced a more continuous degassing process. Upward-migrating juvenile water would then probably freeze in the regolith. If the surficial regolith is relatively fine-grained, ice accumulating in this way could be stable over billions of years (Smoluchowski, 1968). Thus, the trapped water would be largely lithospheric (Sharp, 1974; Soderblom and Wenner, 1978). A combination of origins is also likely, especially given the abundant evidence for fluid release from this permafrost zone. As the dominant sink for Martian water, the ice-rich permafrost serves an equivalent function to the Earth's ocean. Unlike Earth, however, Mars does not have a hydrosphere that dynamically interacts with incoming solar radiation, transferring heat and influencing the atmosphere. The stable, ancient permafrost can only be perturbed by catastrophic events, such as meteor impacts, seismic events, and volcanism.

However it developed, an immense Martian permafrost would profoundly influence the history and distribution of Martian volatiles. Soderblom and Wenner (1978) considered the implications of an immense ice-rich permafrost zone on Mars. They proposed that below the permafrost zone is a second zone in which liquid water is stable and in which it at least temporarily resided. This stratification produced a discontinuity in the Martian crust. Material below the discontinuity underwent diagenetic alteration and cementation; material above remained pristine and fragmented because it was cemented by ice. This discontinuity subsequently controlled the development of the scarps bounding the chaotic and fretted terrains on the planet. Because of diagenetic alteration, the lower zone was resistant, allowing the overlying impact breccia to easily erode except where caprocks of lava locally protected it.

The permafrost would also have functioned as a trap for juvenile volatiles migrating upward from the planet's interior. Thus the concentration of outflow channels just east of the Tharsis region may relate to the immense quantity of water concentrated by volcanic outgassing in that area. Similarly, the association of fluvial-like channels with the Elysium volcanoes (Chapter 4) could also be related to juvenile water accumulations beneath a permafrost cap that was locally ice-saturated. Liquid water beneath the frozen zone might contain dissolved gases such as nitrogen and carbon dioxide. Some carbon dioxide might also occur adsorbed in the regolith (Fanale and Cannon, 1974) or as clathrates. Clathrates are crystallized water structures that incorporate other gas species. Carbon dioxide clathrates are considered to be probable on Mars (Miller and Smythe, 1970).

Through their influence on the hydrologic cycle, the oceans act as both the source and the terminus of rivers on Earth. For the Martian channels this role is taken by the ice-rich permafrost. The different mechanisms that operate in Martian permafrost versus the terrestrial hydrologic cycle can account for many of the differences observed between terrestrial and Martian "rivers." The remaining questions center on the conditions that allowed water to emerge from the permafrost. Could Mars have experienced climatic changes that melted ground ice, or is it even necessary to postulate such changes? What were the release mechanisms for the immense quantities of fluid that are required by the observed morphology of the outflow channels?

CLIMATIC CHANGE

Mars probably had the requisite quantities of water to carve the various channel-like features described in this book. Indeed, it may be a water-rich planet that has been very effective in hiding its aqueous wealth from our peering eyes, as it has hidden other secrets in the past (Chapter 1), but Mars is also extremely cold, and this poses problems for prolonged liquid water flow on its surface. The problem is less extreme for the short-lived outflow events (Chapter 8), but it is critical for the seepage flows envisioned for the small dry valley networks. The widespread distribution of the small dry valley networks requires that they were formed either by rainfall or, more likely, by atmospheric-induced melting of ground ice. Their great age means that the warmer, possibly wetter atmosphere was also very ancient. This could have been the early dense atmosphere discussed above.

The extensive evidence for climatic change on both Mars and Earth (Pollack, 1979) is one of the remarkable coincidences of modern science. It is certainly an attractive hypothesis that the responsible mechanisms are universal in the solar system. However, the search for the mechanisms is complicated by the multiple time scales of terrestrial climatic change. The various mechanisms must reconcile such diverse phenomena as the following:

(1) Isotopic analyses of ancient chert deposits indicate that the Earth generally had a higher temperature 1–3 billion years ago than in the last 600 million years (Knauth and Epstein, 1976).

(2) Macroglaciations affected Earth in the Precambrian (7×10^8 years B.P.), Permo-Carboniferous (3×10^8 years B.P.), and Quaternary (1×10^6 years B.P.).

(3) The spacings of cold maxima inferred from the isotopic character of ocean faunal tests average 23,000, 42,000, and 100,000 years for the Quaternary (Hays, Imbrie, and Shackelton, 1976).

(4) Even the present interglacial period contains closely spaced cold spells, such as occurred from 1450 to 1900 (the Maunder Minimum), 2800 years B.P., 5300 years B.P., and 11,000 years B.P.

The possibility of cyclic climatic change on Mars has theoretical support in the celestrial mechanics of the planet's orbit. The relevant theory for Earth was established by Milankovitch (1941), who demonstrated quantitative changes in various orbital pa-

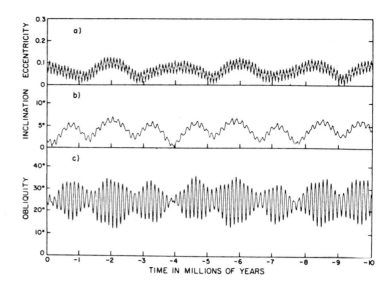

Figure 9.1. Estimated variations in Martian orbital eccentricity (a), orbital inclination (b), and obliquity (c) for the past 10 million years. (From W. R. Ward, 1979, Jour. Geophys. Res., v. 84, p. 237–241, copyrighted by American Geophysical Union.)

rameters of Earth and proposed that they were responsible for Pleistocene climatic change. The three important terrestrial parameters are as follows: (1) Precession of the Earth's spin axis caused by the gravitational attraction of the sun on the Earth's equatorial bulge. This cycle, sometimes termed "precession of the equinoxes," results in a 21,000-year cycle of warmer and colder seasons in each hemisphere. (2) Variations in the Earth's orbital eccentricity, that is, its deviation from a perfectly circular orbit. The amplitude and periodicity of this cycle tends to repeat every 90,000 years. (3) Variations in the obliquity (the tilt between the planet's axis of rotation and the orbital plane). This cycle has an amplitude of 2° and a periodicity of about 41,000 years.

The Milankovitch theory was hotly debated during the decades after its proposal. A major problem was the inability to identify the many cycles in the terrestrial record of glacial advances and retreats. The resolution of this paradox has only recently come from the marine record, where faunal and isotopic changes can be precisely dated and correlated to climatic controls. The correspondence of the marine stratigraphic record to the Milankovitch theory is so close in detail as to constitute a proof (Hays, Imbrie, and Shackelton, 1976).

Insolation changes on Mars are caused by five changing parameters: (1) position of the perihelion (closest approach to the sun), (2) equinox position, (3) orbital eccentricity, (4) orbital plane inclination, and (5) obliquity (Mutch and others, 1976). Probably the most important of these factors is the obliquity, which varies on a cycle of 1.2×10^6 years (W. R. Ward, 1973). W. R. Ward (1979) calculates the relevant variations over the past 10 million years (Figure 9.1). The obliquity oscillation varies from a maximum of 36° to a minimum of 12° with a long-

term average of 24.4°. This is an immense change, and it would exert a profound influence on the polar regions, where annual solar radiation could vary by a factor of two. With relatively low tilt, volatiles might remain frozen at the poles. A larger tilt value might allow more water and carbon dioxide to enter the atmosphere from polar regions because the summer sun would be more effective.

The critical obliquity values for inducing Martian climatic change are not known. W. R. Ward (1979) suggests that an obliquity of 10.8° would drop the carbon dioxide atmospheric pressure to only 0.1 mb, making the wind transport of dust virtually nonexistent. Very high obliquity would have an opposite effect. Although the polar carbon dioxide reservoir is probably far less than the 1 bar envisioned by Sagan, Toon, and Gierasch (1973), there may be an additional input from adsorbed carbon dioxide in the regolith. The critical obliquity value for inducing a major increase in atmospheric pressure is not known. Certainly the present value of about 24° has not been sufficient to induce a dramatic increase in polar evaporation.

The modern insolation cycles of Mars are probably most significant in explaining the more sensitive indicators of climatic change on Mars: the polar layered terrains, the layered terrain of the Valles Marineris, variations in eolian erosion, volatile adsorption by the regolith, and the long-

term composition of the perennial polar caps. The channeling history did not occur on this fine time scale, and the channeling is sufficiently ancient to have involved a primordial atmosphere. Mutch and others (1976) presented thermodynamic arguments that temperature changes resulting from insolation variations are incapable of producing rainfall on Mars. Instead, they favored a model in which a pre-existing dense atmosphere cooled, resulting in the precipitation of rain prior to entrapment of the water in an extensive ice-rich permafrost. In their model the permafrost is the remnant of the ancient atmosphere, and it serves as the major planet-wide sink for water.

Ward, Burns, and Toon (1979) presented an interesting variant of the obliquity hypothesis that could relate to the channeling history. The new theory notes that immense changes in mass distribution have occurred on Mars, particularly associated with the Tharsis region. These changes involve deviations from isostatic adjustment and returns to adjustment. The changes in astronomical perturbations induced by such changes are especially important on Mars because Mars already has a very sensitive set of orbital parameters. Ward, Burns, and Toon speculated that Martian obliquity was profoundly influenced by the growth of the Tharsis bulge. Early in the planet's history the obliquity may have been as low as 9° and as high as 46°. Subsequent adjustments over the planet's history probably shifted it to less variability in its obliquity. Thus, warming in polar areas could have been very pronounced in the ancient epoch of channeling. Later adjustments, including possible polar wandering, might then have yielded the present distribution of relict dry valleys.

Certainly one way to profoundly affect all the planets is through changes in solar output. Current theories of stellar evolution indicate that the sun's luminosity has increased 30–40 percent since the formation of the planets (Bahcall and Shaviv, 1968). Episodic variations in solar luminosity have been proposed on a time scale of a few hundred million years to explain Earth's "megaglaciations" and Martian channels (Hartmann, 1974b). The relevant theory was developed by Öpik (1958, 1965). Öpik hypothesized that disturbances in the radiation transfer rate of the sun's core caused temporary expansions, lowering surface temperature and decreasing luminosity. The low-luminosity periods would last about 6 million years and would be spaced about 250 million years apart.

One possible evidence for changing solar conditions comes from the study of neutrinos, high-energy particles produced by a minor branch of the nuclear cycle that powers the sun. Modern solar theory fails to account for the anomalously low modern abundance of neutrinos. This suggests that the sun may presently be in a core-expansion phase of relatively low solar luminosity (Sagan and Young, 1973). The apparent coincidence of ice-age conditions on Earth and Mars is consistent with the same observation. Sagan and Young (1973) speculated that Earth damps out severe effects of solar luminosity changes because of its permanent oceans and because no significant component of its atmosphere condenses on its soils or at its poles.

The theories of climatic change on Mars provide explanations for the various observed phenomena. What we lack are definitive tests that will tell which explanations are correct. The state of knowledge is not unlike that confronting glacial geologists

several decades ago as they tried to explain terrestrial cycles of climatic change (Imbrie and Imbrie, 1979). The problem was solved only through painstaking stratigraphic analysis on a microscopic scale. Certainly the status of our knowledge about Mars is as good as might be achieved for Earth by the remote viewing of an extraterrestrial intelligence. Ultimate answers will have to await future explorations of Mars.

RELEASE MECHANISMS FOR MARTIAN FLOODS

On Mars the appearance of flowing surface water seems to be either perseverance-limited or release-limited. The former case includes the valley networks (Chapter 4), for which relatively slow release from the Martian permafrost poses few problems. Temperature changes, volcanic heating, and other phenomena can easily release the water; the problem is in maintaining prolonged flows on the planet's surface. Prolonged surface flow requires such assumptions as a warmer, denser, ancient atmosphere (Pollack, 1979), ice-covered rivers (Wallace and Sagan, 1979), or freezing point depressants (Ingersoll, 1970). For the outflow channels the formative mechanism is release-limited. Short-duration floods are less severely constrained even by the present Martian atmosphere (Sharp and Malin, 1975). The major problem is in yielding the immense quantities of water implied by the flow hydraulics of the outflow channels (Chapter 8).

Several investigators have invoked geothermal-permafrost interactions as fluid-release mechanisms. McCauley and others (1972) suggested underground storage of geothermally melted ground ice. But they did not explain how a rapid discharge could be achieved for water that was filling the pores and interstices of subsurface rocks.

Sharp and Malin (1975) were impressed with the apparent localization of Martian channels as they interpreted them from Mariner 9 imagery. They speculated, following Fanale (1971), that parts of the Martian crust may have become temporarily "juicy." The localized juicy spots became seepage zones for lithospheric water. The

seepage zones could have nourished small valley networks, or they could have ponded water for eventual release as a flood. The origin of the juicy spots might be juvenile waters above a crystallizing magma body or melted primordial ground ice. Unfortunately, Sharp and Malin did not address the problems of release mechanics for the flood flows.

Masursky and others (1977) suggested a release mechanism for outflow floods that is analogous to the *jökulhlaups*, or glacial bursts, caused by volcanic eruptions under glaciers, and especially common in Iceland (Thorarinsson, 1957). The chaotic terrain presumably forms when subsurface permafrost is melted by geothermal heat from a deep-seated volcanic center. When the melt zone, really an interstitially stored meltwater lake, expands to reach a topographic slope, the stored water is suddenly released. Carr (1979b) considered this mechanism incapable of achieving the required discharges, and he noted that the mechanism is inconsistent with the lack of evidence for contemporaneous volcanic activity in the outflow channel source regions.

Clark (1978) proposed an ingenious "pressure cooker" for releasing geothermally heated water under pressure. Clark first reviewed evidence from the Viking lander analyses that the Martian surface is probably underlain by materials rich in salts and smectite clays. Besides being a reservoir for planetary water and other volatiles, a regolith rich in these hydrated minerals could be baked by a geothermal source to form a pressurized trapped-ground-water system. The baking effect derives from the poor heat conduction of hydrated salts and clays. Zones of liquid water and ice would build up beneath swollen clays that would prevent upward diffusion of the water. The water would be rich in freezing-point depressants, and pressure would build up beneath the clay cap until relieved by the explosive release of hot water and mud. Unfortunately, the model is not evaluated for potential flow volumes. It appears that the total released fluid is largely determined by the depth of the water-rich zone, certainly no more than a few kilometers.

Large outflow channels appear to require many cubic kilometers of released fluid.

Maxwell and others (1973) suggested that meteorite impact is locally responsible for the release of water from the Martian permafrost. Their model has the attraction of a lithospheric water source. However, it is difficult to accept their identification of certain smooth-rimmed areas of chaotic terrain as impact craters coeval with the outflow (Milton, 1974). Milton (1974) suggested instead that the crater-like shape of some chaos areas results from thick fills of sediment or lava overlying ancient craters. The collapse of these fills would yield the crude crater-like shape and also explain the lack of any indications of impact other than the depression. Even if the impact mechanism is valid, it can apply only to a small number of channels on Mars (Sharp and Malin, 1975).

Schultz (1978) and Schultz and Glicken (1979) analyzed floor-fractured craters on Mars and found them morphologically distinct from lunar analogs. The differences appear to result from crater controlled igneous processes interacting with ice-bearing materials. Schultz and Glicken (1979) presented a model of igneous intrusions developing beneath large craters and thawing crater-contained ice. Release of thawed materials can be relatively slow, to form the small sinuous valleys that appear to emanate from the rim regions of certain ancient Martian craters. Rapid release is also possible, resulting in collapse features. The hypothesis is attractive in accounting for the source of localized high heat flows (crater-controlled igneous processes).

Milton (1974) proposed that ground ice on Mars consisted of a carbon dioxide–water clathrate. This unusual substance dissociates upon release of pressure at temperatures between 0°C and 10°C. Pressure release could come from some tectonic event, triggering the release of carbon dioxide gas and massive amounts of liquid water. The heat of dissociation would be derived from cooling the host rock from its original temperature down to 0°C, at which point the reaction would cease. Catastrophic dissociation of this clathrate presumably occurred during a past Martian epoch when the near-surface temperature was in the proper range. But the optimum temperature range is so narrow and the specific heat contrasts of clathrate and rock are such as to make the mechanism very unlikely (Carr, 1979b).

Peale, Schubert, and Lingenfelter (1975) argued that Milton's clathrate hypothesis was both unlikely and unnecessary to explain the apparent outbursts of flood water from chaotic terrain areas. They pointed out that, at the temperatures and pressures required for carbon dioxide hydrate, liquid water would already exist. This liquid water could easily be tapped by any process that would break the overlying cap of permafrost, as proposed for the origin of the lunar sinuous rilles (Lingenfelter, Peale, and Schubert, 1968; Schubert, Lingenfelter, and Peale, 1970). They also pointed out that the clathrate release mechanism is thermodynamically unlikely, since 150 cal is required to release 1 g of water from clathrate while only 80 cal is required to release 1 g of water liquid from water ice. As water is released from the clathrate, this heat would be obtained not merely from the surrounding rock but also from the recently released water. The result would be an enhanced freezing effect that would lower the discharge of released water.

Carr (1979b) erected the most elaborate fluid-release model for explaining Martian floods. He proposed that the combination of a highly permeable near-surface megaregolith and an ice-rich permafrost constitutes a system of confined aquifers on Mars. The upper seal of these aquifers is a thick permanent ice layer, and the lower seal is created by self-compaction of the rock materials at depths of about 10 km. Where sufficient hydraulic gradients exist and the permeability is continuous, this artesian system could yield high pore pressures at potential discharge zones. Carr found that high potential head differences do indeed exist upgradient from the chaotic terrain zones around the Chryse Basin. The elevation differences imply hydrostatic head differences of 4–5 km, easily capable of exceeding the lithostatic pressure in the chaos regions for an aquifer buried 1–1.5 km deep.

At points where the overlying seal of the confined aquifer is ruptured, the water would gush to the surface under high pressure, and the overlying terrain would collapse to form chaos. By analogy to heat flow equations, Carr (1979b) calculated probable flow rates from the Martian aquifers. For reasonable values of aquifer thickness and permeability, depth of burial and chaos diameter, he calculated discharges of 10^5–10^7 m/sec. The values are perhaps one order of magnitude short of those required by the channel hydraulics (Table 8.1). The result is not disturbing considering the uncertainties involved.

Carr's model has the advantage that the aquifers can be of immense areal extent. Thus there is no problem of sufficient fluid volumes. For extensive aquifers flow will continue for a long period, until the hydraulic gradient around the breakout point is so reduced that the reduced flow rate results in freezing. Moreover, because discharge depends on the diameter of the region over which the water has access to the surface, large chaos zones can yield immense discharges, even given the limited permeability of the aquifers. Chaos zones would enlarge by undermining as the flow emerged. This self-enhancing effect can explain both the huge discharges and the apparent headward growth of outflow channels (Baker, 1977, 1978b). Breakout to the surface probably included disruption of the aquifer in the high-discharge zones, such that part of the porous medium was itself carried along with the flow (Carr, 1979b).

Carr's model explains the preferential location of outflow channels around the Chryse Basin. Because Chryse is one of the lowest regions on the planet, its margins were ideal locations for generating high artesian pressures. Chryse is also younger than the other major Martian depressions, Hellas, Argyre, and Isidis. The older depressions may have formed before the global Martian cooling which Carr believes put the seal of ice over the aquifer systems. The older depressions would therefore have lost their groundwater systems through slow seepage. Chryse, by forming after the sealing, was posed for catastrophic release.

Carr also believes his model provides an explanation for relationships observed in the Valles Marineris (Chapter 2). The transition of the eastern chasmas into chaotic terrain and outflow channels is explained by the eastward regional slope. Artesian pressures would therefore be greatest to the east. Some of the faulting involved in forming the canyons (Blasius and others, 1977) could have disrupted the postulated aquifers in these eastern zones. However, water played an insignificant role in the higher western sections of the Valles Marineris, where artesian pressures would have been low.

Despite the ingenuity of the various proposed fluid-release mechanisms, none provides a full explanation for the Martian outflow source mechanism. This lack of an ultimate answer has its parallel in the debate that developed during the 1920's over J Harlen Bretz' proposed flood origin of the Channeled Scabland (Chapter 7 of this book; Baker, 1978a). It was then concluded by many prominent geologists that the lack of an obvious source for the immense scabland floods should constitute a basis for rejecting the hypothesis. However, the flood origin of the Channeled Scabland was established by the morphological relationships first described by Bretz and inexplicable by any other hypothesis. The story was merely completed when the source of the flooding, glacial Lake Missoula, was subsequently described. Mars seems to pose a similar riddle. We have the morphological evidence, but the source mechanism will be resolved only by new data. Since the critical data involve the Martian subsurface, they will have to come from a future exploration of the planet.

FATE OF THE RELEASED WATER

What happened to the water that burst onto the surface of Mars to shape the outflow channels or which slowly seeped and sapped to form the dry valley networks?

Nearly all the Martian channels terminate at closed depressions. These range in size from small fractures, craters, and troughs for the dry valley networks up to the huge Chryse Basin into which many of the outflow channels debouched their fluid flows. In the present Martian environment the water that reached these depressions would have frozen on its surface to form ice-covered lakes (Wallace and Sagan, 1979). Water from the lake bottom would undoubtedly seep back into the subsurface. Indeed, percolation of water into the subsurface would have been a significant down-channel loss for the postulated seepage flows of the dry valley networks. The volcanic lava flows and impact breccias thought to floor these valleys would probably be very permeable. Only the water which directly vaporized on release, or which sublimated from the lake and river ice, would eventually reach the atmosphere. This would first precipitate temporarily as local snow, but eventually it would become trapped in the polar ice. If a warm, dense atmosphere prevailed during the Martian past, the water could have been repeatedly recycled. The process would be similar to that on Earth, except for the lack of a huge liquid reservoir, the oceans.

Some of the atmospheric water could have dissociated and escaped to outer space (McElroy, 1972; Sagan and Mullen, 1972), but the total for such losses could only have been about 2.5 m planet-wide over 4.6 billion years (Hunten and McElroy, 1970). It is likely that much of the released water found its way back to the subsurface permafrost reservoirs; that which did not constitutes the polar deposits; and but a tenuous vapor occurs in the modern atmosphere. If only those water molecules could tell of their adventures!

CONCLUSION

The early telescopic observers of Mars were fascinated with the Earth-like nature of the planet. Some of the less objective viewers even populated their imaginary Mars with Earth-like creatures and postulated landforms and engineering works exactly as on our own planet. Even the objective analyses that followed the first spacecraft missions to Mars sometimes displayed a form of interplanetary chauvinism, assuming that heavy cratering implied a Moon-like Mars or that networks of valleys meant an Earth-like Mars. However, as the overwhelming data of recent space explorations have been analyzed, the inescapable conclusion is that Mars possesses a character of its own. Indeed the study of Mars is now yielding Mars-analogs for terrestrial processes. The scientific implications are summarized in a review of geomorphological processes on terrestrial planetary surfaces (Sharp, 1980, p. 231):

Planetary exploration has proved to be a two-way street. It not only created interest in Earth-surface processes and features as analogues, it also caused terrestrial geologists to look on Earth for features and relationships best displayed on other planetary surfaces. Impact cratering, so extensive on Moon, Mercury, and Mars, is a well-known example . . . Another is the huge size of features such as great landslides and widespread evidence of large-scale subsidence and collapse on Mars, which suggests that our thinking about features on Earth may have been too small scaled. One of the lessons from space is to "think big."

The Martian channels are, in many respects, uniquely Martian. Neither the early fantasy of excavated canals nor the reasoned analogy to terrestrial rivers has adequately explained all their attributes. However, this book has shown that we have learned a great deal about the channels. We know that some terrestrial processes—sapping, catastrophic flooding, thermokarst—are very closely related to the Martian examples. We know that some terrestrial landscapes have many, though not all, of the attributes of channeled terrains on Mars. Obviously, this is continuing research, and more will be learned, modifying the analysis presented in this book. What I hope has been shown is that the phenomenal spacecraft images of Mars are not simply a tool for solving the mysteries of a far-away planet. In a sense Mars has achieved a kind of retribution for the centuries of misconception by its observers. Now the study of its landscapes is stimulating planetologists to ask new questions about the Earth.

References Cited

Allen, C. C., 1979a, Areal distribution of Martian rampart craters: Icarus, v. 39, p. 111–123.

———, 1979b, Subglacial volcanism: NASA Tech. Memo. 80339, p. 251–253.

———, 1979c, Volcano-ice interactions on Mars: Jour. Geophys. Res., v. 84, p. 8048–8059.

Allen, J. R. L., 1968, Current ripples, their relation to patterns of water and sediment motion: Amsterdam, Elsevier, 433 p.

———, 1971, Transverse erosional marks of mud and rock: Their physical basis and geological significance: Sedimentary Geology, v. 5, p. 167–385.

Anders, E., and Owen, T., 1977, Mars and earth: Origin and abundance of volatiles: Science, v. 198, p. 453–465.

Anderson, D. L., 1972, Internal constitution of Mars: Jour. Geophys. Res., v. 77, p. 789–795.

Anderson, D. M., Gaffney, E. S., and Law, P. F., 1967, Frost phenomena on Mars: Science, v. 155, p. 319–322.

Anderson, D. M., Gatto, L. W., and Ugolini, F. C., 1972, An Antarctic analog of Martian permafrost terrain: Antarctic Jour., v. 7, p. 114–116.

———, 1973, An examination of Mariner 6 and 7 imagery for evidence of permafrost terrain on Mars, in Permafrost: The North American contribution to the second international conference: Washington, D.C., Nat. Acad. Sci., p. 499–508.

Arvidson, R. E., 1972, Aeolian processes on Mars: Erosive velocities, settling velocities, and yellow clouds: Geol. Soc. America Bull., v. 83, p. 1503–1508.

———, 1974, Morphologic classification of Martian craters and some implications: Icarus, v. 22, p. 264–271.

Arvidson, R. E., Goettel, K. A., and Hohenberg, C. M., 1980, A post-Viking view of Martian geologic evolution: Rev. Geophys. and Space Phys., v. 18, p. 565–603.

Arvidson, R. E., Guiness, E. A., and Lee, S. W., 1979, Differential aeolian redistribution rates on Mars: Nature, v. 278, p. 533–535.

Ashton, G. D., 1979, River ice: Am. Scientist, v. 67, p. 38–45.

Bagnold, R. A., 1966, An approach to the sediment transport problem from general physics: U.S. Geol. Survey Prof. Paper 422-I, 37 p.

Bahcall, J. N., and Shaviv, G., 1968, Solar models and neutrino fluxes: Astrophys. Jour., v. 153, p. 113–126.

Baird, A. K., Castro, A. J., Clark, B. C., Toulmin, P., III, Rose, H., Jr., Keil, K., and Gooding, J. L., 1977, The Viking x-ray fluorescence experiment: Sampling strategies and laboratory simulations: Jour. Geophys. Res., v. 82, p. 4595–4624.

Baker, V. R., 1971, Paleohydrology of catastrophic Pleistocene flooding in eastern Washington: Geol. Soc. America Abstracts with Programs, v. 3, p. 497.

———, 1973a, Paleohydrology and sedimentology of Lake Missoula flooding in eastern Washington: Geol. Soc. America Spec. Paper 144, 79 p.

———, 1973b, Erosional forms and processes for the catastrophic Pleistocene Missoula floods in eastern Washington, in Morisawa, M., ed., Fluvial geomorphology: Binghamton, N.Y., Publ. in Geomorphology, State Univ. N.Y., p. 123–148.

———, 1974, Paleohydraulic interpretation of Quaternary alluvium near Golden, Colorado: Quaternary Res., v. 4, p. 95–112.

———, 1977, Viking-slashing at the Martian scabland problem: NASA Tech. Memo. TMX-3511, p. 169–172.

———, 1978a, The Spokane Flood controversy and the Martian outflow channels: Science, v. 202, p. 1249–1256.

———, 1978b, A preliminary assessment of the fluid erosional processes that shaped the Martian outflow channels: Proc. Lunar and Planet. Sci. Conf. 9th, p. 3205–3223.

———, 1978c, Large-scale erosional and depositional features of the Channeled Scabland, in Baker, V. R., and Nummedal, D., eds., The Channeled Scabland: Washington, D.C., NASA Office of Space Science, Planetary Geology Program, p. 81–115.

———, 1978d, Quaternary geology of the Channeled Scabland and adjacent areas, in Baker, V. R., and Nummedal, D., eds., The Channeled Scabland: Washington, D.C., NASA Office of Space Science, Planetary Geology Program, p. 17–35.

———, 1978e, Hydrodynamics of erosion by catastrophic floods: NASA Tech. Memo. 79729, p. 248–250.

———, 1979a, Erosional processes in channelized water flows on Mars: Jour. Geophys. Res., v. 84, p. 7985–7993.

———, 1979b, Cavitation processes in Martian water flows, in Lunar and Planetary Science X: Houston, Lunar and Planetary Inst., p. 57–59.

———, 1979c, Morphology of channels on Mars: NASA Conference Publ. 2072, p. 4–6.

———, 1980a, Geomorphic mapping of dry valley systems on Mars: NASA Tech. Memo. 81776, p. 54–56.

———, 1980b, Some terrestrial analogs to dry valley systems on Mars: NASA Tech. Memo. 81776, p. 286–288.

———, 1980c, Nirgal Vallis: NASA Tech. Memo. 82385, p. 345–347.

———, 1980d, Degradation of volcanic landforms on Mars and Earth: NASA Tech. Memo. 82385, p. 234–235.

Baker, V. R., ed., 1981, Catastrophic flooding: The origin of the Channeled Scabland: Stroudsburg, Pa., Dowden, Hutchinson and Ross, 384 p.

Baker, V. R., and Kochel, R. C., 1978a, Morphological mapping of Martian outflow channels: Proc. Lunar and Planet. Sci. Conf. 9th, p. 3181–3192.

———, 1978b, Morphometry of streamlined forms in terrestrial and Martian channels: Proc. Lunar and Planet. Sci. Conf. 9th, p. 3193–3203.

———, 1979a, Martian channel morphology: Maja and Kasei Valles: Jour. Geophys. Res., v. 84, p. 7961–7983.

———, 1979b, Streamlined erosional forms of Kasei and Maja Valles, Mars, in Lunar and Planetary Science X: Houston, Lunar and Planetary Inst., p. 60–62.

Baker, V. R., and Milton, D. J., 1974, Erosion by catastrophic floods on Mars and Earth: Icarus, v. 23, p. 27–41.

Baker, V. R., and Nummedal, D., eds., 1978, The Channeled Scabland: Washington, D.C., NASA Office of Space Science, Planetary Geology Program, 186 p.

Baker, V. R., and Patton, P. C., 1976, Missoula flooding in the Cheney-Palouse Scabland tract: Terrestrial analogue to the channeled terrain of Mars: Geol. Soc. America Abstracts with Programs, v. 8, p. 351–352.

Baker, V. R., and Pyne, S., 1978, G. K. Gilbert and modern geomorphology: Am. Jour. Sci., v. 278, p. 97–123.

Baker, V. R., and Ritter, D. F., 1975, Competence of rivers to transport coarse bedload material: Geol. Soc. America Bull., v. 86, p. 975–978.

Barnes, H. L., 1956, Cavitation as a geological Agent: Am. Jour. Sci., v. 254, p. 493–505.

Batchelor, G. K., 1967, An introduction to fluid dynamics: London, Cambridge Univ. Press, p. 481–493.

Batson, R. M., Bridges, P. M., and Inge, J. L., 1979, Atlas of Mars: NASA Spec. Publ. 438, 146 p.

Benjamin, T. B., and Ellis, A. T., 1966, The collapse of cavitation bubbles and the pressures thereby produced against solid boundaries: Phil. Trans. Royal Soc. London, Ser. A., v. 260, p. 221–240.

Benson, M. A., and Dalrymple, T., 1968, General field and office procedures for indirect discharge measurements: U.S. Geol. Survey, Techniques Water-Resources Inv., Book 3, Chapter A1, 30 p.

Black, R. F., 1976, Periglacial features indicative of permafrost: Ice and soil wedges: Quaternary Res., v. 6, p. 3–26.

Blackwelder, E., 1934, Yardangs: Geol. Soc. America Bull., v. 45, p. 159–166.

Blasius, K. R., 1976, Topical studies of the geology of the Tharsis region of Mars: Ph.D. thesis, Calif. Inst. Tech., 85 p.

Blasius, K. R., and Cutts, J. A., 1979, Erosion and transport in Martian outflow channels, in Lunar and Planetary Science X: Houston, Lunar and Planetary Inst., p. 140–141.

Blasius, K. R., Cutts, J. A., Guest, J. E., and Masursky, H., 1977, Geology of the Valles Marineris: First analysis of imaging from the Viking 1 orbiter primary mission: Jour. Geophys. Res., v. 82, p. 4067–4091.

Blasius, K. R., Cutts, J. A., and Roberts, W. J., 1978, Large scale erosive flows associated with Chryse Planitia, Mars: Source and sink relationships, in Lunar and Planetary Science IX: Houston, Lunar and Planetary Inst., p. 109–110.

Bradley, W. C., Hutton, J. T., and Twidale, C. R., 1978, Role of salts in development of granitic tafoni, South Australia: Jour. Geology, v. 86, p. 647–654.

Breed, C. S., Grolier, M. J., and McCauley, J. F., 1979, Morphology and distribution of common "sand" dunes on Mars: Comparisons with Earth: Jour. Geophys. Res., v. 84, p. 8183–8204.

Bretz, J H., 1923, The Channeled Scabland of the Columbia Plateau: Jour. Geology, v. 32, p. 139–149.

——, 1924, The Dalles type of river channel: Jour. Geology, v. 32, p. 139–149.

——, 1925, The Spokane flood beyond the Channeled Scabland: Jour. Geology, v. 33, p. 97–115, p. 236–259.

——, 1928a, Bars of the Channeled Scabland: Geol. Soc. America Bull., v. 39, p. 643–702.

——, 1928b, The Channeled Scabland of eastern Washington: Geog. Rev., v. 18, p. 446–477.

——, 1932, The Grand Coulee: Am. Geog. Soc. Spec. Publ. 15, 89 p.

——, 1959, Washington's Channeled Scabland: Wash. Dept. Conserv., Div. Mines and Geology Bull. no. 45, 57 p.

——, 1969, The Lake Missoula floods and the Channeled Scabland: Jour. Geology, v. 77, p. 505–543.

Bretz, J H., Smith, H. T. U., and Neff, G. E., 1956, Channeled Scabland of Washington: New data and interpretations: Geol. Soc. America Bull., v. 67, p. 957–1049.

Carr, M. H., 1974a, Tectonism and volcanism of the Tharsis region of Mars: Jour. Geophys. Res., v. 79, p. 3943–3949.

——, 1974b, The role of lava erosion in the formation of lunar rilles and Martian channels: Icarus, v. 22, p. 1–23.

——, 1975, The volcanoes of Mars: Sci. American, v. 234, p. 32–43.

——, 1979a, Distribution of small channels on Mars: NASA Tech. Memo. 80339, p. 337.

——, 1979b, Formation of Martian flood features by release of water from confined aquifers: Jour. Geophys. Res., v. 84, p. 2995–3007.

——, 1980, The morphology of the Martian surface: Space Science Reviews, v. 25, p. 231–284.

Carr, M. H., Crumpler, L. S., Cutts, J. A., Greeley, R., Guest, J. E., and Masursky, H., 1977a, Martian impact craters and emplacement of ejecta by surface flow: Jour. Geophys. Res., v. 82, p. 4055–4065.

Carr, M. H., Greeley, R., Blasius, K. R., Guest, J. E., and Murray, J. B., 1977b, Some Martian volcanic features as viewed from the Viking orbiters: Jour. Geophys. Res., v. 82, p. 3985–4015.

Carr, M. H., Masursky, H., Baum, W. A., Blasius, K. R., Briggs, G. A., Cutts, J. A., Duxbury, T., Greeley, R., Guest, J. E., Smith, B. A., Soderblom, L. S., Veverka, J., and Wellman, J. B., 1976, Preliminary results from the Viking orbiter imaging experiment: Science, v. 193, p. 766-776.

Carr, M. H., and Schaber, G. G., 1977, Martian permafrost features: Jour. Geophys. Res., v. 82, p. 4039–4054.

Cartwright, K., and Harris, H. J. H., 1976, Ground water at Don Juan Pond, Wright Valley, southern Victoria Land, Antarctica: A probable origin at the base of the east Antarctic icecap: Geol. Soc. America Abstracts with Programs, v. 8, p. 804.

Cess, R. D., Ramanathan, V., and Owen, T., 1980, The Martian paleoclimate and enhanced atmospheric carbon dioxide: Icarus, v. 41, p. 159–165.

Chapman, C. R., 1974, Cratering on Mars: I. Cratering and obliteration history: Icarus, v. 22, p. 272–291.

Chow, V. T., 1959, Open-channel hydraulics: New York, McGraw-Hill, 680 p.

Christensen, P. R., and Kieffer, H. H., 1979, Moderate resolution thermal mapping of Mars: The channel terrain around Chryse Basin: Jour. Geophys. Res., v. 84, p. 8233–8238.

Clark, B. C., 1978, Implications of abundant hygroscopic minerals in the Martian regolith: Icarus, v. 34, p. 645–655.

Clemens, S., 1896, Life on the Mississippi: New York, Harper and Row, 381 p.

Clifford, S. M., and Huguenin, R. L., 1980, The H_2O mass balance on Mars: Implications for a global subpermafrost ground-water flow system: NASA Tech. Memo. 81776, p. 144–146.

Coradini, M., and Flamini, E., 1979, A thermodynamical study of the Martian permafrost: Jour. Geophys. Res., v. 84, p. 8115–8130.

Cutts, J. A., 1973, Nature and origin of layered deposits of the Martian polar regions: Jour. Geophys. Res., v. 78, p. 4231–4249.

Cutts, J. A., and Blasius, K. R., 1979, Martian outflow channels: Quantitative comparisons of erosive capacities for eolian and fluvial models, *in* Lunar and Planetary Science X: Houston, Lunar and Planetary Inst., p. 257–259.

——, 1981, Origin of Martian outflow channels: The eolian hypothesis: Jour. Geophys. Res., v. 86, no. B6, p. 5075–5102.

Cutts, J. A., Blasius, K. R., and Farrell, K. W., 1976, Mars: New data on Chryse Basin land forms: Bull. Am. Astron. Soc., v. 8, p. 480.

Cutts, J. A., Blasius, K. R., and Roberts, W. J., 1978, Chaotic terrain and channels associated with Chryse Planitia, Mars, *in* Lunar and Planetary Science IX: Houston, Lunar and Planetary Inst., p. 206–208.

——, 1979, Evolution of Martian polar landscapes: Interplay of long-term variations in perennial ice cover and dust storm intensity: Jour. Geophys. Res., v. 84, p. 2975–2994.

Cutts, J. A., Roberts, W. J., and Blasius, K. R., 1978, Martian channels formed by lava erosion, *in* Lunar and Planetary Science IX: Houston, Lunar and Planetary Inst., p. 209.

Czudek, T., and Demek, J., 1970, Thermokarst in Siberia and its influence on the development of lowland relief: Quaternary Res., v. 1, p. 103–120.

Dalrymple, G. B., Silver, E. A., and Jackson, E. D., 1973, Origin of the Hawaiian Islands: Am. Scientist, v. 61, no. 3, p. 294–303.

Davis, W. M., 1899, The geographical cycle: Geog. Jour., v. 14, p. 481–504.

Decker, R., and Decker, B., 1981, The eruptions of Mount St. Helens: Sci. American, v. 244, no. 3, p. 68–80.

Dresch, J., 1968, Reconnaissance dans le Lut (Iran): Bull. Assoc. Geogr. Fr., v. 362, no. 3, p. 143–153.

Dunne, T., 1980, Formation and controls of channel networks: Progress in Physical Geography, v. 4, no. 2, p. 211–239.

Einstein, H. A., and Li, H., 1958, Secondary flow in straight channels: Trans. Am. Geophys. Union, v. 39, p. 1085–1088.

Eisenberg, P., and Tulin, M. P., 1961, Mechanics of cavitation, *in* Streeter, V. L., ed., Handbook of fluid mechanics: New York, McGraw-Hill, p. 12.1–12.24.

Evans, D. S., 1976, A fancier of Mars: Science, v. 193, p. 754.

Evans, I. S., 1969, Salt crystallization and rock weathering: A review: Rev. Geomorph. Dyn., v. 19, p. 153–177.

Evans, N., and Rossbacher, L. A., 1980, The last picture show: Small-scale patterned ground in Lunae Planum: NASA Tech. Memo. 82385, p. 376–378.

Fairbridge, R. W., 1972, Planetary spin-rate and evolving cores: Annals N.Y. Acad. Sci., v. 187, p. 88–107.

Fanale, F. P., 1971, History of Martian volatiles: Implications for organic synthesis: Icarus, v. 15, p. 297–303.

——, 1976, Martian volatiles: Their degassing history and geochemical fate: Icarus, v. 28, p. 179–202.

Fanale, F. P., and Cannon, W. A., 1974, Exchange of adsorbed H_2O and CO_2 between the regolith and atmosphere of Mars caused by changes in surface insolation: Jour. Geophys. Res., v. 79, p. 3397–3402.

——, 1979, Mars: CO_2 adsorption and capillary condensation on clays —significance for volatile storage and atmospheric history: Jour. Geophys. Res., v. 84, p. 8404–8414.

Farmer, C. B., Davies, D. W., and LaPorte, D. D., 1976, Mars: Northern summer ice cap—water vapor observations from Viking 2: Science, v. 194, p. 1339–1341.

Farmer, C. B., and Doms, P. E., 1979, Global seasonal variation of water vapor on Mars and the implications for permafrost: Jour. Geophys. Res., v. 84, p. 2881–2888.

Ferrians, O. J., Kachadoorian, R., and Greene, G. W. 1969, Permafrost and related engineering problems in Alaska: U.S. Geol. Survey Prof. Paper 678, 37 p.

Fielder, G., 1963, On the topography of Mars: Astron. Soc. of the Pacific, Publ., v. 75, p. 75–76.

French, H. M., 1976, The periglacial environment: London, Longman, 309 p.

Frey, H., Lowry, B. L., and Chase, S. A., 1979, Pseudocraters on Mars: Jour. Geophys. Res., v. 84, p. 8075–8086.

Fryxell, R., and Cook, E. F., 1964, A field guide to the loess deposits and channeled scablands of the Palouse area, eastern Washington: Wash. State Univ. Lab. Anthropology Rept. Inv. 27, 32 p.

Gatto, L. W., and Anderson, D. M., 1975, Alaskan thermokarst terrain and possible Martian analog: Science, v. 188, p. 255–257.

Gifford, F. A., Jr., 1964, The Martian canals according to a purely aeolian hypothesis: Icarus, v. 3, p. 130–135.

Gilbert, G. K., 1877, Report on the geology of the Henry Mountains: U.S. Geog. Geol. Survey of the Rocky Mountain Region, 160 p.

——, 1890, Lake Bonneville: U.S. Geol. Survey Monograph 1, 438 p.

Glen, J. W., 1955, The creep of polycrystalline ice: Proc. Royal Soc., Ser. A., v. 228, p. 519–538.

Gornitz, Vivien, ed., 1979, Geology of the planet Mars: Stroudsburg, Pa., Dowden, Hutchinson and Ross, 414 p.

Greeley, R., 1971, Observations of actively forming lava tubes and associated structures, Hawaii: Mod. Geology, v. 2, p. 207–223.

——, 1973, Mariner 9 photographs of small volcanic structures on Mars: Geology, v. 1., p. 175–180.

Greeley, R., and Spudis, P. D., 1981, Volcanism on Mars: Rev. Geophys. and Space Phys., v. 19, no. 1, p. 13–41.

Greeley, R., Theilig, E., Guest, J. E., Carr, M. H., Masursky, H., and Cutts, J. A., 1977, Geology of Chryse Planitia: Jour. Geophys. Res., v. 82, p. 4093–4109.

Griggs, G. B., Kulm, L. D., Waters, A. C., and Fowler, G. A., 1970, Deep-sea gravel from Cascadia Channel: Jour. Geology, v. 78, p. 611–619.

Hanks, T. C., and Anderson, D. L., 1969, The early thermal history of the Earth: Phys. of Earth Planet. Interiors, v. 2, p. 19–29.

Hartmann, W. K., 1973, Martian cratering, 4, Mariner 9 initial analysis of cratering chronology: Jour. Geophys. Res., v. 78, p. 4096–4116.

——, 1974a, Geological observations of Martian arroyos: Jour. Geophys. Res., v. 79, p. 3951–3957.

——, 1974b, Martian and terrestrial paleoclimatology: Relevance of solar variability: Icarus, v. 22, p. 301–311.

——, 1977, Cratering in the solar system: Sci. American, v. 236, no. 1, p. 84–99.

Hartmann, W. K., and Raper, O., 1974, The new Mars: The discoveries of Mariner 9: NASA Spec. Publ. SP-337, U.S. Gov. Printing Office, 179 p.

Hays, J. D., Imbrie, J., and Shackleton, N. J., 1976, Variations in the earth's orbit: Pacemaker of the ice ages: Science, v. 194, p. 1121–1132.

Head, J. W., 1976, Lunar volcanism in space and time: Rev. Geophys. and Space Phys., v. 14, p. 265–300.

Head, J. W., Wood, C. A., and Mutch, T. A., 1977, Geologic evolution of the terrestrial planets: Am. Scientist, v. 65, p. 21–29.

Heim, A., 1882, Der Bergsturz von Elm: Zeitschrift geologischen Gesellschaft, v. 34, p. 74–115.

————, 1932, Bergstuz und Menschenleben: Zurich, Fretz and Wasmuth, 218 p.

Helfenstein, P., 1980, Martian fractured terrain: Possible consequences of ice-heaving: NASA Tech. Memo. 82385, p. 373–375.

Hess, S. L., Henry, R. M., Leovy, C. B., Ryan, J. A., and Tillman, J. E., 1977, Meteorological results from the surface of Mars, Viking 1 and 2: Jour. Geophys. Res., v. 82, p. 4559–4574.

Hiller, Konrad, and Neukum, Gerhard, 1980, Time sequence of Martian geologic features: NASA Tech. Memo. 81776, p. 119–121.

Hinds, N. E. A., 1925, Amphitheater valley heads: Jour. Geology, v. 33, p. 816–818.

Hodges, C. A., 1973, Mare ridges and lava lakes: NASA Spec. Publ. 330, p. 31.12–31.21.

Hodges, C. A., and Moore, H. J., 1978, Table mountains of Mars, in Lunar and Planetary Science IX, Houston, Lunar and Planetary Inst., p. 523–525.

————, 1979, The subglacial birth of Olympus Mons and its aureoles: Jour. Geophys. Res., v. 84, p. 8061–8074.

Hoerner, S. F., 1958, Fluid-dynamic drag: Midland Park, N.J., Dr. S. F. Hoerner, 402 p.

Hoffman, J. H., Hodges, R. R., Jr., McElroy, M. B., Donahue, T. M., and Kolpin, M., 1979, Composition and structure of the Venus atmosphere: Results from Pioneer Venus: Science, v. 205, p. 49–52.

Holland, H. D., 1962, Model for the evolution of the Earth's atmosphere, in Petrologic studies: A volume in honor of A. F. Buddington: Boulder, Geol. Soc. America, p. 447–477.

Horton, R. E., 1945, Erosional development of streams and their drainage basins: Hydrophysical approach to quantitative morphology: Geol. Soc. America Bull., v. 56, p. 275–370.

Howard, A. D., 1978, Origin of the stepped topography of the Martian poles: Icarus, v. 34, p. 581–599.

Howard, K. A., and Muehlberger, W. R., 1973, Lunar thrust faults in the Taurus-Littrow region: NASA Spec. Publ. 330, p. 31.22–31.25.

Hoyt, W. G., 1976, Lowell and Mars: Tucson, Univ. Arizona Press, 376 p.

Hsü, K. J., 1975, Catastrophic debris streams (sturzstroms) generated by rockfalls: Geol. Soc. America Bull., v. 86, p. 129–140.

Huguenin, R. L., and Clifford, S. M., 1980, Additional remote sensing evidence for oases on Mars: NASA Tech. Memo. 81776, p. 153–155.

Hunten, D. M., and McElroy, M. B., 1970, Production and escape of hydrogen on Mars: Jour. Geophys. Res., v. 75, p. 5989–6001.

Hutton, J., 1788, Theory of the earth, or an investigation of the laws observable in the composition, dissolution, and restoration of land upon the globe: Royal Soc. Edinburgh Trans., v. 1, p. 209–304.

————, 1795, Theory of the earth with proofs and illustrations: Edinburgh, William Creech; London, Cadell, Jr., and Davies, v. 1, 620 p.; v. 2, 567 p.

Imbrie, J., and Imbrie, K. P., 1979, Ice ages: Solving the mystery: Short Hills, N.J., Enslow Publishers, 224 p.

Ingersoll, A. P., 1970, Mars: Occurrence of liquid water: Icarus, v. 168, p. 972–973.

Ingersoll, A. P., Dobrovolskis, A. R., and Jakosky, B. M., 1979, Planetary atmospheres: Rev. Geophys. and Space Phys., v. 17, no. 7, p. 1722–1735.

Jackson, R. G., 1975, Hierarchical attributes and a unifying model of bed forms composed of cohesionless material and produced by shearing flow: Geol. Soc. America Bull., v. 86, p. 1523–1533.

————, 1976, Sedimentological and fluid-dynamic implications of turbulent bursting phenomenon in geophysical flows: Jour. Fluid Mechanics, v. 77, p. 531–560.

Jamison, D., 1965, Some speculation on the Martian Canals: Astron. Soc. of the Pacific, Publ., v. 77, p. 394–395.

Jones, K. L., 1974, Evidence for an episode of crater obliteration intermediate in Martian history: Jour. Geophys. Res., v. 79, p. 3917–3931.

Jones, K. L., Arvidson, R. E., Guiness, E. A., Bragg, S. L., Wall, S. D., Carlston, C. E., and Pidek, D. G., 1979, One Mars year, Viking lander observations: Science, v. 204, p. 799–806.

Kälin, M., 1977, Hydraulic piping: Theoretical and experimental findings: Canadian Geotechnical Jour., v. 14, p. 107–124.

Karcz, I., 1967, Harrow marks, current aligned sedimentary structures: Jour. Geology, v. 75, p. 113–121.

————, 1968, Fluvial obstacle marks from the wadis of the Negev (southern Israel): Jour. Sed. Petrology, v. 38, p. 1000–1012.

Keller, E. A., and Melhorn, W. N., 1973, Bedforms and fluvial processes in alluvial stream channels, in Morisawa, M., ed., Fluvial geomorphology: Binghamton, N.Y., Publ. in Geomorphology, State Univ. N.Y., p. 253–283.

Kenn, M. J., 1966, Cavitating eddies and their incipient damage to concrete: Civil Engineering and Public Works Review (London), v. 61, p. 404–405.

Kenn, M. J., and Minton, P., 1968, Cavitation induced by vorticity at a smooth flat wall: Nature, v. 217, p. 633–634.

Kieffer, H. H., Chase, S. C., Martin, T. Z., Miner, E. D., and Palluconi, F. D., 1976, Martian north polar summer temperature: Dirty water ice: Science, v. 194, p. 1341–1343.

Klein, H. P., 1979, The Viking mission and the search for life on Mars: Rev. Geophys. and Space Phys., v. 17, no. 7, p. 1655–1662.

Knapp, R. T., Daily, J. W., and Hammitt, F. G., 1970, Cavitation: New York, McGraw-Hill, 578 p.

Knauth, L. P., and Epstein, S., 1976, Hydrogen and oxygen isotope ratios in nodular and bedded cherts: Geochim. Cosmochim. Acta, v. 40, p. 1095–1108.

Kojan, E., and Hutchinson, J. N., 1978, Mayunmarca rockslide and debris flow, Peru, in Voight, B., ed., Rockslides and avalanches, 1, Natural phenomena: New York, Elsevier, p. 315–361.

Komar, P. D., 1979, Comparisons of the hydraulics of water flows in Martian outflow channels with flows of similar scale on Earth: Icarus, v. 37, p. 156–181.

————, 1980, Modes of sediment transport in channelized water flows with ramifications to the erosion of the Martian outflow channels: Icarus, v. 43, p. 317–329.

Kuiper, G. P., 1956, Note on Dr. McLaughlin's paper: Astron. Soc. of the Pacific, Publ., v. 68, p. 219.

Kuiper, G. P., Strom, R. G., and LePoole, R. S., 1966, Interpretation of Ranger records: Calif. Inst. Tech. Jet Propulsion Lab. Tech. Rept. 32-800.

Lachenbruch, A. H., 1962, Mechanics of thermal contraction cracks and ice wedge polygons in permafrost: Geol. Soc. America Spec. Paper 70, 69 p.

Laity, J. E., 1980, Groundwater sapping on the Colorado Plateau: NASA Tech. Memo. 82385, p. 358–360.

Laity, J. E., and Pieri, D. C., 1980, Sapping processes in tributary valley systems: NASA Tech. Memo. 81776, p. 271–273.

Lambert, R. St. J., and Chamberlain, V. E., 1978, CO_2 permafrost and Martian topography: Icarus, v. 34, p. 568–580.

Latham, G. V., Ewing, M., Press, F., Sutton, G., Dorman, J., Nakamura, Y., Toksoz, N., Duennebier, F., and Lammelein, D., 1971, Passive seismic experiment, Apollo 14 preliminary science report: NASA Spec. Publ. 272, p. 133–161.

Laursen, E. M., 1960, Scour at bridge crossings: Jour. Hydraulics Div., Proc. Am. Soc. Civil Engineers, v. 86, no. HY2, p. 39–54.

Lederberg, J., and Sagan, C., 1962, Microenvironments for life on Mars: Proc. Nat. Acad. Sci., v. 48, p. 1473–1475.

Leighton, R. B., and Murray, B. C., 1966, Behavior of carbon dioxide and other volatiles on Mars: Science, v. 153, p. 136–144.

Leopold, L. B., Wolman, M. G., and Miller, J. P., 1964, Fluvial processes in geomorphology: San Francisco, W. H. Freeman, 522 p.

Leovy, C. B., 1979, Global meteorology of Mars: paper presented at the 2nd International Colloquium on Mars, Calif. Inst. Tech., Pasadena, 16 January 1979.

Lingenfelter, R. E., Peale, S. J., Schubert, B., 1968, Lunar rivers: Science, v. 161, p. 266–269.

Lucchitta, B. K., 1978a, Morphology of chasma walls, Mars: U.S. Geol. Survey Jour. Res., v. 6, p. 651–662.

———, 1978b, Survey of cold-climate features on Mars, in Proc. Second Colloquium on Planetary Water and Polar Processes: Washington, D.C., NASA Planetary Geology Program, p. 131–134.

———, 1978c, Landslides in the Valles Marineris, Mars: NASA Tech. Memo. 79729, p. 288–290.

———, 1978d, A large landslide on Mars: Geol. Soc. America Bull., v. 89, p. 1601–1609.

———, 1979, Debris flows on Olympus Mons: NASA Tech. Memo. 80339, p. 34–35.

Lucchitta, B. K., and Anderson, D. M., 1980, Martian outflow channels sculptured by glaciers: NASA Tech. Memo. 81776, p. 271–273.

Lucchitta, B. K., Anderson, D. M., and Shoji, H., 1981, Did ice streams carve Martian outflow channels? Nature, v. 290, no. 5809, p. 759–763.

Lucchitta, B. K., and Klockenbrink, J. L., 1979, Ridges and scarps in the equatorial belt of Mars: NASA Tech. Memo. 80339, p. 31–33.

Lyell, Charles, 1830–1833, Principles of geology, 1st ed.: London, John Murray, v. 1, 1830, 511 p.; Appendices and Index, 109 p.

MacDonald, G. A., and Abbott, A. T., 1970, Volcanoes in the sea: Honolulu, Univ. Press of Hawaii, 441 p.

Malde, H. E., 1968, The catastrophic late Pleistocene Bonneville Flood in the Snake River Plain, Idaho: U.S. Geol. Survey Prof. Paper 596, 52 p.

Malin, M. C., 1974, Salt weathering on Mars: Jour. Geophys. Res., v. 79, p. 3888–3894.

———, 1976, Age of Martian channels: Jour. Geophys. Res., v. 81, p. 4825–4845.

———, 1977, Comparison of volcanic features of Elysium (Mars) and Tibesti (Earth): Geol. Soc. America Bull., v. 88, p. 908–919.

Masursky, H., 1973, An overview of geological results from Mariner 9: Jour. Geophys. Res., v. 78, p. 4009–4030.

Masursky, H., Boyce, J. M., Dial, A. L., Schaber, G. G., and Strobell, M. E., 1977, Classification and time of formation of Martian channels based on Viking data: Jour. Geophys. Res., v. 82, p. 4016–4038.

Masursky, H., Dial, A. L., Jr., and Strobell, M. E., 1980, Martian channels—a late Viking view: NASA Tech. Memo. 82385, p. 184–187.

Matthes, G. H., 1947, Macroturbulence in natural stream flow: Trans. Am. Geophys. Union, v. 28, p. 255–262.

Maxwell, T. A., Otto, E. P., Picard, M. D., and Wilson, R. C., 1973, Meteorite impact: A suggestion for the origin of some stream channels on Mars: Geology, v. 1, p. 9–16.

McCauley, J. F., 1973, Mariner 9 evidence for wind erosion in the equatorial and mid-latitude regions of Mars: Jour. Geophys. Res., v. 78, p. 4132–4137.

McCauley, J. F., Breed, C. S., El-Baz, F., Whitney, M. I., Grolier, M. J., and Ward, A. W., 1979, Pitted and fluted rocks in the western desert of Egypt: Viking comparisons: Jour. Geophys. Res., v. 84, p. 8222–8232.

McCauley, J. F., Carr, M. H., Cutts, J. A., Hartmann, W. K., Masursky, H., Milton, D. J., Sharp, R. P., and Wilhelms, D. E., 1972, Preliminary Mariner 9 report on the geology of Mars: Icarus, v. 17, p. 289–327.

McCauley, J. F., Grolier, M. J., and Breed, C. S., 1977, Yardangs of Peru and other desert regions: U.S. Geol. Survey Interagency Rep., Astrogeology, 81, 177 p.

McElroy, M. B., 1972, Mars: An evolving atmosphere: Science, v. 175, p. 443–445.

McElroy, M. B., Yung, Y. L., and Nier, A. O., 1976, Isotopic composition of nitrogen: Implications for the past history of Mars' atmosphere: Science, v. 194, p. 70–72.

McGill, G. E., 1977, Craters as "fossils": The remote dating of planetary surface materials: Geol. Soc. America Bull., v. 88, p. 1102–1110.

Meier, M. F., 1964, Ice and glaciers: Section 16, in Chow, V. T., ed., Handbook of applied hydrology: New York, McGraw-Hill, p. 16.1–16.32.

Michel, B., 1966, Morphology of frazil ice, in Oura, H., ed., Physics of snow and ice: Sapporo, Japan, Inst. of Low Temperature Science, Hokkaido Univ.

Milankovitch, M., 1941, Canon of insolation and the ice-age problem, translated from German, 1969, U.S. Dept. Commerce and Nat. Sci. Foundation, Publ. TT67-51410/1–2, 484 p.

Miller, S. L., and Smythe, W. D., 1970, Carbon dioxide clathrate in the Martian ice cap: Science, v. 170, p. 531–533.

Milton, D. J., 1973, Water and processes of degradation in the Martian landscape: Jour. Geophys. Res., v. 78, p. 4037–4047.

———, 1974, Carbon dioxide hydrate and floods on Mars: Science, v. 183, p. 654–656.

———, 1975, Geologic map of the Lunae Palus Quadrangle of Mars: U.S. Geol. Survey Misc. Inv. Map I-894

Milton, D. J., Barlow, B. C., Brett, R., Brown, A. R., Glikson, A. Y., Manwaring, E. A., Moss, F. J., Sedmik, E. C. E., Van Son, J., and Young, G. A., 1972, Gosses Bluff impact structure, Australia: Science, v. 175, p. 1199–1207.

Moore, P., 1980, Mars—then and now: Mercury, March–April issue, p. 23–30.

Moore, W. L., and Masch, F. D., 1963, Influence of secondary flow on local scour at obstructions in a channel: Federal Inter-Agency Sedimentation Conf. Proc., U.S. Dept. Agriculture Misc. Publ. 970, p. 314–320.

Morris, E. C., and Underwood, J. R., 1978, Polygonal fractures of the Martian plains: NASA Tech. Memo. 79729, p. 97–99.

Mouginis-Mark, P. J., 1979, Martian fluidized crater morphology: Variations with crater size, latitude, altitude, and target material: Jour. Geophys. Res., v. 84, p. 8011–8022.

Muller, S. W., 1944, Common features in geology and related science in the U.S.S.R. and the U.S.A., in Science and Soviet Russia: Lancaster, Pa., J. Cattell Press, p. 11–17.

Mullineaux, D. R., Hyde, H. J., and Rubin, M., 1975, Widespread late glacial and postglacial tephra deposits from Mount St. Helens volcano, Washington: U.S. Geol. Survey Jour. Res., v. 3, no. 3, p. 329–335.

Mullineaux, D. R., Wilcox, R. E., Ebaugh, W. F., Fryxell, R., and Rubin, M., 1977, Age of the last major scabland flood of eastern Washington, as inferred from associated ash beds of Mount St. Helens set S: Geol. Soc. America Abstracts with Programs, v. 9, no. 7, p. 1105.

Murray, B. C., 1973, Mars from Mariner 9: Sci. American, v. 228, no. 1, p. 49–69.

Murray, B. C., and Malin, M. C., 1973, Polar wandering on Mars? Science, v. 179, p. 997–1000.

Murray, B. C., Malin, M. C., and Greeley, R., 1981, Earthlike planets: Surfaces of Mercury, Venus, Earth, Moon, Mars: San Francisco, W. H. Freeman and Company, 387 p.

Murray, B. C., Soderblom, L. A., Cutts, J. A., Sharp, R. P., Milton, D. J., and Leighton, R. B., 1972, A geological framework for the south polar region of Mars: Icarus, v. 17, p. 328–344.

Mutch, T. A., 1972, Geology of the Moon: A stratigraphic view: Princeton, N.J., Princeton Univ. Press 391 p.

———, 1979, Planetary surfaces: Rev. Geophys. and Space Phys., v. 17, no. 7, p. 1694–1722.

Mutch, T. A., Arvidson, R. E., Head, J. W., Jones, K. L., and Saunders, R. S., 1976, The geology of Mars: Princeton, N.J., Princeton Univ. Press, 400 p.

NASA, 1976, Viking 1, early results: NASA Spec. Publ. 408, 66 p.

Neukum, G., and Wise, D. U., 1976, Mars: A standard crater curve and possible new timescale: Science, v. 194, p. 1381–1387.

Nummedal, D., 1978, The role of liquefaction in channel development on Mars: NASA Tech. Memo. 79729, p. 257–259.

———, 1980, Debris flows and debris avalanches in the large Martian channels: NASA Tech. Memo. 81776, p. 289–291.

Nummedal, D., Gonsiewski, J. J., and Boothroyd, J. C., 1976, Geological significance of large channels on Mars: Geol. Romana, v. 15, p. 407–418.

Offen, G. R., and Kline, S. J., 1975, A proposed model of the bursting process in turbulent boundary layers: Jour. Fluid Mechanics, v. 70, p. 209–228.

Öpik, E. J., 1958, Solar variability and paleoclimatic changes: Irish Astron. Jour., v. 5, p. 97–109.

———, 1965, Climatic change in cosmic perspective: Icarus, v. 4, p. 289–307.

———, 1966, The Martian surface: Science, v. 153, p. 255–265.

Owen, T., and Biemann, K., 1976, Composition of the atmosphere at the surface of Mars: detection of argon-36 and preliminary analysis: Science, v. 193, p. 801–803.

Pardee, J. T., 1942, Unusual currents in glacial Lake Missoula, Montana: Geol. Soc. America Bull., v. 53, p. 1569–1600.

Parker, G. G., 1963, Piping, a geomorphic agent in landform development of the drylands: International Assoc. of Scientific Hydrology Publ. 65, p. 103–113.

Patton, P. C., and Baker, V. R., 1976, Morphometry and floods in small drainage basins subject to diverse hydrogeomorphic controls: Water Resources Res., v. 12, p. 941–952.

———, 1978a, New evidence for pre-Wisconsin flooding in the Channeled Scabland of eastern Washington: Geology, v. 6, p. 567–571.

———, 1978b, Origin of the Cheney-Palouse Scabland Tract, in Baker, V. R., and Nummedal, D., eds., The Channeled Scabland: Washington, D.C., NASA Office of Space Science, Planetary Geology Program, p. 117–130.

———, 1979, The Cheney-Palouse Scabland Tract: NASA Tech. Memo. 80339, p. 341–343.

Peale, S. J., Schubert, G., and Lingenfelter, R. E., 1975, Origin of Martian channels: Clathrates and water: Science, v. 187, p. 273–274.

Pechmann, J. C., 1980, The origin of polygonal troughs on the northern plains of Mars: Icarus, v. 42, p. 185–210.

Péwé, T., 1975, Quaternary geology of Alaska: U.S. Geol. Survey Prof. Paper 835, 145 p.

Pieri, D. C., 1976, Martian channels: Distribution of small channels in the Martian surface: Icarus, v. 27, p. 25–50.

———, 1978, Small channels on Mars from Viking orbiter: NASA Tech. Memo. 79729, p. 267.

———, 1979, Global distribution of Martian valley systems: NASA Tech. Memo. 80339, p. 353–356.

———, 1980a, Geomorphology of Martian valleys: NASA Tech. Memo. 81979, p. 1–160.

———, 1980b, Martian valleys: Morphology, distribution, age, and origin: Science, v. 210, p. 895–897.

Pieri, D. C., Malin, M. C., and Laity, J. E., 1980, Sapping: Network structure in terrestrial and Martian valleys: NASA Tech. Memo. 87116, p. 292–294.

Pieri, D. C., and Sagan, C., 1978, Junction angles of Martian channels: NASA Tech. Memo. 79729, p. 268.

———, 1979, Origin of Martian valleys: NASA Tech. Memo. 80339, p. 349–352.

———, 1980, Low energy cavitation in Martian floods: NASA Tech. Memo. 82385, p. 355–357.

Pollack, J. B., 1979, Climatic change on the terrestrial planets: Icarus, v. 37, p. 479–553.

Pollack, J. B., and Black, D. C., 1979, Implications of the gas compositional measurements of Pioneer Venus for the origin of planetary atmospheres: Science, v. 205, p. 56–59.

Pollack, J. B., and Yung, Y. L., 1980, Origin and evolution of planetary atmospheres: Ann. Rev. Earth and Planet. Sci., v. 8, p. 425–487.

Prandtl, L., 1949, Essentials of fluid dynamics: London, Blackie and Son Ltd., 452 p.

Prandtl, L., and Tietjens, O. G., 1934, Applied hydro- and aeromechanics: New York, Dover, 311 p.

Prior, D. B., and Coleman, J. M., 1978, Submarine landslides on the Mississippi River delta-front slope: Geoscience and Man, v. 19, p. 41–53.

Prior, D. B., Coleman, J. M., and Garrison, L. E., 1979, Digitally acquired undistorted side-scan sonar images of submarine landslides, Mississippi River delta: Geology, v. 7, p. 423–425.

Rayleigh, Lord, 1917, On the pressure developed in a liquid during the collapse of a spherical cavity: Phil. Mag., v. 34, p. 94–98.

Reimers, C. E., and Komar, P. D., 1979, Evidence for explosive volcanic density currents on certain Martian volcanoes: Icarus, v. 39, p. 88–110.

Richardson, L. F., 1920, The supply of

energy from and to atmospheric eddies: Proc. Royal Soc., Series A, v. 97, no. 686, p. 354–373.

Richardson, P. D., 1968, The generation of scour marks near obstacles: Jour. Sed. Petrology, v. 38, p. 965–970.

Richmond, G. M., Fryxell, R., Neff, G. E., and Weiss, P. L., 1965, The Cordilleran ice sheet of the northern Rocky Mountains and related Quaternary history of the Columbia Plateau, in Wright, H. E., and Frey, D. G., eds., The Quaternary of the United States: Princeton, N.J., Princeton Univ. Press, p. 217–230.

Ringe, D., 1970, Sub-loess basalt topography in the Palouse Hills, southeastern Washington: Geol. Soc. America Bull., v. 81, p. 3049–3060.

Ringwood, A. E., 1978, Water in the solar system, in McIntyre, A. K., ed., Water, planets, plants and people. Canberra, Australian Acad. Sci., p. 18–34.

———, 1979, Composition and origin of the earth: Abstracts of the Interdisciplinary Symposia, International Union of Geodesy and Geophysics XVII General Assembly, Canberra, Dec. 1979, p. 16.

Roberts, W. J., and Cutts, J. A., 1978, Shapes of streamlined islands in Martian outflow channels, in Proc. Second Colloquium on Planetary Water and Polar Processes: Washington, D.C., NASA Planetary Geology Program, p. 146–150.

Rubey, W. W., 1955, Development of the hydrosphere and atmosphere with special reference to the probable composition of the early atmosphere: Geol. Soc. America Spec. Paper 62, p. 631–650.

Sagan, C., 1971, The long winter model of Martian biology: A speculation: Icarus, v. 15, p. 511–514.

———, 1973a, The cosmic connection: New York, Doubleday, 274 p.

———, 1973b, Sandstorms and eolian erosion on Mars: Jour. Geophys. Res., v. 78, p. 4117–4122.

———, 1977, Reducing greenhouses and the temperature history of Earth and Mars: Nature, v. 269, p. 224–226.

Sagan, C., and Mullen, G., 1972, Earth and Mars: Evolution of atmospheres and surface temperature: Science, v. 177, p. 52–55.

Sagan, C., and Pollack, J. B., 1966, On the nature of the canals of Mars: Nature, v. 212, p. 117–121.

Sagan, C., Toon, O. B., and Gierasch, P. J., 1973, Climatic change on Mars: Science, v. 181, p. 1045–1049.

Sagan, C., and Young, A. T., 1973, Solar neutrinos, Martian rivers, and Praesepe: Nature, v. 243, p. 459–468.

Schaber, G. G., Horstman, K. C., and Dial, A. L., Jr., 1978, Lava flow materials in the Tharsis region of Mars: Proc. Lunar and Planet. Sci. Conf. 9th, p. 3433–3458.

Scheidegger, A. E., 1973, On the prediction of the reach and velocity of catastrophic landslides: Rock Mechanics, v. 9, p. 231–236.

Schmincke, H. U., 1967, Stratigraphy and petrography of four upper Yakima Basalt flows in south-central Washington: Geol. Soc. America Bull., v. 78, p. 1385–1422.

Schonfeld, E., 1976, On the origin of the Martian channels: EOS, v. 57, p. 1948.

———, 1977, Martian volcanism, in Lunar Science VIII: Houston, Lunar Science Inst., p. 843–845.

Schubert, G., Lingenfelter, R. E., and Peale, S. J., 1970, The morphology, distribution, and origin of lunar sinuous rilles: Rev. Geophys. and Space Phys., v. 8, p. 199–224.

Schultz, P. H., 1978, Martian intrusions: Possible sites and implications: Geophys. Res. Letters, v. 5, no. 6, p. 457–460.

Schultz, P. H., and Glicken, H., 1979, Impact crater and basin control of igneous processes on Mars: Jour. Geophys. Res., v. 84, p. 8033–8047.

Schumm, S. A., 1970, Experimental studies on the formation of lunar surface features by fluidization: Geol. Soc. America Bull., v. 81, p. 2539–2552.

———, 1974, Structural origin of large Martian channels: Icarus, v. 22, p. 371–384.

———, 1977, The fluvial system: New York, John Wiley, 338 p.

Schumm, S. A., and Shepherd, R. G., 1973, Valley floor morphology: Evidence of subglacial erosion: Area, v. 5, p. 5–9.

Scott, D. H., 1978, Mars, highlands-lowlands: Viking contributions to Mariner relative age studies: Icarus, v. 34, p. 479–485.

Scott, D. H., and Carr, M. H., 1978, Geologic map of Mars: U.S. Geol. Survey Misc. Geol. Inv. Map I-1083.

Scott, D. H., Schaber, G. G., Horstman, K. C., and Dial, A. L., Jr., 1979, Lava flows of Tharsis Montes: NASA Tech. Memo. 80339, p. 237–238.

Selby, M. J., and Wilson, A. T., 1971, The origin of the Labyrinth, Wright Valley, Antarctica: Geol. Soc. America Bull., v. 82, p. 471–476.

Sharp, R. P., 1968, Surface processes modifying Martian craters: Icarus, v. 8, p. 472–480.

———, 1973a, Mars: Troughed terrain: Jour. Geophys. Res., v. 78, p. 4063–4072.

———, 1973b, Mars: Fretted and chaotic terrains: Jour. Geophys. Res., v. 78, p. 4073–4083.

———, 1973c, Mars: South polar pits and etched terrain: Jour. Geophys. Res., v. 78, p. 4222–4230.

———, 1973d, Mass movements on Mars, in Moran, D. E., Slosson, J. E., Stone, R. O., and Yelverton, C. A., eds., Geology, seismicity, and environmental impact: Assoc. Eng. Geologists Spec. Publ., p. 115–122.

———, 1974, Ice on Mars: Jour. Glaciology, v. 13, p. 173–185.

———, 1978, Martian sedimentation, in Fairbridge, R. W., and Bourgeois, J., eds., The encyclopedia of sedimentology: Stroudsburg, Pa., Dowden, Hutchinson and Ross, p. 475–479.

———, 1980, Geomorphological processes on terrestrial planetary surfaces: Ann. Rev. Earth and Planet. Sci., v. 8, p. 231–261.

Sharp, R. P., and Malin, M. C., 1975, Channels on Mars: Geol. Soc. America Bull., v. 86, p. 593–609.

Sharp, R. P., Soderblom, L. A., Murray, B. C., and Cutts, J. A., 1971, The surface of Mars 2: Uncratered terrains: Jour. Geophys. Res., v. 76, p. 331–342.

Shen, H. W., 1971, Scour near piers, in Shen, H. W., ed., River mechanics: Ft. Collins, Colo., H. W. Shen Publisher, p. 23.1–23.25.

Shepherd, R. G., 1972, A model study of river incision: M.S. thesis, Colorado State Univ., Fort Collins, 135 p.

Shepherd, R. G., and Schumm, S. A., 1974, An experimental study of river incision: Geol. Soc. America Bull., v. 85, p. 257–268.

Shoemaker, E. M., 1962, Interpretation of lunar craters, in Kopal, Z., ed., Physics and astronomy of the Moon: New York, Academic Press, p. 283–360.

Shreve, R. L., 1966a, Statistical law of stream numbers: Jour. Geology, v. 74, p. 17–37.

———, 1966b, Sherman landslide, Alaska: Science, v. 154, p. 1639–1643.

———, 1968, The Blackhawk landslide: Geol. Soc. America Spec. Paper 108, 47 p.

———, 1975, The probabilistic topologic approach to drainage-basin geomorphology: Geology, v. 3, p. 527–529.

Simons, D. B., and Richardson, E. V., 1966, Resistance to flow in alluvial channels: U.S. Geol. Survey Prof. Paper 422-J, 61 p.

Smith, D. G., 1979, River ice processes: Thresholds and geomorphic effects in northern and mountain rivers, in Coates, D., and Vittek, J. D., Geomorphic thresholds: London, George Allen and Unwin, p. 323–343.

Smoluchowski, R., 1968, Mars: Retention of ice: Science, v. 159, p. 1348–1350.

Smyth, J. R., Huguenin, R., and McGetchin, T. R., 1978, Composition of Martian primary lavas—convergence of model results, in Lunar and Planetary Science IX: Houston, Lunar and Planetary Inst., p. 1077–1079.

Snyder, C. W., 1979, The planet Mars as seen at the end of the Viking Mission: Jour. Geophys. Res., v. 84, p. 8487–8519.

Soderblom, L. A., Kreidler, T. J., and Masursky, H., 1973, Latitudinal distribution of a debris mantle on the Martian surface: Jour. Geophys. Res., v. 78, p. 4117–4122.

Soderblom, L. A., and Wenner, D. B., 1978, Possible fossil H_2O liquid-ice interfaces in the Martian Crust: Icarus, v. 34, p. 622–637.

Soderblom, L. A., West, R. A., Herman, B. M., Kreidler, T. J., and Condit, C. D., 1974, Martian planetwide crater distributions: Implications for geologic history and surface processes: Icarus, v. 22, p. 239–263.

Spry, A., 1962, The origin of columnar jointing, particularly in basalt flows: Geol. Soc. Australia Jour., v. 8, p. 191–216.

Squyres, S. W., 1978, Martian fretted terrain: Flow of erosional debris: Icarus, v. 34, p. 600–613.

———, 1979, The distribution of lobate debris aprons and similar flows on Mars: Jour. Geophys. Res., v. 84, p. 8087–8096.

Strahler, A. N., 1957, Quantitative analysis of watershed geomorphology: Trans. Am. Geophys. Union, v. 38, p. 913–920.

Swanson, D. A., 1967, Yakima Basalt of the Tieton River area, south-central Washington: Geol. Soc. America Bull., v. 78, p. 1077–1110.

———, 1973, Pahoehoe flows from the 1969–1971 Mauna Ulu eruption, Kilauea volcano, Hawaii: Geol. Soc. America Bull., v. 84, p. 615–626.

Swanson, D. A., and Wright, T. L., 1978, Bedrock geology of the northern Columbia Plateau and adjacent areas, in Baker, V. R., and Nummedal, D., eds., The Channeled Scabland: Washington, D.C., NASA Office of Space Science, Planetary Geology Program, p. 37–57.

Taylor, S. R., 1975, Lunar science: A post-Apollo view: New York, Pergamon, 372 p.

Theilig, E., and Greeley, R., 1978, Episodic channeling and layered terrain on Mars: Implications for ground ice, in Proc. Second Colloquium on Planetary Water and Polar Processes: Washington, D.C., NASA Planetary Geology Program, p. 151–157.

———, 1979, Plains and channels in the Lunae Planum–Chryse Planitia Region of Mars: Jour. Geophys. Res., v. 84, p. 7994–8010.

Thompson, D. E., 1979, Origin of longitudinal grooving in Tiu Vallis, Mars: Isolation of responsible fluid-types: Geophys. Res. Letters, v. 6, p. 735–738.

Thorarinsson, S., 1957, The jokulhlaup from the Katla area in 1955 compared with other jokulhlaups in Iceland: Reykjavik, Iceland, Mus. Natur. Hist., Misc. Paper 18, p. 21–25.

Thornbury, W. P., 1969, Principles of geomorphology: New York, John Wiley and Sons, 594 p.

Tomkeieff, S. I., 1940, The basalt lavas of the Giant's Causeway district of northern Ireland: Bull. Volcanologique, v. 6, p. 90–143.

Toulmin, P., Baird, A. K., Clark, B. C., Keil, K., Rose, H. J., Christian, R. P., Evans, P. H., and Kelliher, W. C., 1977, Geochemical and mineralogical interpretation of the Viking inorganic chemical results: Jour. Geophys. Res., v. 82, p. 4625–4634.

Trevena, A. S., and Picard, M. D., 1978, Morphometric comparison of braided Martian channels and some braided terrestrial features: Icarus, v. 35, p. 385–394.

Trimble, D. E., 1950, Joint controlled channeling in the Columbia River basalt: Northwest Sci., v. 24, p. 84–88.

———, 1963, Geology of Portland, Oregon, and adjacent areas: U.S. Geol. Survey Bull. 1119, 119 p.

Tsoar, H., Greeley, R., and Peterfreund, A. R., 1979, Mars: The north polar sand sea and related wind patterns: Jour. Geophys. Res., v. 84, p. 8167–8180.

U.S. Geological Survey, 1976, Topographic map of Mars: U.S. Geol. Survey Misc. Inv. Map I-961.

Veverka, J., and Sagan, C., 1974, McLaughlin and Mars: Am. Scientist, v. 62, p. 44–53.

Voight, B., and Pariseau, W. G., 1978, Rockslides and avalanches: An introduction, in Voight, B., ed., Rockslides and avalanches, 1, Natural phenomena: New York, Elsevier, p. 1–67.

Wade, F. A., and deWys, J. N., 1968, Permafrost features on the Martian surface: Icarus, v. 9, p. 175–185.

Waitt, R. B., Jr., 1980, About forty last-glacial Lake Missoula jokulhlaups through southern Washington: Jour. Geology, v. 88, p. 653–679.

Wallace, D., and Sagan, C., 1979, Evaporation of ice in planetary atmospheres: Ice-covered rivers on Mars: Icarus, v. 39, p. 385–400.

Walther, J., 1891, Die Denudation in der Wüstend ihre geologische Bedeutung: Abh. math-phys. Kl.d. Kön. Sächsischen Ges. d. Wiss., v. 16, p. 345–570.

Ward, A. W., 1979, Yardangs on Mars: Evidence of recent wind erosion: Jour. Geophys. Res., v. 84, p. 8147–8166.

Ward, A. W., McCauley, J. F., and Grolier, M. J., 1977, The yardangs at Rogers Playa, California: Geol. Soc. America Abstracts with Programs, v. 9, p. 1216.

Ward, W. R., 1973, Large-scale variations in the obliquity of Mars: Science, v. 181, p. 260–262.

———, 1979, Present obliquity oscillations of Mars: Fourth-order accuracy in orbital e and I: Jour. Geophys. Res., v. 84, p. 237–241.

Ward, W. R., Burns, J. A., and Toon, O. B., 1979, Past obliquity oscillations of Mars: The role of the Tharsis Uplift: Jour. Geophys. Res., v. 84, p. 243–259.

Washburn, A. L., 1973, Periglacial processes and environments: New York, St. Martin's, 320 p.

———, 1980, Geocryology: New York, John Wiley and Sons, 406 p.

Waters, A. C., 1960, Determining direction of flow in basalts: Am. Jour. Sci., v. 258-A, p. 583–611.

Weihaupt, J., 1974, Possible origin and probable discharges of meandering channels on the planet Mars: Jour. Geophys. Res., v. 79, p. 2073–2076.

Wellman, H. W., and Wilson, A. T., 1965, Salt weathering, a neglected geological erosive agent in coastal and arid environments: Nature, v. 205, p. 1097.

Wells, R. A., 1979, Geophysics of Mars: Amsterdam, Elsevier, 678 p.

Wentworth, C. K., 1928, Principles of stream erosion in Hawaii: Jour. Geology, v. 36, p. 385–410.

———, 1943, Soil avalanches on Oahu, Hawaii: Geol. Soc. America Bull., v. 54, p. 53–64.

Williams, P. F., and Rust, B. R., 1969, The sedimentology of a braided river: Jour. Sed. Petrology, v. 39, p. 649–679.

Winkler, E. M., 1975, Stone: Properties, durability in Man's environment: New York, Springer-Verlag, 230 p.

Wise, D. U., Golombek, M. P., and McGill, G. E., 1979, Tharsis province of Mars: Geologic sequence, geometry, and a deformation mechanism: Icarus, v. 38, p. 456–472.

Wu, S. S. C., 1979, Mars photogrammetry: NASA Tech. Memo. 80229, p. 432–435.

Yung, Y. L., and Pinto, J. P., 1978, Primitive atmosphere and implications for the formation of channels on Mars: Nature, v. 273, p. 730–732.

Zisk, S. H., and Mouginis-Mark, P. J., 1980, Anomalous region on Mars: Implications for near-surface liquid water: Nature, v. 288, no. 5787, p. 735–738.

Index

Aa lava flows, 53
Absolute ages, 17
Accretion, 175–179
A-frame images, 7
Age
    of channels, 38
    establishment of, 17
    of geomorphic surfaces, 21
    of landslides, 100
Alas, 91–93
Alba Patera, 21, 25, 52
Albor Tholus, 24
Al Qahira Vallis, 36, 72
Amphitheater-headed valleys, 64, 85
Anastomosis
    in Channeled Scabland, 147, 157
    on Mars, 47
Antediluvian epoch, 39
Antoniadi, E. M., 5, 6
Appollinarsis Patera, 24–25
Aram Chaos, 108
Ares Vallis, 5, 36, 44–45, 108–117
    age of, 38
    flow parameters of, 162
    morphology of, 108–117
Argon, 105, 175, 178, 179
Argyre impact basin, 14, 75
Argyre valleys, 73–75
Aromatum Chaos, 50
Arsia Mons, 22, 24, 88–90, 100
Artesian Lake, 159
Ascraeus Mons, 22, 24
Atmosphere
    in early dense phase, 178–179
    of Earth, 105
    evolution of, 105–107, 175–179
    of Mars, 105
Auqukuh Vallis, 36, 40
Aureole, 17

Back-wearing, 91, 94
Bahram Vallis, 70–72
Bamberg, 18
Barnard, E. E., 4
Bars, 46
    in Channeled Scabland, 140,
        148–152

in Martian outflow channels, 46,
    111, 125, 128–129
Bed forms, 42, 44–46, 127, 140,
    147–160, 166–167
Bed shear, 158, 160
B-frame images, 7
Bonneville flooding, 142–143, 152,
    166
Bonneville, Lake, 141–142
Borealis Chasma, 32
Braiding (of streams), 46–47, 118,
    147–148, 152
Bretz, J Harlen, 14, 140, 142, 160, 184
Broad channels, 34. *See also* Outflow
    channels
Burroughs, Edgar Rice, 4
Butte-and-basin topography, 151,
    156–158

Canali, 4–5, 13
Canals, 4–5, 13
Canyons, 26–28, 32
Capri Chasma, 28–29
Carbonaceous chondrites, 175–177
Cataracts, 152–158, 160, 167
Catastrophic flooding, 47, 111
    in Channeled Scabland, 140–160
    mechanics of, 161–174
    processes of, 161–174
    release mechanisms of, on Mars,
        182–184
Cavitation, 162, 170–174
Ceraunius Tholus, 79, 81
Channeled Scabland, 14, 140–160,
    163–164, 184
Channels
    ages of, 38–39
    classification of, 34–37
    definition of, 34, 41
    discovery of, 7
    distribution of, 37–38
    fretted, 40–41
    lava, 52–55
    location of, 36
    mystery of, 13
    outflow. *See* Outflow channels
    runoff, 41
    See also Valley networks; *names*
        of individual channels
Chaotic terrain, 17, 29–30, 43, 50,
    82–83, 92, 110–112
Cheney-Palouse Scabland tract, 115,
    118, 144–146, 148–151, 170
Chryse Planitia, 9, 10
    age of, 21, 38
    description of, 108
    map of, 109
    photo of, 125, 128, 129
    relation of, to confined aquifers,
        183–184

scabland of, 45
    streamlined uplands of, 44,
        112–118
    thermokarst of, 92–93
Chryse Trough, 108
Chutes (hillslope), 92, 94
Classification of Martian channels,
    34
Clathrate, 180, 183
Clay minerals, 105
Climatic change, 181–183
Clouds, 3
Colorado Plateau, 83–85
Columbia Plateau
    basalts of, 26
    flooding of, 144–145
Columbia River Basalt Group, 25
Composite volcanoes, 24
Cooling-contraction cracks, 90
Confined aquifers, 183
Coprates Chasma, 26, 28
Crab Creek, 144
Crater age determinations, 39
Crater counting, 21, 38–39
Crater curves, 39
Crater densities, 17
Crater number, 21, 38
Crater production rates, 21–22
Craters, 17–22, 33
    degradation of, 18
    erosion of, by floods, 115–117,
        126–127
    exhumed, 138
    flat-floored, 130–131
    lunar, 18
    Martian, 18
    morphology of, 18–21
    rampart, 18–21
Creep, 96
Current ripples, giant, 140, 148, 156,
    158–160
Cydonia region, 91

Debris aprons, 94–98
Debris blankets, 41
Debris fan, 95, 128
Debris flows, 51, 95
Debris lobes, 92
Deep sea channels, 51
Demos, 3
Desertism, 13
Deuteronilus Mensae, 96
Deuteronilus-Protonilus region, 40,
    96
Diluvian epoch, 39
Distribution of channels on Mars,
    37–38
Drag, 167–168
Drainage basin, 56
Drainage density, 56

Dry Falls, 152, 154, 163
Dry Valley, 56
Dunes, 32–33, 48
Duricrust, 11

Earth
    absolute ages for, 17
    volatile evolution and, 176–177
Earthflows, 49–52
Eccentricity of orbit, 181
Echus Chasma, 132–133
Ejecta, 18, 20–21
    action of fluid flows on, 115
    blankets of, 18, 20
    fluidization of, 21
Elysium Fossae, 80–81
Elysium Mons, 24, 79–80
Elysium region
    channels of, 75, 78–81
    volcanoes of, 24–25
Endogenic channels, 34
English schoolboy experiment, 5
Eolian processes, 48–49, 104
Equifinality, 14
Erosional landforms
    in Channeled Scabland, 147–158
    in Martian outflow channels,
        126–127
Escarpments, 29, 127
Etched terrain, 33
Excavated channels, 37
Exogenic channels, 34
Exospheric escape, 180
Expansion bars, 46, 111

Flooding, 47, 54, 161–174. See also
    Catastrophic flooding
Fractured plains, 90–91
Frazil, 117
Frenchman Springs cataract, 153
Fretted channels, 34, 40–41
Fretted terrain, 29–31, 92, 96–97
Frost, 12
Frost caps, 31
Frost shattering, 94
Froude number, 162, 173

Galileo, 17
Gangis Chasma, 100–102
Gas-exchange experiment (GEX),
    11–12
Gelifluction, 96
Geological Society of America, 142
Geomorphic maps, 116–123,
        128–129, 135–136, 150–151
Geomorphology, 14–15, 161
Gilbert, Grove Karl, 14
Glaciation, 88–89
Glaciers, 51–52
Gosses Bluff, 115

Grand Coulee, 154, 157, 163
Greenhouse effect, 178–179
Ground ice, 88, 106
Ground resolution, 7

Hadriaca Patera, 21
Haleakala volcano, 84–86
Hall, Asaph, 3–4
Hanging valleys
    in Moses Coulee, 157
    in Nirgal Vallis, 64, 69
Hawaiian volcanoes, 24–25, 52–54,
    84–87
Headcuts, 44–45
Heavily cratered uplands, 17
Hecates Tholus, 24–25, 78–79,
    82–83
Hellas Planitia, 18, 73
Hellas valleys, 73–75
Herschel, William, 3
High-water marks, 144–145
Hilly and cratered terrain, 126
Horseshoe-vortex system, 166
Housden, C. E., 13
Hrad Vallis, 92
Hudson cataract, 153
Huo Hsing Vallis, 36, 40
Hutton, James, 14
Huygens, Christian, 3
Hydaspis Chaos, 30, 108
Hydraulic calculations, 161–163

Iani Chaos, 110
Ice. See Ground ice
Ice covers (on rivers), 161, 171, 173
Ice drives, 171–172
Ice jams, 171
Iceland, 182
Impact basins, 18
Inclination of orbit, 181
Infrared thermal mapper (IRTM), 9,
    111
Inner channels
    in Channeled Scabland, 153–158
    on Mars, 44–45, 111, 125, 136
    origin of, 165, 167
Insolation changes for Mars, 181
Intermediate channels, 34
International Astronomical Union, 7
Io, 15
Ius Chasma, 26, 73, 100

Jökullhlaup, 145, 182
Junction angle, 56
Juventae Chasma, 118, 120

Kasei Vallis, 36, 40
    age of formation of, 21, 38
    morphology of, 132–139
    origin of, 132

streamlined form morphometry in,
    169–170
and thermal inertia, 111–112
Kaupo Gap, 86
Kepler, Johannes, 3
Kohala volcano, 85
Kolk, 164–166

Labeled release experiment (LR),
    11–12
Ladon Vallis, 36, 57–58
Landslides, 88, 92, 98–104, 112
Lava channels, 52–55. See also Vol-
    canic channels
Lava fans, 24
Lava flows, 53–54
Lava tubes, 52–54
Lemniscate, 115, 168
Lenore Canyon, 163
Lind Coulee, 159
Liquefaction, 49
Loess, 142–144, 148, 151
Longitudinal bars
    in Channeled Scabland, 152
    on Mars, 118, 128–129
Longitudinal grooves
    in Channeled Scabland, 151,
        154–158, 166–167
    on Mars, 44–45, 116, 125, 134
    origin of, 166–167
Longitudinal vortices, 166
Long Lake, 156
Lowell, Percival, 4–5, 13
Luminosity of the sun, 182
Lunae Planum, 40
    age of, 21
    channel erosion in, 118, 126–127
    plateau material in, 126
    thermokarst in, 92–93
Lyell, Charles, 14

Ma'adim Vallis, 36, 41, 72
McLaughlin, Dean B., 4, 7
Macroturbulence, 164–167, 174
Maja Vallis, 10, 36, 118–132
    age of, 21, 38
    flow parameters of, 162
    maps of, 118–119
    streamlined form morphometry in,
        169–170
    temporal relationships in, 131–132
Malaspina Glacier, 100
Mamers Valles, 36–37, 39, 99
Mangala Vallis, 7–8, 36, 46–47
    age of, 38
    braided reach of, 46–47
    flow parameters of, 162
Manning equation, 161
Mare-like ridges, 126–127. See also
    Wrinkle ridges

Margaritifer Sinus region, 60–64
Mariner 4 mission, 6
Mariner 6 mission, 6
Mariner 7 mission, 6
Mariner 9 mission, 7, 22
Marineris, Valles. See Valles
    Marineris
Marlin, 159
Mars
    ages of geomorphic surfaces on, 21
    comparison of, to other planets,
        14–15
    crater density on, 22
    craters of, 18–21
    geologic map of, 15–17
    geologic mapping units for, 17
    megaregolith of, 21
    names for features on, 7
    seasonal wave of darkening on, 4
    variable dark regions of, 4
    weather of, 11
    See also Atmosphere; Catastrophic
        flooding; Channels; Craters;
        Geomorphic maps; names of in-
        dividual features and locations;
        Permafrost; Valley networks
Mars atmospheric water detector
    (MAWD), 9
Maui, 84–86
Maumee Vallis, 36, 120–132
Mauna Iki, 54
Mauna Loa, 22, 53
Maunder, E. W., 5
Maunder Minimum, 180
Mayunmarca landslide, 102
Megaregolith, 21, 26, 106, 175, 183
Melas Chasma, 26, 28, 95
Mercury, 15, 176–178
Mid-channel bars, 125, 128–129
Milankovitch theory, 181
Mississippi Delta, 51
Missoula flooding, 142–147, 161–163
Missoula, Lake, 47, 118, 141–142, 145,
    184
Moberg, 26
Moon
    age curve for, 17
    crater degradation on, 18
    crater production curve for, 21–22
    craters of, 18
    megaregolith of, 106
    regolith of, 21
    sinuous rilles on, 53–54
    volatiles on, 176–178
Moraines, 88–89
Moses Coulee, 156–157
Mount St. Helens, 142–143
Mudflows, 51

Nanedi Vallis, 36, 70
Nilosyrtis region, 41, 96–98
Nirgal Vallis, 36, 41, 64–70
Nitrogen, 105–106, 178
Noctis Labyrinthus, 26–27
Nontronite, 104

Obliquity of orbit, 181–182
Olympus Mons, 21–24, 26, 100,
    103–104
Ophir Chasma, 74
Outflow channels
    ages of, 38–39
    description of, 108–139
    discharge of, 162, 183
    erosional processes of, 161–174
    origin of, 41–52
    source and sink relationships of, 48
Outgassing, 175–178

Pahoehoe lava flows, 53
Paleohydrology, 143–147
Palm Lake, 157
Palouse Falls, 153
Palouse Formation, 142, 146–147
Palouse-Snake divide, 149, 163
Parana Vallis, 36, 62–63
Patera, 25
Patterned ground, 11, 90–92
Pavonis Mons, 22, 24
Pendant bars
    in Channeled Scabland, 151–152
    on Mars, 128
Pend Oreille, Lake, 142
Periapsis, 7
Periglacial landforms, 88–98
Perihelion, 181
Permafrost
    definition of, 88
    development of ice-rich zone of, on
        Mars, 29–30, 105–107, 175–176,
        179–180
    distribution of, 106
    evidence for, 92
    geothermal melting of, 182
    origin of, 105–107
    thickness of, 106, 175
Phobos, 3–4
Piping, 81, 83
Pitted terrain, 33
Plains, 17
Planetesimals, 177
Plateau material, 126
Plate tectonics, 178, 179
Polar caps, 31–33
Polar terrains, 31–33
Polygonal fractures, 90–92, 137
Portland, 142–143
Postdiluvian epoch, 39
Potholes, 156

Precession, 181
Protonilus Mensae region, 30
Protonilus region, 41, 96–97
Pyrolytic release experiments (PR),
    11–12

Quaternary geology, 142–143
Quick clays, 49

Rampart craters, 18–21
Rathdrum Prairie, 145
Ravi Vallis, 36, 49–50, 108, 110
Red Deer River, 172
Regolith, 21
Resurfacing processes, 21
Relative ages, 17
Relative release factors, 178
Residual uplands, 112–118, 126
Reull Vallis, 36, 74–75
Reynolds number, 164–168, 170–171
River ice processes, 170
Rock glaciers, 94–98
Rock Lake, 157
Roller, 165
Runoff channels, 34, 36, 41. See also
    Valley networks

Sacra Fossa, 40
Salt crusts, 104–105
Salt weathering, 105, 136, 139
Samara Vallis, 36, 58
Sapping, 39, 41, 81–85, 92, 98
Scabland topography
    in Channeled Scabland, 142–150,
        152–158
    on Mars, 9, 44–45, 112, 114,
        116–117
Schiaparelli, Giovanni, 3–5, 13
Scour marks
    in Channeled Scabland, 166
    on Mars, 44, 126
    origin of, 166
Secchi, Pietro Angelo, 4, 13
Secondary craters, 18
Seepage gullies, 73
Shalbatana Vallis, 36, 108–109, 112
Sherman glacier, 100
Sherman landslide, 102
Shield volcanoes, 22–24, 52–54,
    84–87
Simud Vallis, 36, 108–109
Sinuous rilles, 53–54
Slope gullies, 42
Small channels, 34. See also Runoff
    channels; Valley networks
Soil, 104
Solar output variations, 182
Solifluction, 92
Spokane, 142
Spreading failures, 49

Spur-and-gully topography, 92, 94–95, 138
Stickney, 4
Streamlined hills (Channeled Scabland), 148–152
Streamlined uplands (Mars), 111–118, 126–129, 134–135
Streamlining, 115, 167–170, 174
Stream power, 158, 160
Stripes, 92
Swift, Jonathan, 3

Table mountains, 26
Talus cone, 95
Talus slope, 95
Telford–Crab Creek Scabland, 147
Tephra, 143
Tharsis, 22, 24, 181
Tharsis bulge, 22
Thermal inertia, 111
Thermocirque, 91–93
Thermokarst, 91–92, 125, 139
Tithonium Chasma, 26–27, 100
Tiu Vallis, 30, 36, 51, 108–109, 113, 116–117
T-Tauri phase, 179
Tyrrhena Patera, 25
  age of, 21
  valleys of, 75–77

Uniformitarianism, 14
Utopia Planitia, 9
Uzboi Vallis, 36, 64–69

Valles Marineris, 26–29, 33
  landslides in, 98, 100–103
  origin of, 184
  slope morphology of, 94–95
  stratigraphy of, 28
  tectonic activity in, 28
Valley networks, 56–87
  origin of, by sapping, 81–87
  slope of, 73
  small, 57–64
  longitudinal, 64–72
Valleys, 17, 56–87
Vedra Vallis, 36, 120, 128
Venus, 179
Vidicon, 7
Viking mission, 9
  biology experiment of, 11–12
  lander of, 10
  landing sites of, 9–12, 21
Visual imaging subsystem (VIS), 9
Volatiles
  evolution of, on Mars, 176–180
  origin of, 105–107
Volcanic channels, 34
Volcanic valleys, 84–87

Volcanoes, 17, 22–26, 33
Volcano-ice interactions, 26
Vortex, 165–167
Vortex cavitation, 172
Voyager mission, 15

Wake vortex, 166
Washtucna Coulee, 149
Water
  amount of, on Mars, 175–180
  and indicators of water-related processes on Mars, 178
  fate of, 184
  properties of, 161, 171
  source of, for channels, 105–107
Weathering processes, 104–105
West Bar, 159
West Potholes cataract, 153, 155
Wind, 48–49, 104
Wrinkle ridges
  and channeling processes, 126–127
  origin of, 25–26

Yakima Basalt, 142, 164
Yardangs, 48–49, 112
Yuty, 20, 21